Effective Advocacy

Lessons from East Asia's Environmentalists

Mary Alice Haddad

The MIT Press
Cambridge, Massachusetts
London, England

The open access edition of this book was made possible by generous funding from Arcadia—a charitable fund of Lisbet Rausing and Peter Baldwin.

Open access edition funded by the National Endowment for the Humanities. Any views, findings, conclusions, or recommendations expressed in this book do not necessarily represent those of the National Endowment for the Humanities.

NATIONAL
ENDOWMENT
FOR THE
HUMANITIES

The MIT Press would like to thank the anonymous peer reviewers who provided comments on drafts of this book. The generous work of academic experts is essential for establishing the authority and quality of our publications. We acknowledge with gratitude the contributions of these otherwise uncredited readers.

This book was set in Stone Serif and Stone Sans by Westchester Publishing Services. Printed and bound in the United States of America.

Library of Congress Cataloging-in-Publication Data

Names: Haddad, Mary Alice, 1973- author.
Title: Effective advocacy : lessons from East Asia's environmentalists / Mary Alice Haddad.
Description: Cambridge, Massachusetts : The MIT Press, [2021] | Series: American and comparative environmental policy | Includes bibliographical references and index.
Identifiers: LCCN 2020027086 | ISBN 9780262542357 (paperback)
Subjects: LCSH: Environmentalism--East Asia. | Environmental policy--East Asia.
Classification: LCC GE199.E17 H34 2021 | DDC 333.7095--dc23
LC record available at https://lccn.loc.gov/2020027086

10 9 8 7 6 5 4 3 2 1

For Tammer and Reja, who constantly give me hope and inspiration. May they leave their children a better world than the one they are inheriting from me.

Contents

Series Foreword

Public participation and advocacy are core concepts in democratic theory and Western democracy. Most agree that they are essential components of decision-making at all levels of government. There is some disagreement, however, concerning the appropriate form and extent of public participation and advocacy in government decision-making.

Although levels of public participation in the US have declined over time, opportunities for civic involvement have significantly increased in environmental and natural resource policymaking. A large segment of major legislation involving such policy issues, for example, has included requirements of some form of citizen participation and input. More recently, US environmental and natural resource agencies have formulated innovative ways to encourage and facilitate civic engagement in rulemaking and other aspects of decision-making.

Generally speaking, most of the literature on public participation and advocacy focuses on the nature and efficacy of such activity and the extent to which it actually influences environmental policy in meaningful ways in Western democracies, mainly those in the US and Europe. The many ways citizens can become involved and lobby for effective environmental policymaking and the wide availability of empirical data on public participation in environmental politics and policy make this an attractive focus for researchers. Lately, however, a growing number of scholars have begun to analyze citizen involvement and advocacy in other parts of the world by adopting novel and creative methodological approaches.

Mary Alice Haddad has written a fascinating and outstanding book that examines the activities of environmental advocates in East Asia, particularly in countries that pose serious impediments to political advocacy. Specifically, in China advocacy has been significantly stifled by the government, in South Korea and Taiwan there exist significant legal restrictions, and in Japan political advocacy is legally and culturally discouraged. Such obstacles to public involvement are not at all unique to these four countries.

Despite the existence of these obstacles, however, the governments of these nations are among the world's most innovative in terms of environmental policy development. Japan has been leading the world in high emissions standards for decades, China has become the world's largest producer of photovoltaic panels and the world leader in renewable energy, and South Korea and Taiwan have both embarked on major green initiatives that involve not just green business development but also new national parks, widespread energy conservation, and comprehensive recycling efforts. For Haddad, East Asia presents an important opportunity to examine a set of countries that are puzzling: Their political opportunity structures are such that one might expect few, if any, efforts to improve environmental quality. Yet she finds that governments in the region are among the world's most innovative in terms of proenvironmental policy development.

Haddad's in-depth investigation addresses two main questions: Which advocacy strategies are the most successful in persuading citizens, governments, and businesses to alter their behavior, and why are some strategies more successful than others in the four countries she studies? To answer these questions, she analyzes the strategies that are proving to be effective even under the hostile political conditions faced by advocates in East Asia in the hope of yielding insights into how advocacy can be effective elsewhere in the world. The goal of her book is to discover which advocacy strategies work and why. Among other things, *Effective Advocacy* concludes that while societies in East Asia have unique cultural histories and political contexts that affect the way that advocacy strategies are implemented, the strategies that are most common and effective in East Asia are also common and effective in other parts of the world.

At a time when environmental policies are increasingly seen as controversial, and new and alternative approaches are being implemented widely, we especially encourage studies that assess policy successes and failures, evaluate new institutional arrangements and policy tools, and clarify new directions for environmental politics and policy. The books in this series are written for a wide audience that includes academics, policymakers, environmental scientists and professionals, business and labor leaders, environmental activists, and students concerned with environmental issues. We hope they contribute to the public's understanding of environmental problems, issues, and policies of concern today and also suggest promising actions for the future.

Sheldon Kamieniecki, University of California, Santa Cruz
Michael Kraft, University of Wisconsin–Green Bay
Coeditors, American and Comparative Environmental Policy series

Acknowledgments

Writing a book takes a long time, and I have been helped by hundreds of people who have inspired, challenged, and encouraged me along the way. Although it is impossible for me to acknowledge all of the individuals and institutions that have assisted in the process of bringing this book to fruition, I would like to recognize a few of the most important. I begin by recognizing the institutions that supported this research through grants, fellowships, and residencies, which enabled me to travel and live in East Asia to conduct interviews, visit project sites, and learn about how advocates across the region have been changing their communities and countries for the better.

This project began with a seed grant from Wesleyan University that enabled me to attend a fascinating and eye-opening faculty seminar at the Studienforum Berlin called "Germany as a Model? The Environmental and Energy Strategy." It served as a kind of boot camp in environmental politics and policy, and I also met a number of faculty who would become important sources of information and inspiration, including Carol Hager, who would ultimately coedit *NIMBY Is Beautiful: Cases of Local Activism and Environmental Innovation around the World* (2015) with me. Wesleyan University continued to support me with numerous project and other grants over the years, including a very helpful fellowship in the College of the Environment's think tank, which introduced me to future coauthor Helen Poulos and funded a book workshop for *NIMBY Is Beautiful*.

The bulk of the research for *Effective Advocacy* was conducted during a very eventful year of research in the field during 2010–2011. The East Asian Institute's Fellows Program on Peace, Governance, and Development in East Asia funded research trips to Taipei, Beijing, and Seoul in 2010. Those trips included very helpful seminars in each city that offered me a chance to workshop my early ideas and facilitated connections to important informants in the three countries that would significantly enhance my research. My family and I lived in Tokyo for several months in early 2011, hosted by the Faculty of Law at Keio University as part of an Abe Fellowship sponsored by the Japan Foundation Center for Global Partnership (CGP) and the Social Science Research Council. The CGP staff deserve special recognition for the compassion and support that they offered me and my family as we had to make a premature return to the United States after the disaster of 3/11 made life in Tokyo too difficult, and I had to rapidly adjust my research plans for the year to accommodate that change. CGP also supported the "Catalyst Workshop on Policymaking in East Asia" at Wesleyan University in 2017, which facilitated an intellectual breakthrough in my own work and contributed to the publication of a number of colleagues' research as well.

The Japan-US Friendship Commission and the National Endowment for the Humanities' Fellowship for Advanced Social Science Research on Japan supported significant research and writing during a sabbatical in 2014–2015. The bulk of *Effective Advocacy* was written during the 2018–2019 academic year, where the Universities Service Centre, Chinese University of Hong Kong, provided a welcoming and supportive academic home for me while my family and I spent the year in Hong Kong.

Although a book may have a single author, it is the result of the intellectual contributions of dozens of people over the course of a decade. These contributions range from offhand comments in the hallway of a conference that facilitated theoretical breakthroughs to lengthy, thoughtful comments on draft manuscripts or at public presentations. I would like to recognize participants of a few particularly helpful gatherings: the 2009 faculty seminar hosted by the Studienforum Berlin; the National Bureau of Asian Research's 2013 Pacific Energy Summit; the Japan External Trade Organization workshop in 2015; the 2016 Conference on the Environment and Environmentalism in East Asia at the Banff Centre; the 2017 Catalyst Workshop on Policymaking in East Asia at Wesleyan University; the 2017 conference "Civil Society versus the State? Emergent Trajectories of Civic Agency in East Asia in Comparative and Transnational Perspective" at the University of Zürich; a presentation of my research at the Institute for the Environment, Hong Kong University of Science and Technology; the 2018 workshop at the Institute of Social Science, University of Tokyo; and the 2019 meeting of the Israeli Association for Japanese Studies in Jerusalem.

Of all these gatherings, perhaps the group that has proved to be the most influential on my research has been the US-Japan Network for the Future, funded by the Japan Foundation CGP and the Maureen and Mike Mansfield Foundation, which has not only put together a number of exceptionally fruitful conferences and workshops but also created a network of supportive colleagues who were all willing to offer helpful advice and useful resources anytime I needed them.

In addition to collective feedback, a number of individuals have offered important support at critical junctures. It is impossible to acknowledge them all, but I would like to recognize a few who have been particularly helpful. I met Miranda Schreurs at the 2009 faculty seminar in Berlin, and she has regularly offered feedback in the ten years since, including exceptionally helpful comments on the full manuscript as a reviewer for MIT Press. Stevan Harrell, who, along with Ashley Esarey and Johanna Lewis, would become coeditor with me of *Greening East Asia: The Rise of the Eco-developmental State* (forthcoming), has been especially helpful, since working together on that project enabled me to fix theoretical weaknesses I needed to address in *Effective Advocacy*.

In Japan, Professor Hironori Hamanaka not only repeatedly offered numerous insights of his own on how environmental advocacy worked in Japan, he also provided numerous helpful introductions to other exceptionally useful informants. In China, Ted Pflaker did the same, generously sharing his time, insights, and introductions. Jennifer Holdaway and Shawn Shieh similarly imparted important insights gained from decades working with Chinese civil society actors. In South Korea, Kwang-Yeong Shin at Chung-Ang University, whom I met in Zurich, was exceptionally generous with his time, insights, and contacts. My Wesleyan colleague Hyejoo Back has also been incredibly helpful, introducing me to important informants, sharing her insights into her home country, and even offering quick responses to desperate translation questions. Beth Clevenger at MIT Press has been the best editor that I have ever worked with, patiently shepherding this project through the review and publication process and offering helpful feedback and advice along the way.

In addition to financial and intellectual support, I benefitted greatly from the able assistance from a number of exceptionally talented research assistants. Gena Yoo, Charlie Chung, Guangshuo Yang, Haru Mitani, Somin Lee, Bohao Zhao, Julia Jonas-Day, Julia Michaels, Elijah Stevens, Melody

Chen, James Hall, Zhuo Chen, Bryan Chong, and Zhaoyu Sun all contributed in vital ways to this project. They coded newspaper articles written in languages I didn't speak, learned new statistical programs, ran countless statistical analyses, and hunted down obscure government data. Along the way they helped me vet and shape my ideas, finding the aspects of my study that were the most relevant and interesting and identifying which were intellectual dead ends best abandoned.

Researching and writing these kinds of books is not just an intellectual exercise conducted in an academic ivory tower somewhere; it involves personal engagement and connection with numerous people, their stories, and their lives. From a more personal standpoint, I want to acknowledge my deep gratitude to the Vandenbrink and Shioji families, both of whom provided housing and much-needed emotional support (which also translated into research support) when my family was displaced during the 3/11 disaster in Japan. Sam Paik and Pheobe Shin offered me a warm welcome, provided connections to relevant informants, and helped me sort out tangled and confused ideas about how politics worked in South Korea during my visits to Seoul. In Hong Kong, the American Women's Association's environmental group provided helpful opportunities to "do" some environmental advocacy while I lived in the city. The AWA Globe Paddlers helped make sure that my year in Hong Kong was not only professionally productive but also full of comradery, laughter, and buff biceps, generating what I hope to be many lifelong friendships.

Finally, none of this would have been possible without the support of my husband, Rami, who endured his wife's erratic and frequently inconvenient travel schedule. He quit jobs and took overseas assignments to support my sabbaticals abroad, all while working with me to keep lively boys healthy and happy in their ever-changing living situations. Those boys, Tammer and Reja, were tiny preschoolers when this project began and now are strapping teenagers, both of whom are considerably taller than I am. Their joy for life has kept me going through the more challenging parts of this project, and my wish for them to pass on to their children a better planet than they are inheriting from me has been a driving motivation for this research. It is to them that I dedicate this book.

1 Introduction: Environmental Advocacy and Policymaking

As sea levels rise and pollution spreads, people around the world are organizing to advocate for a better environment. They are protesting in the streets, lobbying politicians, filing public interest lawsuits, writing policy briefs, forming green companies, participating in local cleanup campaigns, installing eco-art, and engaging in many other activities that are all designed to change behavior and policy in ways that will improve environmental outcomes in communities around the world. Which of these advocacy strategies are the most successful in persuading citizens, governments, and businesses to change their behavior? Why are some strategies more successful than others? *Effective Advocacy* sets out to answer these vital questions by investigating the success of environmental advocates in East Asia.

East Asia has seen an "economic miracle" that has lifted the most people out of poverty in the shortest period of time in history, but that same "miracle" has also generated unprecedented levels of pollution. Indeed, East Asia's pollution now threatens the ecology and livability of the entire planet, so its environmental problems have become the world's environmental problems. The premise of this book is that its environmental solutions might serve as solutions for the world as well.

East Asia is a region of developmental states that have strong ties to business and are oriented toward economic development.[1] It is also a region of poorly institutionalized advocacy organizations; its civic organizations tend to be small, locally based, and volunteer run. Although most countries in the region have established green parties, their representation in national legislatures is insignificant.[2] Furthermore, political advocacy is severely repressed China,[3] faces significant legal restrictions in South Korea and Taiwan,[4] and has been legally and culturally discouraged in Japan.[5]

And yet Japan has been leading the world in high emissions standards for decades, China has become the world's largest producer of photovoltaic panels and the world leader in renewable energy, and South Korea and Taiwan have both embarked on major green initiatives that involve not just green business development but also new national parks, widespread energy conservation, and comprehensive recycling efforts. Therefore, East Asia presents the opportunity to examine a set of countries that are puzzling: their political opportunity structures are such that we might expect poor environmental policy, but we find that governments in the region are among the world's most innovative in terms of proenvironmental policy development. East Asia's environmental advocates must be doing something right.

The task of this book is to discover which of their advocacy strategies work and why. We will study the strategies that are proving to be effective even under the hostile political conditions faced by advocates in East Asia in the hope that we might garner insights into how advocacy can be effective elsewhere in the world. This chapter begins with an overview of the advocacy and policymaking literature. It will then provide an explanation of the research design and logic of the book and conclude with a brief overview of the rest of the volume.

Environmental Advocacy and Policymaking

Environmental advocacy has a long, global history. From early conservation organizations such as the Plumage League in Europe (founded in 1889), the Sierra Club in the United States (founded in 1892), and the Wild Bird Society of Japan (founded in 1934) to advocacy groups with broad policy agendas such as the Natural Resources Defense Council (founded in 1970) and Greenpeace (founded in 1971) to more recent, digital-based organizations such as 350.org (founded in 2007) and Youth Strike for Climate (founded in 2018), individuals have joined together locally, nationally, and, increasingly, internationally to protect and improve the earth's environment. Scholars studying environmental politics have tended to look at the political dynamics of their efforts from two different perspectives: the grassroots level, studying environmental movements as a kind of social movement, and the elite level, studying environmental policy as a type of public policy.

Much of the social movement literature has been primarily concerned with democratization movements and regime change.[6] When social movement theories have been used to examine environmental politics, they have focused on how citizens mobilize their resources (economic, social, and political) to take advantage of political opportunities to promote their environmental causes. This strand of scholarship emphasizes advocacy strategies that are intended to mobilize publics who can then pressure elite actors (governmental and corporate) to improve their environmental policy and behavior.

One of the most common strategies studied by this branch of the literature is public protest. In public protests, advocates gather large numbers of people together in a public place in order to draw attention to their issue and pressure policymakers to develop better environmental policy.[7] In democratic systems, elected officials are commonly the target of the protests, where protesters threaten the electoral prospects of individual politicians and their parties if demands are not heeded.[8] In nondemocratic countries, public demonstrations are a nonelectoral method for citizens to voice concerns. While protesters may not be threatening the electoral prospects of authoritarian leaders, their potent dissatisfaction can undermine the legitimacy and credibility of political leadership, which even authoritarian leaders are keen to avoid.[9]

Around the world, public protests are usually peaceful, but sometimes they turn violent. Violent environmental protests offer similar political benefits to peaceful protests—raising the profile of a particular environmental issue and undermining the legitimacy of the leaders who have failed to resolve the issue peacefully—but at a very high cost. Commonly protesters are thrown in jail, but sometimes there can be mass violence with thousands killed at once when well-resourced companies work with corrupt local officials to crush dissent. Violent protests tend to be more common in undemocratic countries and in places where the protesters are socially and economically marginalized (e.g., indigenous communities, ethnic minorities, rural residents).[10]

A similarly contentious but more law-abiding strategy used by advocates is lawsuits. Filing a legal challenge against a polluting company, local government, or even national government is a lawful method for those seeking change. In democratic and undemocratic countries alike, environmental lawsuits are usually lost, but they offer another opportunity to raise the profile of the advocates' issues on the public agenda. Furthermore, in the rare case when plaintiffs do win their case, the win can be a very big one. Companies—not just the polluting company but all companies with similar profiles—can be forced to accept responsibility for environmental damage. Local and national governments can be required to enhance their enforcement of environmental regulations. Successful and even unsuccessful lawsuits can encourage lawmakers to enact proenvironmental legislation, benefiting many more people than just the original victims.[11] Perhaps even more importantly, lawsuits can shift the discourse surrounding elite and popular understandings about the environment and the responsibilities of government, corporations, and individuals to protect it.[12]

Critical to the effectiveness of both protests and lawsuits, whether they are used alone or in combination with one another, is engagement with the media. Fundamentally, both protests and lawsuits are intended to attract the attention of policymakers, and the media is ultimately the source of that attention. The constraints of the media often shape the timing, location, and content of protests, while important procedural moments in legal proceedings (e.g., when the lawsuit is filed, when major figures appear in court to testify, when the verdict is rendered) can create natural opportunities for activists to draw media attention to their issue.[13]

Turning toward less contentious modes of grassroots advocacy, scholars have also studied the role of nonpartisan environmental education. Environmental education, especially among children, is seen as a vital foundation for all other forms of advocacy. Children need to be exposed to the natural environment and given opportunities to enjoy it. They need to be taught how to behave in environmentally responsible ways and that they have a duty to care for the environment. Citizens who have an appreciation for nature are more likely to act in ways to protect and promote it.[14]

Another mode that is directed at emotional engagement of citizens and leaders is the extensive use of art and cultural symbols to give meaning to environmental issues and make those issues relevant to people's everyday lives. Whether this is achieved by having celebrities perform music at an event to support a cause[15] or offering spectacular visuals to make visible environmental beauties and horrors to those who cannot witness them firsthand,[16] the emotive power of art can be used to engage, enrage, and energize a previously passive population.[17] As a result, art can be a powerful tool for activists seeking change.[18]

Finally, one of the most potent forms of grassroots environmental advocacy is scaling direct local action—finding a specific location, implementing a positive, proenvironmental change there, and then working with others to disseminate the successful change to other localities. Local environmental projects require that activists execute concrete projects in actual places; they must do more than just talk about ideals and problems. Frequently, in order to make the projects successful, they must work closely with local authorities and collaborate with other related entities. Perhaps surprisingly, it is often the case that the local authorities who might have initially resisted local environmental projects become their greatest advocates.[19] When local environmental organizations are able to network together, they can spread the benefits of a small, local improvement over a much larger area, sometimes even changing policy at the national level and beyond.[20]

Shifting focus from scholarship emphasizing citizen-based, grassroots forms of advocacy to research examining environmental policy itself, we find several different strands of literature. Perhaps the most robust area of research has been on regulatory design, where scholars and practitioners work to generate new knowledge about the types of environmental regulations, as well as the process and governance structures most likely to generate the desired policy outcomes.[21]

However, scholars and advocates have long recognized that no matter how perfect a proposed regulation or policy is, it will never be adopted or implemented unless it gets the attention of policymakers.[22] Therefore, significant research has also been devoted to examining the best way for advocates to influence policy elites—to get their attention and convince them to design policies that favor the advocates' goals.[23] As the environmental agenda has become a global one, many of these elite-related strategies are global, involving influence brokers who move across national boundaries. The influence of international networks operates in several different ways: International actors (e.g., global nongovernmental organizations [NGOs], foreign governmental and nongovernmental actors, multinational corporations) can pressure domestic elites to change national laws to be more compatible with globally accepted norms.[24] Domestic elites can work with and through international networks to pressure policymakers in their own governments.[25] At the same time, numerous governmental, nongovernmental, and corporate actors are constantly seeking to reshape global discourse, agendas, and models in ways that are compatible with their own local and global goals.[26]

Collectively, the scholarship on environmental advocacy and policymaking has generated valuable new knowledge about how to design effective environmental policy, the processes through which it is developed, and the mechanisms through which advocates can exert influence on that process. Thus far, however, most studies have remained relatively narrow, focusing largely on the environmental politics in democratic countries located in western Europe and North America or international organizations or multinational corporations that have headquarters located in those two regions.

When scholars have studied environmental politics in East Asia, they have tended to focus on single countries, seeking to understand the environmental politics of China,[27] Japan,[28] South Korea,[29] or Taiwan.[30] To the extent that single books have tried to cover the region as a whole, they are generally edited volumes where individual chapters cover individual countries. While these volumes do draw commonalities across the countries within the region, they share an underlying assumption that East Asian politics is fundamentally different from politics found in Europe and North America. The authors frequently assume that because of those differences, any insights garnered from studying the politics of the region, while important, may have limited value in helping to understand politics in other parts of the world.[31]

Research Design

Effective Advocacy takes a fundamentally different perspective. As discussed at the beginning of this chapter, East Asia represents a particularly "hard" case in which to study effective environmental advocacy: while some governments in the region are more democratic than others, they all have a multidecade history of following a developmental state model that favors business and is hostile to citizen-based advocacy.[32] As a result, while citizens may be very active at the local level, civil society in general and the advocacy sector in particular tend to be underdeveloped across the region. Concerned citizens face significant legal and social constraints when they seek to form nonprofit and advocacy organizations, so there are very few national, nonprofit organizations with large budgets and numerous professional staff members anywhere in the region.[33] Those organizations that do register as official nonprofit organizations and hire staff often take the form of government-organized NGOs, which have most of their funding or leadership or both coming from the government, or they find themselves working in close collaborative relationships with the state rather than as autonomous, independent organizations that challenge and confront government regularly.[34]

There are some political advantages to the developmental state when it comes to making environmental policy. The largest is perhaps that when proenvironmental policy can be tailored to promote business interests, such as the many "green growth" initiatives found across the region, the resulting policies can be powerful, widespread, and quickly implemented.[35] On the flip side, when action to protect the environment goes against corporate interests or threatens political elites by raising the concerns of marginalized peoples, change is particularly hard. Although the region may be a leader in green energy and technology, East Asia is a global laggard in biodiversity preservation and environmental justice, and it continues to struggle with industrial pollution.[36]

Overall, advocates in East Asia have found it particularly difficult to persuade governments, businesses, and individuals to change their behaviors in ways that benefit the environment because of significant legal, political, and social barriers inhibiting political advocacy. The premise of this book is that it is useful to examine environmental advocates in East Asia to garner insights into how advocacy can be effective under difficult conditions. The assumption of the volume is that strategies that are effective in East Asia can be adapted to fit nearly any region of the world, since advocates in most places have it easier than they do in East Asia.

Effective Advocacy assumes that every community has unique political dynamics that have been shaped by its own particular culture, history, legal system, economic structure, and other factors. However, just as the experience of North America and Europe can generate insights into the nature of social movements and the tactics of policy entrepreneurs that can be applied to other contexts, lessons from East Asia's environmental advocates can be useful for policymakers, advocates, and citizens around the world as well. In fact, *Effective Advocacy* asserts that the insights gained from East Asia's environmentalists are likely to be more useful and more generalizable than those gained by looking at advocates located in Europe and North America because the strategies that work across East Asia do not require democracy, a free press, a rich advocacy sector, or a culture of national political engagement by citizens. If an advocacy strategy is effective across East Asia, it should work almost anywhere.

Although East Asia has experienced some of the worst pollution in world history,[37] in the last few decades it has become a world leader in many areas of environmental policymaking. It has some of the most stringent emission standards, highest use of renewable energy, extensive recycling infrastructure, and has been expanding its greenspace and protected natural areas. Despite their probusiness and antiadvocate orientations, governments in the region are among the world's most innovative in terms of proenvironmental policy development. Surely we can learn something useful from the environmental advocates in the region about how they can be effective in such difficult circumstances.

Thus, this study of effective advocacy is rooted in East Asia and will examine the issue of environmental policy in particular. By limiting the issue focus to the environment, I am able to control for the fact that the political dynamics of other policy areas might be very different. One would not expect the same configuration of interests and actors in the health care or foreign policy arenas to be the same as in the environment issue area. While, as will be discussed more in chapter 3 and in the conclusion, I suspect that the lessons from East Asia's environmentalists should have widespread application to other policy areas, this study focuses exclusively on the environment to limit the confounding factors that might influence advocacy effectiveness.

Similarly, by restricting the qualitative research to China, Japan, South Korea, and Taiwan, I was able to control a number of factors that might otherwise cause variation among the cases, allowing me to focus on the factors in which I am most interested. All four places have a somewhat related cultural background rooted in rice cultivation and Confucianism. All four places experienced a period of "high-growth," rapid industrial development. All four places have export-oriented industrial structures. All four places have been dominated by probusiness governments for most of the past fifty years. All four have legal environments that restrict political advocacy, and none of them have green parties that are viable in national-level politics. All four have relatively small and underinstitutionalized advocacy sectors. All four have experienced significant, intense levels of pollution that resulted from their rapid development. All four have seen citizens and their organizations demand that their governments address the pollution problems, and governments and businesses in all four places have responded, developing significant and progressive environmental policies to address concerns raised by the public. Therefore, the four places have a number of similarities on variables likely to influence the ability of advocates to be effective in generating proenvironmental changes in policy and

behavior among governments, businesses, and citizens, which makes it useful to consider them as "similar" as well as "hard" cases.

There are, of course, a number of ways in which the places are different. The
most obvious is size—mainland China is the most populous country (1.4 billion people) and has the largest gross domestic product (at purchasing power
parity) ($25 trillion) on the planet, while Taiwan is comparatively tiny with
only 24 million people and a GDP (PPP) of $1.2 trillion. While the govern-
ment on the mainland has been led by the Chinese Communist Party for the
last 70 years, Taiwan's head of state is chosen through a democratic electoral
process which has resulted in a regular change in ruling power. And yet, the
governments of both the People's Republic of China (China) and the Republic
of China (Taiwan) officially view themselves as belonging to "One China."

I do not take a stand on that very sensitive political topic. As will become
obvious in the coming chapters, while there are some loose linkages between
the environmental politics on the mainland and that on the islands of Taiwan, for the most part they can be considered separately. Their citizens
vote in different elections; their political parties are different; their environmental organizations are different; their main universities and think tanks
are different; their most influential businesses are different; their elected
officials are different; their bureaucratic structures are different; and their
relationships to international organizations are different.

Therefore, for rhetorical ease, the Republic of China will be referred to as
Taiwan throughout this book and will be discussed as a "place," "country,"
"polity," and "society." None of these word choices are meant to imply
anything about Taiwan's national sovereignty, which is not particularly relevant for this study. What matters for this study is that both China and Taiwan have functioning governments that make policies for their respective
territories, and those governments respond to their local civil societies and
business communities when making environmental policy.

Another important difference among these four countries, one that
helps ensure variation on a key variable that might affect advocacy effectiveness, is their very different levels of experience with democracy. Japan
is the oldest democracy in the region and was one of the first nonwhite,
non-Christian, non-Western democracies in the world. Since 1947 it has
had a democratic constitution that guarantees its citizens equality under
the law; freedom of expression, assembly, and religion; and due process.

It also guarantees human rights, protection from discrimination, freedom of movement, the right to a "minimum standard of wholesome and cultured living," and academic freedom.[38] Japan's democracy is not just a paper democracy. In the more than seventy years since its constitution came into force, it has developed a robust set of democratic institutions, values, and practices at both the elite and grassroots levels of society that continue to evolve as Japan's political culture changes over time.[39]

South Korea and Taiwan are newer democracies. They both experienced lengthy and sometimes brutal occupations by Japan (South Korea from 1910 to 1945 and Taiwan from 1895 to 1945). Following Japan's defeat, they suffered destructive civil wars and found their prewar territories split into two parts politically, with a communist party gaining influence over one section of territory and a nationalist or democratic party occupying another section. In both places, that political division remains in place and continues to be a defining issue in each country's politics. Their domestic civil wars put them on the front line of the global Cold War, which was partly responsible for protracted periods of martial law and military governments that directed export-led rapid industrial development during the 1970s and 1980s. Economic growth led to the expansion of the middle class and calls for more political freedom in the 1980s. Both countries engaged in relatively peaceful democratization processes in the late 1980s.[40] South Korea revised its constitution and held its first democratic election in 1987. In Taiwan, martial law was lifted in 1987, and the constitution was revised to allow for free elections in 1991.

South Korea and Taiwan are both considered to be part of the third wave of democracy, which spread across the globe and coincided with the breakup of the Soviet Union in the late 1980s.[41] Although they have struggled with some undemocratic practices such as high-level corruption and "blacklisting" political opponents,[42] their citizens enjoy equal protection under the law and freedom of speech and assembly, and they have highly competitive elections that enable citizens to hold their leaders directly accountable.

Finally, mainland China is not democratic—its constitution uses the phrase "dictatorship of the proletariat" to describe its political system. While China's constitution grants its citizens equality under the law, freedom of the press, and freedom of assembly and religion, those individual rights are second to the rights of the state (Article 51). Similarly, although elections are frequently competitive, candidates have been preselected to limit voter choices to those acceptable to the Chinese Communist Party.[43] The traditional press is highly restricted, and while significant freedom is allowed on the internet, it too is frequently censored.[44] The NGO and civil society sector in China has been growing more diverse and more robust in recent decades, and the legal structure is becoming more sophisticated to cope with this expansion. During the 2000s the nonprofit and civil society sector expanded, with a proliferation of organizations and an increasing sophistication in the legal structure in which they operate. During the 2010s the growth and diversification of civil society stalled, and in some cases was reversed, as the state increased its political control over activists and their organizations.[45]

Thus, a focus on the northeast Asian region allows me to control for a number of variables that are likely to influence the frequency and effectiveness of advocacy strategies—their cultural background, export-oriented industrial policy, probusiness governments, high-speed economic growth, intense pollution problems, and small and legally constrained political advocacy sectors. And yet the four places have very different levels of experience with democracy.

Effective Advocacy is concerned with the strategies that activists of all kinds can use to inspire (and sometimes compel) proenvironmental behavior change on the part of governments, businesses, and citizens. It takes a very broad view of what "counts" as advocacy. For the purposes of this book, environmental advocacy is an organized, collective effort to promote behavior or policy change in proenvironmental ways.

Thus, individual, private environmental efforts are not included under this definition. For example, an individual who decides to compost may be involved in environmental action, but for the purpose of this book, the individual's action would not constitute environmental advocacy. However, if an individual joins his or her neighbors to promote community composting and organizes a community garden event to encourage members of a community to compost, that action would be considered advocacy since the effort is an organized, collective action to promote proenvironmental behavior change. This broad definition is intended to capture a culturally diverse set of advocacy efforts and strategies in order to investigate how advocacy goals, strategies, and efficacy might vary by region and regime type.

This definition of advocacy is consistent with most of the literature on advocacy, but it includes a broader range of actors than are sometimes considered to be advocates by political scientists. In the popular press, *advocacy* is a fairly general term that can be employed by a wide range of people. In his book *Advocacy: Championing Ideas and Influencing Others* (2012), John Daly emphasizes the importance of communication strategies. In his words, "If new ideas are to gain the attention and support of decision makers, they must be touted in memorable and persuasive ways."[46] For Daly, advocacy is not just about influencing public policy but also about the process of getting one's idea adopted by others, whether they are corporate CEOs, school principals, or national legislators.

Although he does not generally use the word *advocacy*, John Kingdon's pathbreaking work on policymaking, *Agendas, Alternatives, and Public Policies* (1984) describes the interaction of a wide range of actors—including bureaucrats, politicians, business leaders, grassroots activists, scientists, and public intellectuals—as they collectively make and implement public policy. In his model, these different actors, all of whom have different stakes in the outcome, struggle to get their preferred policy alternative onto the public agenda, into the mix of alternatives from which decision makers are choosing, selected as the policy to be adopted, and eventually implemented.

Expanding on Kingdon's model, Paul Sabatier specifies this process further by developing his concept of "advocacy coalitions," which are "composed of people from various organizations who share a set of normative and causal beliefs and who often act in concert. At any particular point in time, each coalition adopts a strategy(s) envisaging one or more institutional innovations which it feels will further its policy objectives."[47] For Sabatier, the people involved in these coalitions are as diverse as those in Kingdon's model—they can be grassroots activists, academics, oil tycoons, bureaucrats, and everyone in between. As would be expected, the kinds of policies for which these coalitions might be advocating are as varied as their memberships. Proenvironment coalitions of grassroots activists, scientists, global NGOs, and supportive politicians and bureaucrats might advocate on behalf of strict emissions standards and subsidies for clean energy, while those policies would be opposed by advocates on the other "side," who might consist of oil company executives, a different set of scientists, and another group of politicians and bureaucrats. All the actors on both sides are considered to be advocates, working to convince the other side that they are right and their preferred policies should be adopted.

This book takes a similarly broad view of who "counts" as an advocate. Essentially, anyone who is participating in an organized, collective effort to promote behavior or policy change in proenvironmental ways counts as an advocate and is considered to be engaging in advocacy. These people might be grassroots activists. They might be working inside global corporations to make change from within, or individual entrepreneurs building green businesses and seeking to change corporate culture and finance systems. They might be bureaucrats or political leaders. The only thing that is required is that they are working with others in an organized effort to promote proenvironmental behavior and policy change.

The research presented here is based on two primary sources: fieldwork conducted in East Asia during short and medium-length research trips between 2010 and 2019, and two original databases—one of environmental organizations in the region and another of environmental events that occurred around the world. More details about the interviews, as well as the databases and statistical techniques used to analyze them, can be found in appendix A.

As discussed earlier, the research design for this investigation was inductive—I sought to discover the answer to a puzzle: Why are East Asia's environmentalists so successful even while working under such hostile political conditions? My hope was that the answers to that question not only would generate insights into the nature of East Asia's environmental politics but would also offer lessons that advocates elsewhere could utilize in their own political contexts.

My methodology for selecting interview subjects was a combination of snowball (using one interview to generate additional interviews) and diversification (purposefully seeking out as diverse a range of actors related to environmental politics as I could reach). In particular, I endeavored to speak with government officials, advocates, business leaders, journalists, academics, artists, and grassroots volunteers in each country. I tried to connect both with people who were on the proenvironmental advocating side (e.g., NGO leaders) and with people who were the targets of that advocacy (e.g., government officials, business leaders). Some people (e.g., academics) were easier for me to access than others (e.g., Chinese officials), but in the end, I was able to talk with more than one hundred people from very different backgrounds, giving me a broad perspective on environmental advocacy in each country. Appendix A offers an overview of my interview subjects.

My interviews ranged from single, short meetings when I was able to catch someone I wanted to reach while he or she was at a conference or coming to or from the office, to multihour interviews conducted multiple times over the course of several years as I made return visits to high-quality sources to get updates and seek the names of additional people to contact. For the most part, my conversations lasted about an hour and were usually conducted in the person's office or at a local coffee shop where I was able to take notes as we talked.

The flow of our conversations generally followed the same pattern. I would begin by introducing myself and my project—investigating the seeming paradox of hostile advocacy conditions but comparatively good environmental policy outcomes in East Asia and seeking to discover "strategies that work." After a few descriptive questions about the person and his or her organization and position, I would ask short, open-ended questions designed to encourage my interlocutors to think creatively and in detail about the modes of advocacy that they see operating in their country, which ones they thought were particularly effective, and why they thought those strategies worked. Example questions include, "What do you think that advocates in [relevant country] do that is most effective?" "Why do you think that strategy works?" and "Can you give an example?"

I would close the interview by asking whether I could acknowledge them by name or if they wished to be anonymous, thanking them, and asking if there was anyone else whom they thought I should talk with about my project. Not all interview subjects who indicated a willingness to be named are actually named in this study. Political conditions have shifted since the time I conducted the interviews, so I have exercised my own judgment about the potential for personal harm when quoting directly. In all cases I have erred on the side of anonymity if I thought that anything they might have said to me could cause them difficulty. The safety of my interlocutors has been prioritized over all else.

In 2010 I set out to begin my fieldwork. It should be noted that what was supposed to be about a year in the field was interrupted rather abruptly by Japan's triple disaster when the March 11, 2011, earthquake in Tohoku triggered both a tsunami and a nuclear disaster. I was staying in Tokyo at the time with my family. We evacuated first to Kobe for a few weeks before returning to the United States.

In spite of the disruption, I was able to conduct significant fieldwork over the course of the year, speaking to more than sixty people across all four countries. To my astonishment, what had started as a project to document and investigate how democracy influenced advocacy strategies and effectiveness across East Asia turned into something rather different when

I discovered that some of my fundamental assumptions were wrong or at least inadequate.

Based on my own and other authors' previous research, I had assumed that a country's experience with democracy should shape the kinds of strategies that its advocates employed, as well as their effectiveness. My question at the outset of this research was not so much whether democracy mattered, but how. I was curious to shed more light on the ways that advocates in mature democracies (Japan) were able to utilize a wider range of tools than those in newer democracies (South Korea and Taiwan), who in turn had more options than those in nondemocratic states (China). I wanted to discover which tools were available and effective in Japan that were not in South Korea and Taiwan or in China.

However, as I talked with my diverse set of interlocutors across all four countries, they did not indicate significant differences in effective advocacy strategies across countries. No matter whom I talked with in whatever country, they all kept giving me the same set of advocacy strategies that they thought were effective:

1) Cultivate policy access. Cultivate and empower friends who have policy access. For example, put a former Ministry of Environment official on the board of your NGO, or help the midlevel bureaucrat who knows nothing about the environment attend an international environmental conference.

2) Make it work locally. Successfully implement an environmental solution locally and then disseminate that success to other localities.

3) Make it work for business. Proenvironmental solutions that are also profit making (or cost reducing) are easy win-win-wins for advocates.

4) Educate. At the grassroots and at the elite levels, advocates can promote positive environmental outcomes by helping people understand environmental problems and the solutions that can address them.

5) Engage the heart. Use art (such as photography, documentary films, sculpture, public art, and music) to attract attention, help people to care about the environment, and inspire them to act.

6) Think outside the box—be a game changer. A small number of proenvironmental advocates have innovative ideas that change the entire landscape of advocacy for everyone. These game changers can be exceptionally effective advocates.

It did not matter whether I was talking with a volunteering housewife in Tokyo, a Green Party activist in Taipei, a businessman in Seoul, or a professor in Beijing. While not every person I spoke with listed all six of these strategies, most people listed three or four, and these same six strategies showed up again and again in my conversations. In my interviews I could identify no pattern suggesting that there were substantial differences across the East Asian countries or according to the person's perspective (e.g., between NGO advocates and government officials). Tellingly, hardly anyone suggested that protests were particularly effective, and very few mentioned lawsuits. Media coverage and campaigns were discussed, but usually in conjunction with other strategies, not on their own. Thus, to my shock, the strategies that have garnered the most attention by environmental politics scholars were generally not mentioned by my interlocutors as particularly effective.

There were definitely some differences related to democracy that were in line with what I expected before starting my fieldwork—activists in China reported significantly more repression than those in the other countries, and those in South Korea and Taiwan exhibited more caution and concern about political retaliation than those in Japan. However, the variation seemed to be entirely about the level of repression and the repercussions of failed advocacy: Activists in Japan who failed went home and had a beer before returning to the effort another day. Those in South Korea and Taiwan who failed were sometimes put on a government blacklist that would limit their access to funding or policymaking, but they too were also able to go home and have a beer and try again another day. In contrast, activists who failed in China were occasionally thrown in jail, were put under house arrest, or disappeared. Very few were able to continue their advocacy after it had failed unless they left the country and persisted from abroad. Thus, the consequences of advocacy failure varied, but in terms of my core question—effective advocacy strategies—there was remarkable consistency.

This finding was not consistent with my expectations, so I sought some verification from outside my interview sample. I created a database of environmental organizations in the region and coded their activities. While this would not allow me to test the question of strategy effectiveness, I hoped it would enable me to verify (or call into question) the finding from my interviews that the strategies that environmental organizations in Japan, South Korea, Taiwan, and China were relatively similar. More about the methods related to this database can be found in appendix B.

As will be discussed more in the next chapter, the quantitative analysis of the activities of organizations supported the reports of my interview subjects. Environmental organizations in China, Japan, South Korea, and Taiwan employed similar advocacy strategies—they did not vary by level of democracy. Thus, one of the fundamental assumptions on which I had based this research was called into question: it may be that the most effective advocacy strategies are not particularly affected by regime type or length of experience with democracy. Perhaps effective advocacy strategies are effective no matter where you are.

Once again, my qualitative research would not be sufficient to test the broader applicability and effectiveness of these advocacy strategies. Rather than looking at the strategies that organizations were using—as explained by their websites—I sought to discover which strategies were being used on the ground in actual events, and which of those strategies seemed to be most effective.

Therefore, I created a new database based on cases of environmental advocacy gathered from media reports from around the world. From a pool of 3,390 environmental advocacy events in the five years from January 1, 2005, through December 31, 2009, I randomly selected 200 cases to investigate which strategies had been employed and whether the advocates had been effective. While certainly not a perfect test (challenges related to the study are discussed in more detail in the next chapter and in appendix B), the plausibility test appeared to confirm what was reported in my interviews and also the findings of the first quantitative study of organizations. Essentially, the strategies that East Asian environmental advocates were using to generate positive environmental outcomes under hostile political conditions were quite common around the world—everyone was using them. It turned out that effective advocacy strategies are effective no matter where you are.

As I was engaged in conducting and analyzing the large-N data, I returned to East Asia several more times to reinterview some people and to seek out new perspectives. I was more focused during these follow up conversations. I knew which strategies were effective; what I wanted to know was why. When I returned to East Asia for interviews in 2015, 2018, and 2019, I asked again about effective strategies (after all, the political situation in all of these countries was very different in 2018–2019 from how it had been in 2010–2011, so things might have changed), but this time I pressed harder on the why question. Why do you think that strategy was effective? How was it linked to other strategies? What was the process through which the advocacy strategy generated policy or behavior change?

My goal was to discover a common thread across all of the strategies. I sought to develop a coherent understanding of the political process surrounding environmental advocacy that could help explain why these particular strategies were effective under such a wide variety of political conditions. My intention was to develop a model of policymaking that could offer greater insight into the process through which advocates influence policy. The Connected Stakeholder Model (CSM), presented in chapter 3, is the result of these deliberations. The model offers a new conceptual framework for understanding the policymaking process. It helps explain why these particular advocacy strategies are effective, as well as why they are effective everywhere, not just in democracies.

In brief, the CSM posits that the key to understanding policymaking is to recognize that stakeholders involved in the policymaking process are not institutionally bound individuals with single interests. Rather, they are individuals who are connected to diverse and multiple networks, which enable them to develop complex ideas about the policies they are developing. The networks connected to the process are more important than the particular individuals, more important than the institutions from which the individuals come, and more important than the formal decision-making structure itself. Rather than focusing narrowly on the institutions and interests that are "at the table," scholars and advocates should pay more attention to the diverse networks linked to policymakers, including the informal channels of influences that these networks create to those "outside the room."

Overview of the Book

This chapter has presented the intellectual foundations on which the remaining chapters will build. Chapter 2 presents a "big picture" of advocacy around the world, examining how the various strategies that were investigated throughout the research are used by advocates around the world. It demonstrates empirically through statistical analyses that the strategies identified by my East Asian interlocutors are both common and effective everywhere in the world. It also digs into, but does not resolve, the complex ways that regime type influences advocacy success, and the consequences of advocacy failure.

Chapter 3 introduces the CSM as a conceptual framework for understanding how policies are made around the world. The model helps explain why certain advocacy strategies are more effective than others. It also helps us understand why some strategies, such as protests and lawsuits, might not be particularly effective when measured according how often their advocates succeed in gaining their stated goals but can sometimes be exceptionally effective in generating widespread cultural and political change.

Chapters 4–8 discuss five of the most effective advocacy strategies in turn: cultivate policy access, make it work locally, make it work for business, engage the heart, and be a game changer. Examples from all four of the East Asian polities will be used to illustrate and explain how these strategies work and why each is effective. The goal of these chapters is to provide lessons from East Asian advocates from which others can learn.

I do not have a chapter for the very important "educate" strategy, which involves environmental education at the grassroots level, as well as policy papers designed to educate at the elite level. The reason I have not dedicated a chapter to this strategy is not because it is unimportant, but rather because it has been extensively studied. Other scholars have already documented the critical role that environmental education plays in environmental advocacy.[48] Even in an East Asian context, there has already been considerable research on environmental education and its role in advocacy at the grassroots and elite levels.[49] In short, I did not believe that the strategy was neglected in the current literature, and I could not discern a distinct "lesson" from East Asia's environmentalists that would provide added value to my readers. As a result, although I discuss the "educate" strategy throughout the book in conjunction with other strategies, it does not have a separate chapter, even though it is a vitally important and effective strategy.

The concluding chapter of the book will bring all of these findings together and discuss their implications for advocates, policymakers, and scholars, as well as suggest some directions for further research.

2   Environmental Advocacy Strategies That Work

This chapter examines the full range of advocacy strategies found in the advocacy literature discussed in the previous chapter, offering an overview of environmental advocacy around the world. It will use data from an original dataset of environmental events from 2005 to 2009 to create a portrait of environmental activity during that period. The first section will present a brief description of the data gathered and the analysis performed. The second section will use these data to describe the issues that garnered the most activity, the locations where events occurred, and the advocacy strategies employed. The third section will delve into the key questions at the center of this book, examining which advocacy strategies were most effective and how their rates of utilization and effectiveness were affected by regime type or geographic region. The fourth section will dig a bit deeper to try to understand the ways in which regime type affects advocacy strategies, paying particularly close attention to how it may influence the incidence of violence. The chapter will conclude with some broader thoughts about how these strategies may be connected to one another to set the stage for the rest of the book.

Environmental Advocacy around the World

This chapter draws primarily on analyses of an original database of environmental advocacy events from around the world. To create the database, Factiva's major news and business publications[1] were searched for articles related to the environment that were published between January 1, 2005, and December 31, 2009. These five years were chosen because they are recent enough to be able to capture advocacy strategies used in contemporary environmental politics, the focus of this study, and they are old enough that there would be a good chance that the outcome (success or failure) of the advocacy could be determined. The search generated 3,567 relevant articles with 177 duplicates, for a final pool of 3,390 articles. Articles were then randomly selected until the dataset contained 200 cases of environmental advocacy.

The dependent variable in the analysis was success. An advocacy effort was coded as successful if the goal of the action as articulated by the advocates was achieved. Thus, a public protest to close a factory was coded as a success if the factory was closed; a local clean-the-river event was coded as a success if the river was cleaned. The outcome was coded as a failure if the advocacy did not result in the desired outcome. Using the prior examples, if the factory was not closed, or if the clean-the-river event was canceled, it would be coded as a failure. Success was coded as mixed if the advocacy effort was partially successful, such as if the factory was closed for a while but then reopened after some cleanup had occurred, or if the clean-the-river event was originally scheduled to clean three riverbanks but only cleaned one. If the outcome could not yet be determined—for example, if discussions about factory closure were ongoing, or if the clean-the-river event was rescheduled but had not yet occurred—then the outcome was coded as undetermined.

Although it is intuitively obvious that some advocacy goals are harder to achieve and more important than others, I found it impossible to develop a reliable measure of advocacy impact. While the impact and importance of a national carbon tax is clearly larger than a pop-up exhibit of eco-art, most cases in the dataset were somewhere in between these two extremes. Is the successful prevention of the construction of a single new petrochemical facility more important than the successful creation of a regional watershed management plan? The closest I could come to measuring the scale of the impact was to measure the scope of the advocacy.[2] Scope was measured according to whether the advocacy was directed at the local, regional, national, or global level. It was coded as local if the goal was specific to a particular community—for example, closing a local power plant or conducting a local river cleanup effort. The advocacy was coded as regional if it included multiple communities, such as a watershed protection effort. It was coded as national if the goal was nationwide, such as a new national regulatory standard. It was coded as global if it crossed national boundaries, such as a multinational effort to preserve international fisheries.

For most advocacy events in the dataset, the goals of the advocates were clear and quite specific, as in the foregoing examples of closing a factory or cleaning a local river. However, for a small subset of events (7 percent), the stated goals of the advocates were broad—for example, "improve understanding about climate issues." In those few cases, the event was coded as successful if the advocacy took place and people participated.

I recognize that this definition of success is very limited. In many cases, the ultimate goal of advocates is not so much to win a particular advocacy effort but rather to draw attention to an issue, shift the narrative frame to one that better favors advocates, or contribute to changing political culture. As Daniel Gillion persuasively argues in his *Political Power of Protest* (2013), the goal of protesters is to "set the political agenda and to focus governmental attention on a grievance that requires redress."[3] Thus, while this dataset codes for the "success" of a particular advocacy effort (e.g., halt construction on a new chemical-processing facility) based on whether the advocates achieved their stated objective in the specific advocacy event, the "real" goal of the advocates may have been more about setting political agendas and focusing government attention than about achieving a specific outcome in one particular event. Similar to protests, lawsuits also often have a clear, specific official goal (i.e., to win the lawsuit), but the "real" goal may be much broader and longer term.[4] For advocates employing both of these strategies (as well as others), the result of a particular action is much less important than these broader shifts. Because the time horizon of this study made it too difficult to code consistently for these longer-term effects, they are not included. This limitation of the study is discussed in greater detail in the concluding section of the chapter.

Each case was coded for the strategies that were present in the advocacy effort. I included all the strategies discussed in chapter 1 that are generally identified as effective in the social movement, policymaking, and East Asian literatures: protests, lawsuits, media campaigns, letter writing, lobbying, cultivating connections with policymakers, work for business, public education, policy papers, art, and local projects. Additionally, a number of control variables were also included: level of democracy, issue type, scope of advocacy, whether it was an event related to a NIMBY (not in my backyard) issue, and whether violence occurred at the event. Finally, I coded which actors were involved in the event (grassroots nongovernmental organizations [NGOs], businesses, government, international organizations), as well as which actors initiated the advocacy effort. For more details about how these variables were defined and coded, see appendix B.

This chapter focuses on identifying patterns in the types of environmental advocacy around the world and determining which strategies are the most effective. Additionally, I am interested in whether regime type or geographic region affects either the types of advocacy or its effectiveness. A number of statistical techniques were employed to carry out the investigation. In brief, descriptive statistics were used to determine the prevalence of different strategies and their relative success rates. Then, in order to gain greater analytic leverage on which strategies were more successful, I also employed ordinary least squares (OLS) regressions and recursive partitioning. More details about the statistical techniques used in the analyses can be found in appendix B. I will begin with the descriptive findings about strategy frequency and then present the statistical analyses aimed at determining efficacy.

Patterns of Environmental Advocacy around the World

Figure 2.1 shows that pollution remains the top environmental advocacy issue around the world, with more than 60 percent of all environmental advocacy events in the dataset, and more than half of all the cases in each region, having something to do with pollution. Figure 2.1 also shows that there is some regional variation—for example, while 76 percent of advocacy events in the Americas were related to climate change, only 36 percent of

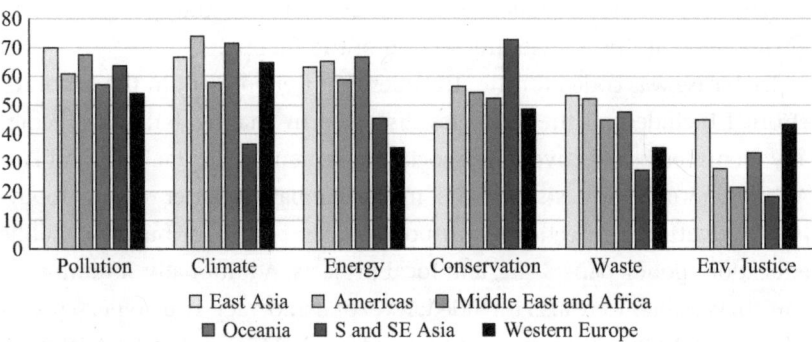

Figure 2.1
Advocacy issues around the world by issue and region, percent of events in each region addressing each advocacy issue.

those in East Asia were connected to that issue. However, overall, there was general consistency across the globe concerning issues, with pollution and climate issues garnering considerably more attention than environmental justice almost everywhere.

This relatively high level of consistency is not just found in terms of the issues that advocates engage; it is also true for the advocacy strategies that they employ, as illustrated in figure 2.2. Nearly everywhere, the most common advocacy strategies were working locally and cultivating a connection to policymakers, and the least common strategies were protests, art, and lawsuits.

The outliers in these data are the unusually high rates of protest and lawsuits in South and Southeast Asia. Although there may be something about that region that generates particularly high rates of protest and lawsuits (or at least particularly high rates of media coverage about protests and lawsuits), it is quite likely that the results are more a result of the relatively small number of overall cases in the dataset from that region. As discussed earlier and described in more detail in appendix B, the environmental events in this dataset were randomly selected. This means that they did not have an even distribution across the different regions. Figure 2.3 shows the number of environmental events in each region and their ratio within the

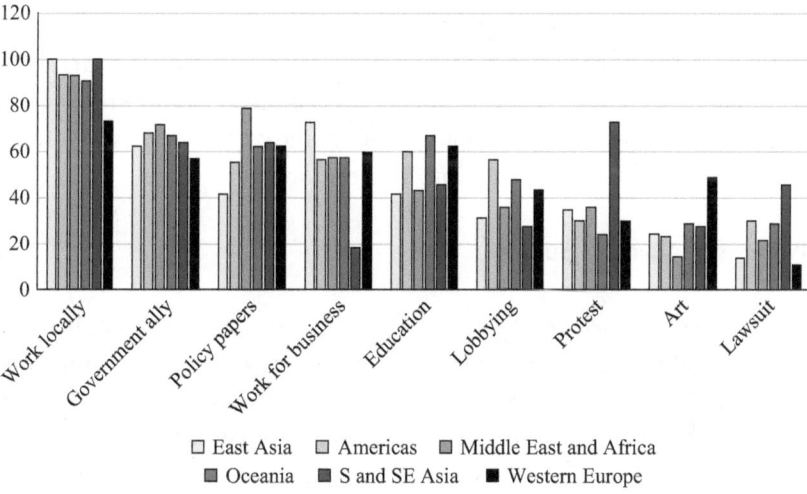

Figure 2.2
Strategy present rate by region, percent of events in each region that utilize each advocacy strategy.

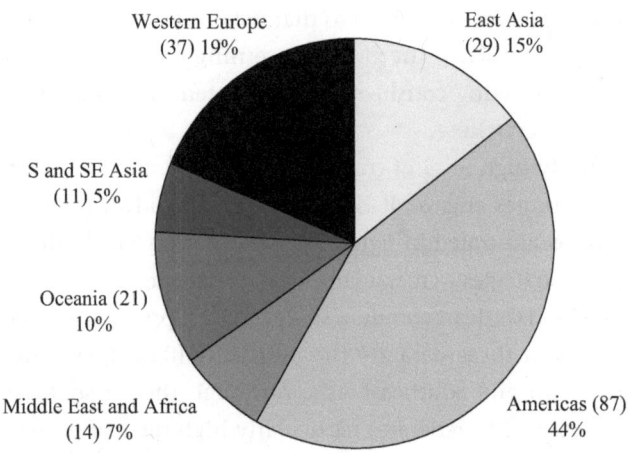

Figure 2.3
Distribution of cases by region.

two-hundred-case dataset. Thus, even though the Americas had more than three times as many protests as South and Southeast Asia during the time period (twenty-six compared with eight), because there were so many more cases coming from the Americas as compared with South and Southeast Asia (eighty-seven vs. eleven), the number of protests was a much smaller percentage of the whole. With a small total number of cases coming from the region, the differences of only a couple of cases can greatly affect the percentages (e.g., if the random sample had captured just three fewer cases of protest, the percentage of South and Southeast Asian cases involving protests would have dropped from 73 percent to 45 percent).

Patterns of Advocacy Success

Thus far, we have seen that the types of issues about which people become engaged and the advocacy strategies that they employ are remarkably similar everywhere in the world. What about efficacy? Are some strategies more effective in some places than in others? To investigate what factors affect the efficacy of different advocacy strategies, I'll utilize slightly more sophisticated statistical methodology.

First, a few more descriptive statistics. Not surprisingly, advocates will have an easier time being successful with some strategies than with others.

Achieving success for an environmental artist—who merely wants to engage the attention of the public and expose them to a different way of thinking and viewing the world—will be considerably easier than for an activist who has taken a giant multinational corporation to court for violating local environmental protection laws.

Figure 2.4 indicates the percentage of cases using each advocacy strategy that succeeded, failed, or had mixed outcomes. It is worth noting that every single strategy succeeded more often than it failed—even lawsuits, which had the highest failure rate of all the strategies, succeeded more often than they failed. Relatedly, for all of the more confrontational forms of advocacy— lobbying, policy papers, lawsuits, and public protests—the most common outcome was mixed, in which the advocates gained some, but not all, of what they wanted.

Figure 2.5 illustrates that success rates also varied by issue area. Once again, it becomes clear that most environmental advocacy, irrespective of issue area, is at least partially successful. Somewhat surprisingly, advocacy concerning waste issues, which can often be highly contentious NIMBY-type politics,[5] was remarkably successful, gaining outright successes 63 percent of the time. Not surprisingly, environmental justice issues, which frequently involve communities that are politically marginalized and are often threatening to ruling political elites,[6] were the least successful, although even these advocates experienced far more successful and mixed results than outright failures.

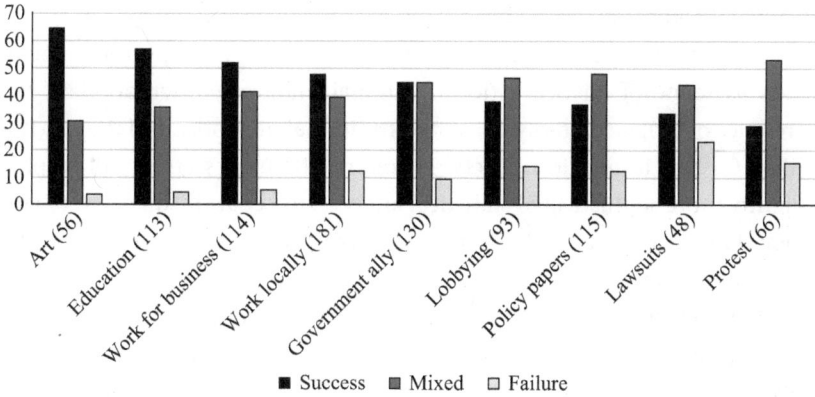

Figure 2.4
Advocacy outcomes (percentage) by strategy type (total cases in parentheses).

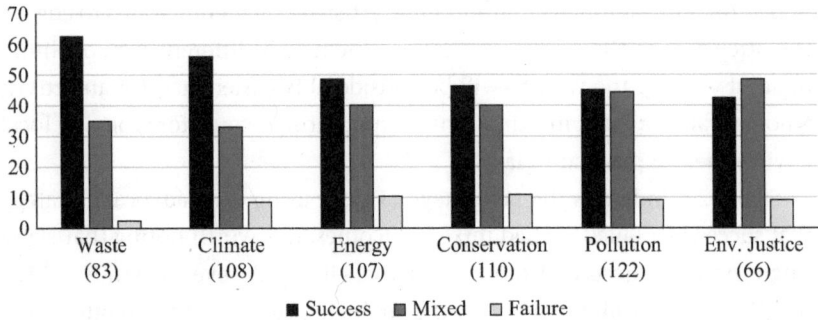

Figure 2.5
Success rates by issue area (total cases in parentheses).

These descriptive statistics help us understand some valuable information about advocacy. First, in contrast to the expectations of much of the literature, advocacy strategies do not appear to vary much by region. Second, in support of the expectations of the literature, more collaborative strategies tend to be more successful (at least when measured by the crude metric of "Did advocates get what they asked for?") than confrontational strategies. Finally, the literature is unclear about what the expectations are for overall success rates, but the numbers from this study strongly suggest that advocacy is frequently successful in achieving at least some gains. Advocates often win outright victories, gaining exactly what they ask for (more than 60 percent in terms of waste issues, and more than 40 percent for all of the other issues), and rates of complete failure are quite low— below 10 percent across all issue areas.

However, these descriptive statistics do not have any control variables such as scope of advocacy (local, regional, national, international), the presence of violence, start year, or organizational networks, and they do not allow for any interaction effects. Therefore, the next section will employ slightly more sophisticated statistical analyses to get at the question of efficacy. Table 2.1 presents the results of the OLS regression of the strategies and control variables.

These results show that two strategies, art and education, have statistically significant positive relationships with successful outcomes. Additionally, two issue areas are strongly related to the likelihood that an advocacy effort will be successful: climate change and waste. One strategy, protest, is negatively related to success, and the presence of violence is also negatively

Table 2.1
OLS regression of advocacy strategies and success.

	Estimate	Std. error	t-value	Pr $(>|t|)$
(Intercept)	−7.205	8.918	−0.808	0.420
Government ally	0.020	0.103	0.194	0.847
Work locally	−0.145	0.180	−0.804	0.423
Local network	0.163	0.129	1.264	0.208
International network	−0.095	0.107	−0.891	0.374
Probusiness	−0.071	0.106	−0.672	0.502
Art	**0.288**	**0.108**	**2.667**	**0.008**
Education	**0.313**	**0.096**	**3.257**	**0.001**
Policy paper	−0.178	0.111	−1.604	0.111
Protest	**−0.267**	**0.133**	**−2.011**	**0.046**
Lawsuit	−0.037	0.133	−0.278	0.781
Lobby	−0.115	0.102	−1.130	0.260
Letters	0.027	0.123	0.219	0.827
Media	−0.098	0.102	−0.963	0.337
Violence	**−0.428**	**0.192**	**−2.226**	**0.027**
Regime type	0.080	0.073	1.085	0.279
Scope	0.066	0.046	1.450	0.149
Start year	0.003	0.004	0.760	0.449
Conservation	0.043	0.100	0.425	0.671
Pollution	−0.028	0.094	−0.293	0.770
Energy	0.056	0.100	0.564	0.574
Development	−0.094	0.099	−0.953	0.342
Climate change	**0.182**	**0.100**	**1.817**	**0.071**
Waste	**0.326**	**0.097**	**3.374**	**0.001**
Environmental justice	0.016	0.103	0.159	0.874
Other	0.217	0.644	0.337	0.737

Note: $R^2 = 0.372$; adjusted $R^2 = 0.274$; $N= 192$. These results do not include the four cases with undetermined outcomes or the four cases where the regime type could not be coded either because the cases began before Freedom House coded the country or because the advocacy effort was transnational and involved countries with different Freedom House scores.

associated with advocacy success. Surprisingly, regime type did not have a statistically significant relationship with success. The same model was run replacing the aggregated "freedom" variable with separate measures for civic rights and political rights with no appreciable difference in the results.

The OLS regression tests all the variables as if they are operating independently, but since many advocacy efforts include the use of multiple strategies, I also used the random forest classification algorithm (CTree)[7] in order to gain greater insight into the conditions under which different strategies might be more effective, as well as their relationship to one another. The algorithm performs recursive partitioning of the dataset into successively more homogeneous groups, sequencing the variables hierarchically to predict successful and failed outcomes. Ideal groups would be entirely homogeneous. Figure 2.6 illustrates the CTree output for all of the predictor variables utilized in the foregoing OLS regression, using the 196 cases with known outcomes. Failed cases are indicated by a –1 value, mixed cases with a 0 value, and successful cases with a +1 value.

The recursive partitioning tree is much more informative than the OLS output. It indicates that the most important factor for determining advocacy outcome is scope: failure is much more likely when activists are advocating at a local level (scope $\leq$ 3). About half of activists working at the local level use policy papers as one of their advocacy strategies. If they combine policy papers with grassroots education, they are likely to gain at least partial success. However, if they use a policy paper and fail to engage in grassroots education, they are likely to fail in the end. Indeed, this is the only node—local advocacy, use of policy paper, no grassroots education—where failure is more likely than success. For every other node, success (full or partial) is more likely.

For advocacy conducted at the regional, national, or global level (scope > 3), the deciding factor is whether the advocacy is engaging with a waste-related issue. Nonwaste issues generated mixed results, while waste issues tended not to fail. How successful advocates were depended on their use of protest and grassroots education. Cases with protests did not fail but tended to generate more mixed resolutions than total victories. Cases without protests and without grassroots education efforts did better than those with protests, but advocates who combined a nonprotest strategy with grassroots education cases were far more successful than either of the other groups. Once again, regime type was not a significant predictor variable.

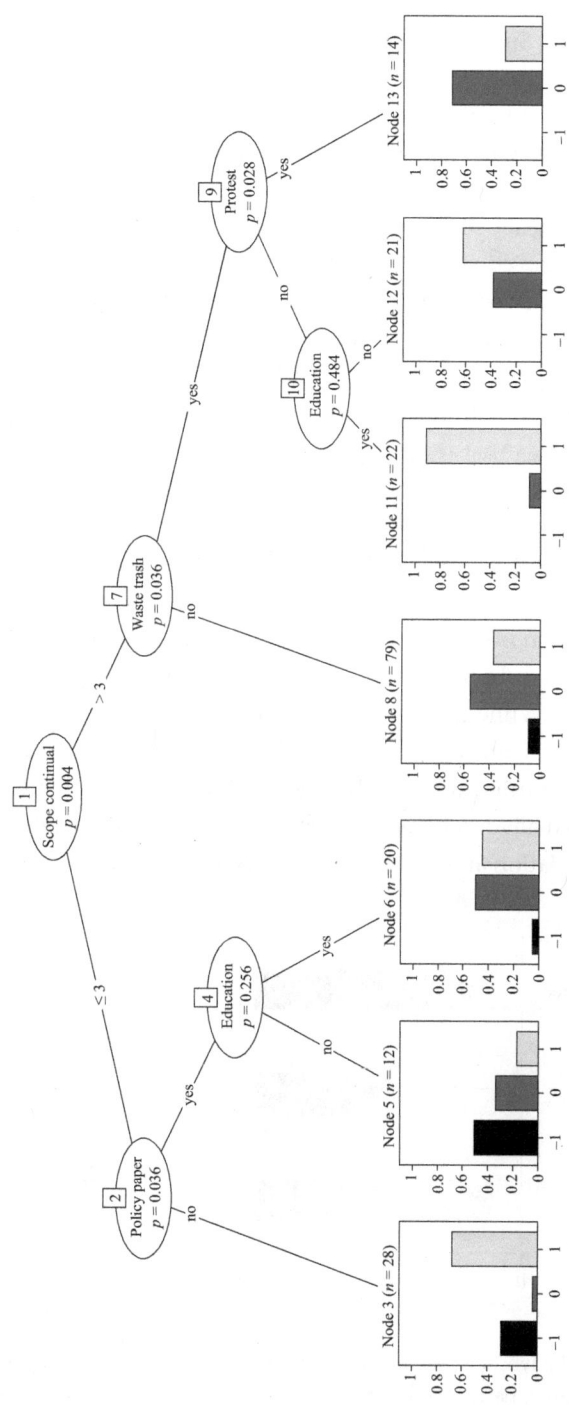

Figure 2.6

Recursive partitioning of advocacy strategy and context variables (-1 = failure, 0 = mixed, 1 = success).

Regime Type and Successful Advocacy

The descriptive statistics indicate that advocacy strategies do not vary much by region, and the OLS regressions and recursive partitioning lend further evidence to suggest that regime type is not a statistically significant factor determining advocacy success. Indeed, it appears that strategy selection and success are influenced primarily by local factors. These results support the findings of Russell Dalton and colleagues in their 2003 study of environmental organizations in which they wrote, "Within each nation, the environmental movement is so diverse that national-level opportunity structures have little influence on the participation patterns of environmental groups."[8]

And yet the idea that regime type does not matter for advocacy success goes against all instincts, scholarly and otherwise. Digging into the data a bit more, we can see that regime type does appear to affect the likelihood of success, even if it does not register as statistically significant when other factors are taken into account. Figure 2.7 illustrates the outcomes for the 192 cases in which both regime type and outcome could be determined.

The relationship between regime type and advocacy outcomes appears to be nonlinear. While the chances of complete success rise with the level of freedom in the country, the chances of failure are about the same in free and not free countries, with partly free countries having the greatest proportion of failed advocacy cases. When these data are broken down more, the complex relationship between regime type and advocacy outcome is

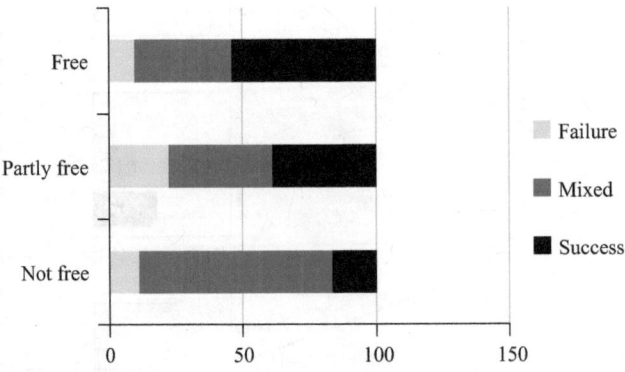

Figure 2.7
Regime type and advocacy outcome (percentage). Of the 192 cases, 156 were in free countries, 18 were in partly free countries, and 18 were in countries that are not free.

further illuminated. Figure 2.8 illustrates the percentage of advocacy outcomes according to Freedom House's political rights (e.g., right to vote, right to join a political party, the existence of competitive elections) and civil liberty (e.g., freedom of expression and belief, right to free association, rule of law) ratings (1 is most free, 7 is least free).

Civil liberty appears to have the expected linear relationship with success—the percentage of successful cases tends to rise as civil liberties rise. However, political rights do not appear to be related to success in any linear way—successful, mixed, and failed cases are almost randomly distributed. Thus, it appears that for environmental advocates, having free elections and active political parties matters much less than securing the rights of a free press and equal protection under the law.

Failure is more common in countries that fall in the midrange of both civil liberties and political rights than in countries on either the democracy or autocracy end of the spectrum. Mixed outcomes do not appear to follow any kind of pattern across regime types. These results suggest that political regime context affects advocacy outcomes in complex, nonlinear ways.

Digging further, the findings here suggest that while regime type may not affect success, it does appear to affect failure—in particular, the consequence of failure. This study was designed to examine the factors that help determine success and distinguish among success, mixed success, and

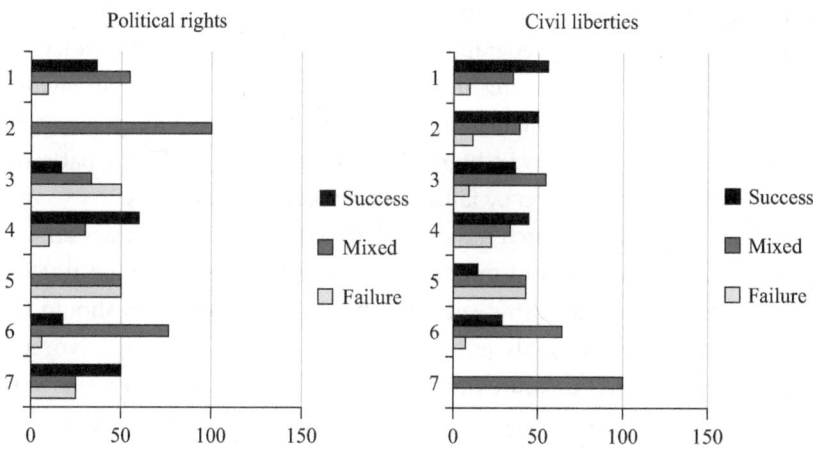

Figure 2.8
Advocacy outcome by political rights and civil liberty rating (percentage of cases).

failure. It was not designed to study failure, so there were not enough cases of complete failure to make nuanced distinctions among them.

However, if we look at the small subset of cases (thirteen of two hundred) in which there was violence associated with the environmental advocacy, there is a strong indication that while regime type might not affect advocacy success rates, it may have an important influence on the consequences of advocacy failure. In terms of overall numbers, environmental advocacy in undemocratic and partly democratic countries was disproportionately likely to involve violence: 22 percent of the environmental advocacy efforts in unfree countries and 11 percent of the cases in partly free countries turned violent, in contrast to only 4 percent of advocacy efforts in free countries.

Because of the relatively small number of failed cases in this dataset, it is impossible to draw definitive conclusions about the influence of regime type on the consequences of advocacy failure, but they do offer tentative support for other research that has studied violence and environmental advocacy. One study by Helen Poulos and Mary Alice Haddad (2016) suggests that grassroots environmental protests in democratic countries are much less likely to turn violent than those in nondemocratic countries.[9] Global Witness, a global NGO that tracks violence against environmental activists, found that 2015 was the most violent year on record, with nearly two hundred people being killed for trying to protect their communities against various forms of resource extraction. The Global Witness map of deaths highlights strong geographic and political patterns to the violence, with more violence occurring in undemocratic countries.[10] Thus, while regime type and geographic region may not help us understand patterns of advocacy success, they may be highly influential in explaining the patterns and consequences of advocacy failure, and violence in particular.

Finally, while this research has demonstrated that some advocacy strategies are more common and more effective than others, it should not be interpreted to mean that strategies that have low success rates should be abandoned. While the analyses here measured the scope of the advocacy effort, they could not measure the impact of the outcome. It may very well be that the types of advocacy that are less successful in terms of frequency, which was the measure here, are more important in terms of overall impact.

For example, the statistical evidence presented here suggests that legal advocacy and public protests are neither common nor particularly effective in gaining outright victories anywhere. However, it may be that one out-right victory can more than make up for the cases of defeat. Furthermore, as Austin Sarat and Stuart Scheingold (2006) have demonstrated in the US, Patricia Steinhoff and colleagues (2014) have shown in Japan for legal advo-cacy, and Daniel Gillion (2013) has demonstrated for protests, even when they fail or only partially succeed, these forms of advocacy can often serve important organizing and public-awareness-raising functions that can con-tribute to a broader advocacy goal even when they have failed to achieve the immediate goal. Therefore, this research should not be interpreted to mean that strategies with low success–failure ratios are ineffective. Rather, their frequency and effectiveness should be placed in an appropriate context.

Conclusion

The research presented here has three important findings: (1) Most envi-ronmental advocacy is successful. (2) Neither advocacy strategy choice nor strategy success rates vary significantly by region. And (3) regime type has a complex relationship with advocacy. These findings challenge several com-mon assumptions in social movement and public policy literatures.

First, to my surprise, most environmental advocacy efforts are successful. While the environmental advocacy efforts in this dataset were quite wide ranging—including activities as far removed from each other geographically and politically as a project to promote green hotels in Vermont, to violent antipollution protests in Henan, China, to environmental education efforts in Gambia and many in between—I did not expect success rates to be so high. My own daily reading of the news is full of doom and gloom about the devastation that climate change is causing and will wreak on the planet. These results clearly show that I have been missing a broader picture that includes a lot of positive action at the grassroots level in which individu-als, communities, businesses, and local and sometimes even national gov-ernments are actively engaged in numerous proenvironmental activities all around the world.

Second, these results challenge the assumption that advocacy strategies and their success rates are particularly linked to political culture. Certainly, the exact form that any given advocacy effort takes will be culturally rooted in the location where it is taking place. For example, while it might be common to have martial arts demonstrations as part of public protests in

Taiwan,[11] you would be very unlikely to see martial arts performed at a climate rally in Germany. However, although they may be engaging in slightly different implementations of the strategy, advocates in both places are using public protest as an advocacy strategy. Thus, the idea that somehow "Western" cultures are more likely to employ confrontational strategies such as filing lawsuits, organizing protests, and lobbying politicians,[12] while "Eastern" countries tend to employ more "embedded" collaborative strategies such as cultivating government allies, writing policy papers, and engaging in grassroots education,[13] is false. Advocates across the globe all employ a wide variety of advocacy strategies, and they do so at roughly similar rates.

Furthermore, successful strategies are similarly successful everywhere. Asking a former Ministry of Environment official to be on the board of your environmental NGO is as effective in Bonn and Buenos Aires as it is in Beijing. Cultivating government allies gives you policy access that is useful in any political context in which you live. Similarly, grassroots education that teaches children to appreciate and value the planet on which they live is as important in Bangalore as it is in Cincinnati. Advocates filing legal lawsuits against polluters and those taking to the streets to oppose the building of a new polluting facility in their neighborhood are going to face opposition and difficulties no matter where they are.

Finally, these results complicate the assumed relationship between regime type and political advocacy. While some scholars[14] have also found that regime type does not affect advocacy patterns, much of the literature on environmental politics assumes that regime type will affect the types and efficacy of advocacy efforts.[15] The findings here suggest that it is not the case that advocacy is more common or more generally successful in democratic states. Perhaps even more surprising, it does not appear that advocacy strategies vary by regime type—advocates in democratic countries are not, in fact, more likely to use lawsuits or protests to promote their environmental causes.

Additional analysis of the data has revealed that democracy has complex, nonlinear relationships with advocacy. When we separate out political and civil liberties, it appears that civil liberties are much more important in ensuring the ability of advocates to promote their causes utilizing safe, legal, nonviolent means. Advocates whose civil liberties were protected, who enjoyed freedom of the press, freedom of movement, and equal legal protection, had greater chances of success than those whose civil rights were not protected.

While the small number of outright failures in the dataset does not allow for a rigorous analysis, the distribution of violence among those cases leads to findings that are deeply troubling, although not particularly surprising: advocates in nondemocratic contexts who engage in failed advocacy are much, much more likely to face violence than their counterparts in democracies. This suggests that the factors that influence failure, and especially the consequence of failure, may not be the same factors that affect success. Success and failure may not lie on a continuum.

Rather, "mixed outcome" may be the equivalent of "0," and the dynamics that move an advocacy effort in the positive direction from 0 (mixed) to +1 (complete success) may be quite different from the factors that move an advocacy effort from 0 (mixed) to –1 (complete failure). Just as it is the case that the factors that lead to "failed states" in international relations are not merely the inverse of those that lead to "sustainable development," scholars of social movements and public policy should perhaps be thinking about success and failure as two rather different phenomena instead of as merely parts of the same continuum.

3   The Connected Stakeholder Model: How Advocates Influence Policy

Effective Advocacy has shown that advocates across the world are utilizing a remarkably similar set of strategies to promote proenvironmental policy and behavior across a diverse range of political contexts. My research has revealed that six strategies are particularly effective: make friends on the inside, make it work locally, make it work for business, engage the heart, educate, and be a game changer. At first glance, it may appear that these are a strange set of strategies. How do they fit together? Why would they work to generate positive change in so many different political contexts? In order to help explain why these strategies are so effective across such diverse cultural and political contexts, this chapter introduces a new conceptual framework for understanding the policymaking process.

In brief, the Connected Stakeholder Model (CSM) posits that the key to understanding policymaking is to recognize that stakeholders involved in the policymaking process are not individuals championing a single, institutionally determined interest in a battle of ideas. Rather, they are complex individuals who likely belong to several different institutions and have diverse interests. Their perspective on any given policy issue will be deeply influenced by the people to whom they are connected through a diverse set of personal and professional networks. Furthermore, these connections enable them to develop complex and nuanced ideas about the issues, which help inform their policymaking.

In this model, the networks—both formal and informal—are the most important. Their scale and diversity are what determines the content and quality of the policy outcomes. Policymaking in this model is not a game played by a set of individual players all trying to win, nor is it a contest between two teams seeking to triumph over each other. Rather, it is a process through which stakeholders with multiple interests are connected to one another through complex networks. These individuals and their networks then influence the people in positions of power who are making public policy. To lay the intellectual groundwork for the CSM, we return to the foundations of current advocacy and policy literatures.

Many current theories and models about policymaking are based on assumptions rooted in democratic political theory, especially the idea that policy is made when multiple stakeholders with differing interests compete. Perhaps most cogently articulated by Robert Dahl in *Polyarchy (1971)*, the basic assumption is that when relevant stakeholders promote their interests through a free and fair political process, policies that benefit the majority and, hopefully, protect the minority emerge. In this conceptualization of politics, there is a relatively clear distinction between the public and private spheres, where the essence of politics consists of actors in the latter sphere trying to influence those in the former.[1] Polyarchy cannot be maintained if one side dominates the other. If that occurs, the polyarchy, which helps to ensure a political process that will be beneficial to the public good, will dissolve into an autocracy, anarchy, oligarchy, or another political system where a small minority benefits at the cost of the majority, or the majority benefits to the detriment of the minority.

Of primary importance for ensuring that polyarchy does not devolve into one of the less optimal political systems are the institutions governing the behavior of political actors in the system. A commonly adopted definition for institutions is the one posited by Douglass North: "Institutions are the rules of the game in a society, or more formally, are the humanly devised constraints that shape human interaction. ... They are perfectly analogous to the rules of the game in a competitive team sport. That is, they consist of formal written rules as well as typically unwritten codes of conduct that underlie and supplement formal rules, such as not deliberately injuring a key player on the opposing team."[2] Given this commonly adopted definition of institutions, it is not surprising that pluralist-based models focus primarily on the constraining capacity of institutions and assume that the actors who are interacting within them are cooperative with teammates, competitive toward rivals, and neutral (or suspicious) of referees.

Moving from the realm of general theory to the concrete investigation of policymaking, John Kingdon articulates in more detail the relationships among advocates, policymakers, and the institutions that constrain them.

In his influential *Agendas, Alternatives, and Public Policies* (1984), he argues that there are three main streams in the policymaking process—problems, policies, and politics—and that they all flow simultaneously through any given political system. Furthermore, he suggests that the policymaking process can be conceptualized as occurring in four stages: setting the agenda, specifying the alternative policy options, deciding among the alternatives, and implementing that decision. Advocates enter the process at one or more of these stages and engage with all three streams. Political entrepreneurs take advantage of political opportunities and frame their desired outcome in ways that make it a policy solution for current problems decision makers are trying to solve.

From these general building blocks, scholars examining elite-level politics have tended to focus on the choices that individuals and organizations make in their efforts to obtain policy outcomes that maximize the activists' preferences. Some scholars focus on advocates' efforts to influence the policy agenda;[3] others examine alternative specification[4] or implementation and compliance.[5] Many scholars have studied the role of political entrepreneurs and the ways that they work to influence the policymaking process.[6]

Another group of scholars has focused less on the choices of individual actors, instead scrutinizing the institutional constraints that shape both the choices available and the process through which advocates must operate. Since institutional constraints vary considerably by level of government, these scholars tend to examine the institutional influences on policymaking at different levels of governance: local,[7] national,[8] or international.[9] At all levels of government, research has focused on strategies that are aimed at influencing the institutional environment in which elite actors operate, such as altering market incentives or directly influencing policymakers through lobbying.[10] Figure 3.1 offers a visualization of ideal-typical, pluralist-based, multistakeholder policymaking.

Assumptions of Pluralist-Based, Multistakeholder Policymaking Models

1. Policy actors are known. Influential policy actors are relatively few and can be clearly identified. They include politicians, bureaucrats, businesses, nongovernmental organization (NGO) activists, grassroots activists, and the various organizations that gather these actors together.

There are other actors who may be significantly involved in influencing policy outcomes (e.g., scientists, journalists), but they are generally
considered to be acting on behalf of the key actors, or merely providing
technical information, and are not usually viewed as independent actors
themselves.

2. Policy actors have narrow, hierarchically organized interests. This is not
   to claim that the actors have narrow interests generally (surely a Greenpeace activist cares about clean air and clean water and endangered species), but rather to say that in any given policy negotiation, each actor
   is focused on a narrow set of interests that are usually identifiable based
   on the actor's institutional role. Each actor is trying to maximize his or
   her interests in any given policy negotiation. This idea can be easily conceptualized using the commonly utilized term *stakeholder*: for any given
   policy negotiation, each actor has a single, identifiable "stake" for which
   he or she is fighting as a highest priority.

3. Actors participating in the policymaking process emerge largely because of
   their institutional roles. Good policy decision-making includes "multiple
   stakeholders" in the process in order to represent a wide range of society's
   interests and increase the opportunity to develop policy that is beneficial
   to the public good. When each stakeholder fights for his or her stake, multiple perspectives can be heard and an optimal policy can be developed.

4. Some of the actors in the policymaking process are more political than
   others. It is expected that business, advocacy NGOs, and citizen group
   actors will work hard to promote their own interests in the course of
   policy discussions. In contrast, bureaucrats are often portrayed as facilitators, keeping the peace among the competing interests, listening to
   their different viewpoints, and trying to develop a policy that offers the
   highest public good.[11] Similarly, academics are frequently asked to lend
   their technical expertise to policymaking, and may be supporting other
   key actors (e.g., a business or NGO), but are not generally considered to
   be independent political actors.

5. Policy is the outcome of competing interests. The policymaking process
   is fundamentally one where multiple actors promote different interests,
   and the policy outcome that emerges is the result of that competition.

6. The policy outcome that emerges from this process is rigid. Actors worked
   hard to incorporate their interests into the policy, so they will take steps

to ensure that the outcome is "locked in" and resistant to post hoc negotiations.

In this conceptualization of the policymaking process, any given policy advocate (in this illustration, one located in a citizen group) is connected to a single member of the policymaking process who is the stakeholder representing the interest of that person or group.

Building on the idea of multistakeholder policymaking, policy scholars recognized that those who are influencing the policy process are not just linked to decision makers; they are also connected to one another through a variety of networks. Policy network theory strove to understand how these networks operated and how they influenced policymaking. Early

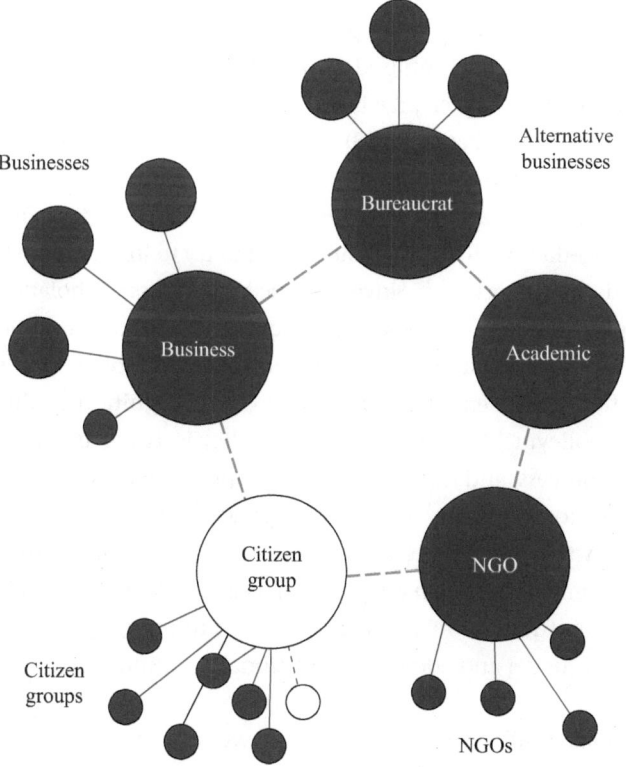

Figure 3.1
Ideal-typical, pluralist-based, multistakeholder model. The small white dot is the advocate, the large white dot is their policymaking representative, and the gray dots are others.

developers of the concept of a policy network tended to examine "iron triangles" and similarly narrow and fixed networks that helped explain why policy was slow to change and why certain actors were able to protect their interests.[12] Subsequent iterations of the theory sought to categorize different types of networks. Categories might vary according to their power relations, functions, or size.[13] Other typologies, such as the influential one developed by David Marsh and Roderick Rhodes Marsh (1992), separated "policy communities" from "issue networks" along several dimensions, such as type of interest, number of participants, and distribution of resources.[14] Moving beyond typologies, scholars such as Hugh Compston (2009) further specified the ties that connected different members inside a network (e.g., resource interdependencies) and articulated specific pathways through which changes in resources, preferences, or rules would then generate predictable changes in policies.[15]

One of the most widely utilized theoretical frameworks that uses policy networks to explain policy change is the Advocacy Coalition Framework (ACF) developed by Paul Sabatier (1988). First in an article and then in an edited volume, Sabatier and his colleagues have argued for a theoretical framework in which advocates are not working alone to change policy but rather are working together within a policy subsystem to try to influence policy change in their desired direction.[16] Since its conceptualization, scholars have fruitfully applied the idea of advocacy networks and advocacy coalitions to the study of environmental politics within and across national boundaries.[17]

The theoretical benefits of the ACF are its recognition of the dynamic nature of policymaking, its emphasis on policy learning as part of the policymaking process, and its inclusion of multiple types of actors coming from different sectors (government, NGO, corporate, academic). Because the ACF emphasizes the importance of coalitions of actors working together in a kind of "team," rather than assuming that each stakeholder is disconnected from one another, the ACF builds on the assumptions of the basic multistakeholder model and adds some additional assumptions.

Assumptions of Advocacy Coalition Framework Policymaking Models

The following assumptions are shared with the basic multistakeholder model:

1. Actors are known.
2. Actors have clearly identifiable, prioritized interests.

3. Actors emerge because of their institutional role.

The following assumptions are modified by ACF models:

4. While some actors are understood to be more political than others, it is recognized that no actor is neutral. Bureaucrats and academics are generally not neutral but rather part of one or another coalition and will support their team in negotiations.

5. Policy outcomes are the result of a competitive process. The ACF contends that there are different coalitions (teams) that are competing rather than individual stakeholders, but the policy process is still conceptualized as a competitive one in which one team will win and another lose.

6. Policy outcomes may be rigid or flexible. The rigidity of a policy is part of the negotiation process.

In this conceptualization of the policymaking process, any given policy advocate (in this illustration, one located in a citizen group) is connected to a few participants in the process. The advocate is connected to an advocacy coalition, which is connected to a few policymakers, and they will then represent the interests of the coalition in policy negotiations.

The basic multistakeholder model and the ACF both emerged from research based primarily on the policymaking processes of western Europe, North America, and international organizations that are headquartered in those regions. While some scholars have been very successful at applying these theoretical approaches to East Asian contexts,[18] others have found that many of the basic assumptions of the Western-based models are not valid for other regions of the world.

For example, many East Asian societies lack clear distinctions among state, societal, and business actors. Scholars examining politics in East Asia often use the term *embedded* to describe the close personal and institutional ties between nonstate organizations and the government and the informal mechanisms through which different actors interact.[19] Scholars of the Middle East, who also struggle with conceptualizing more complex public-private relations, utilize "state-in-society" approaches in order to capture the more porous and dynamic nature of state-society relations in that region.[20]

Similar to those employing the ACF, scholars of East Asia have also noted the importance of networks of advocates working together to promote environmental outcomes. Like the coalitions in the ACF, these networks contain individuals who come from different sectors. Unlike in the

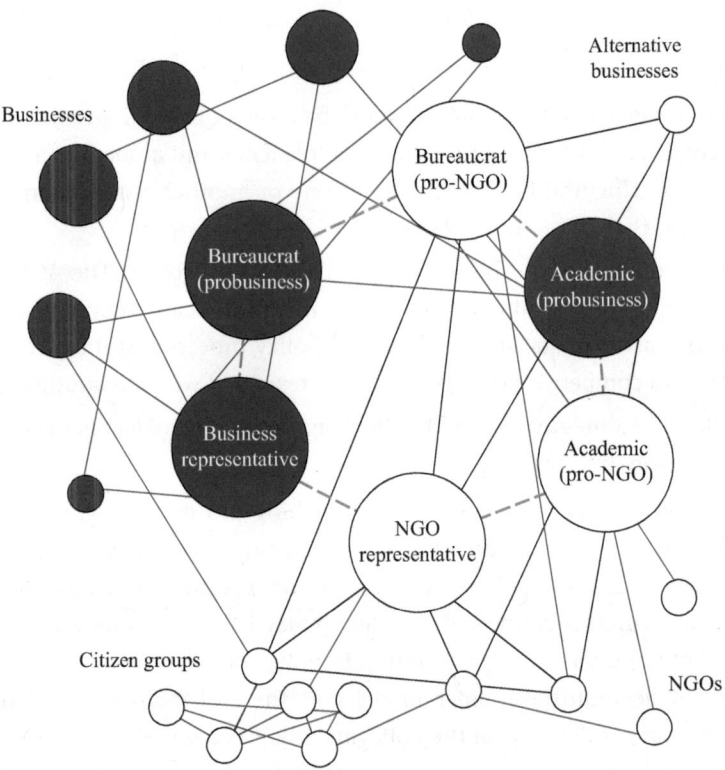

Figure 3.2
The Advocacy Coalition Framework (ACF). The white dots represent advocacy coalition members. The darker dots represent people not connected to the advocate's
coalition.

ACF conceptualization, actors in East Asian networks commonly belong to
multiple sectors simultaneously (e.g., a retired bureaucrat who sits on the
board of an NGO and is the founder of a business). This means that rather
than having narrow, hierarchically organized interests, East Asian policy
actors are usually assumed to have multiple, diverse interests simultaneously.[21] Furthermore, in East Asian scholarship, there is a strong emphasis
on the informal and personal nature of the networks rather than their formal institutional ties. The informal and personal nature of the networks
allows them to be highly flexible, able to change with altering political
circumstances, and also enables them to activate people outside any given

policy subsystem. Spouses, children, school friends, and others are important members of the networks, not just those who share a common belief or policy goal.[22]

As a result, scholars studying environmental politics and policymaking in East Asia have additional advocacy strategies that they investigate, many of which take place through some form of "embedded activism,"[23] where activists pursue their goals utilizing close personal ties with officials and informal channels between nonstate organizations and the government.[24] In addition to governmental and advocacy actors, this literature highlights the role of government-organized NGOs[25] and emphasizes the importance of informal personal connections to policymakers and informal institutional arrangements.[26]

As well as acknowledging the diversity of networks that individual policymakers and policy influencers have, scholars of East Asia also commonly recognize that "the state" cannot be taken as a uniform actor with a single perspective or set of interests. Indeed, identifying the diverse interests and intragovernmental dynamics that can be found within different divisions of central and national governments and between central governments and their counterparts in localities has been one of the region's most valuable contributions to broader scholarship about governance and politics outside the region. Concepts such as the "developmental state," pioneered by Chalmers Johnson's study of Japan,[27] and "fragmented authoritarianism," introduced by Kenneth Lieberthal and Michel Oksenberg in their study of China,[28] have been productively used by scholars for decades to understand politics inside and outside the region.[29]

While East Asian scholars recognize the importance of policy subsystems to bureaucratic politics, the ways in which they are used in ACF analyses frequently do not apply well to understanding policy advocacy in an East Asian context. Environmental (and other) advocacy organizations tend to be small, volunteer run, and involved in a diverse set of issues that do not fall into a single policy subsystem.[30] Furthermore, frequent job rotation is a common practice for East Asian bureaucracies, and civil servants are reassigned to a different division every two or three years in order to develop broad expertise within a given ministry or local government.[31] This means that while an advocate may form a deep connection with a particular civil servant in a policy subsystem of great interest to the advocate, it is almost guaranteed that any particular government official will experience a job rotation that removes him or her from the relevant policy subsystem before the policy of interest to the advocate comes to fruition. This means that the core ACF expectation that a relatively stable set of actors can work in a policy subsystem for a sustained period of time does not hold in most East Asian contexts.

In the East Asian context, networks are more commonly formed between people based on shared experience—they went to the same university, attended the same conference, worked on a common project—rather than because they are part of a single policy subsystem where actors have shared beliefs. Networks are maintained for social reasons and also because it is impossible to know ahead of time who in one's network will be in a position to be helpful at some future moment. One would never expect the members of a college football club to share a common set of beliefs. Nor would one necessarily expect everyone attending a conference to have similar beliefs—corporate exhibitors at a climate conference likely have different beliefs than municipal officials or NGO activists, even if they all find attending the conference useful. And yet, the shared experience offers the opportunity to make personal connections that can be tapped into later. As we will see, whereas ACF identifies shared beliefs as a fundamental source of strength to its advocacy coalitions, the CSM presented here posits that diversity of members and beliefs can contribute to the flexibility, strength, and effectiveness of the networks that advocates utilize to influence policy.

In Chinese there is a specific word to describe an individual's personal network—*guanxi*, commonly translated in English as "social network" or "connections." *Guanxi* is built as a kind of social capital by cultivating relationships with and doing favors for those in your network, with the expectation that those favors can be repaid at some future point in time. The exchange of favors need not be direct. A favor done by your old college roommate for your boss's nephew can be added to your store of *guanxi*. The concept is central to any understanding of Chinese culture and is frequently studied as an important factor in China's political economy. A common phrase about a political, economic, or social problem is, "She used *guanxi* to solve the problem," which means that she reached out to the people in her network to find people in their network who were in a position to solve her problem. The study of the use of *guanxi* has been important in the study of Chinese politics and, especially, business.[32]

Although we do not have an equivalent word in English, the concept is entirely common to American and European experiences. Individuals who are described as having "powerful friends" or who "know all the right people" are people with "a lot of *guanxi*." In earlier centuries, fraternity organizations, eating clubs, country clubs, and the like were designed to create and foster professionally relevant social networks. In the twenty-first century, companies such as LinkedIn and Facebook have used the digital mapping of these social and professional networks as the foundation of their multibillion-dollar business models.

One way that social scientists around the world have tried to map and measure these *guanxi* connections is through social network analysis.[33] Beginning in the field of sociology, social network analysis sought to reveal and describe the social ties that connected people and discover how these connections influenced individual behavior and social world.[34] The idea of network analysis then branched out to many fields, finding especially fertile ground in the computer engineering and life sciences. Furthermore, the rise of social media and the expansion of "big data" have combined with the development of sophisticated statistical techniques to enable not just the mapping of networks but also the testing of how various network features (e.g., centrality, connectivity) affect the behavior of people in the network.[35]

CSM borrows some of the concepts of social network theory when it describes the position of people within a policy-related network. In particular, some people act as "nodes" within a network, meaning multiple people are able to connect to others because of their relationship with the highly connected person. To the extent that these people are connected not just to a lot of people (picture the hub of a bicycle wheel) but also to multiple, diverse networks (picture the city hubs in a map of airline routes), they are more likely to have access to diverse perspectives and also be influential in the policymaking process.

The key point here is that for the CSM, policy-relevant networks are built through personal relationships made among individuals who may or may not share belief systems and may or may not be involved in the same policy subsystem. Unlike the ACF conceptualization, which has networks forming because of common beliefs and goals and bounded by policy subsystems that remain relatively stable over many years, in this model, networks are formed because of social, political, and economic relationships, and while the personal connections are maintained, their strength and relevance to any given policy subsystem shift over time. Thus, the CSM of policymaking combines the insights of the East Asian politics literature with policy models that were developed based on the North American and European experience to create a model that can help explain effective advocacy across the entire world.

Assumptions of the CSM (Network-Based, Multi-interest Policymaking)

1. Key actors are not always known. Many important policy actors are predictable and known, such as government bureaucrats, businesses, NGO activists, grassroots activists, and the various organizations that gather these actors together. Some actors, however, might be highly influential but less visible since they may not be obvious stakeholders. For example, journalists, academics, celebrities, and even artists can exert an independent influence on the policymaking process. An exposé news story, academic study, art installation, or documentary film can influence policymaking depending on the content and timing of the work. These contributions are not exogenous to the policymaking model but rather are incorporated into it through the influence of networks.

2. Actors have multiple, diverse interests that generally cannot be hierarchically organized. It will commonly be the case that key actors "wear many hats" simultaneously and have multiple connections across diverse sectors and institutions that they build and maintain (e.g., a former ministry official who is also the founding director of an influential NGO, or an NGO advocate whose husband is the president of an energy corporation, or an academic who is on the board of an NGO and serves on government advisory committees and runs his or her own for-profit consulting firm). Therefore, it is assumed that policy actors have multiple and diverse interests, and it will generally be impossible to infer the exact nature of the actor's interests based on his or her known institutional position.

3. Actors who are nodes in multiple networks will commonly find themselves in influential positions with respect to the policymaking process. Their role will not be primarily derived from their institutional roles but rather from their connections to diverse networks. Rather than the policymaking process being conceptualized as one in which individual stakeholders meet and compete for their stake, it is conceptualized as a group of individuals who are connected to diverse stakeholders coming together to discuss policy. Thus, the most influential people in the

process will likely be those who are nodes in diverse networks of stakeholders, not those who represent a particular stake.

4. All actors involved in policymaking are assumed to be political. Because
   all actors, whether they are businesspeople, NGO advocates, bureaucrats,
   or academics, will be approaching the policymaking process with an eye
   toward improving outcomes for members of their networks (and themselves), they will all be playing a political role in the process.

5. Policy is the outcome of personal negotiations among multiple actors
   with complex and diverse interests. It is not a competition. While policymaking frequently requires that there be people, causes, and organizations that "win" and those that "lose," with any specific policy, the
   conceptualization is more one of negotiation than competition, with
   an emphasis on areas of shared and multiple gains rather than outright
   victories.

6. Policy outcomes will be designed to be flexible. It is assumed by all
   participants that implementation will not proceed exactly as planned
   and that some actors will be ignored while some interests are under-
   represented. Because policymakers value building and maintaining their
   networks with one another over the outcome of any particular policy
   negotiation, policy outcomes will be crafted to be flexible to account for
   new knowledge and changing circumstances.

In sum, networks are the most important part of the policymaking
process—not the particular individuals at the table, not the institutions that
they come from, not the institutional constraints of the policy process. The
number, size, power, and diversity of the networks connected to the policymaking process will be the key to determining the policy outcome—its
shape and its efficacy. The people matter, but an actor's networks matter
more than his or her institutional role, technical knowledge, or financial
resources. When the people sit at the table to negotiate a policy, they will
be thinking about the technical details of the policy, but they will also be
seeking to benefit those connected to their entire network matrix. In theory, crafting policy that contributes to the collective improvement of the
network resources of those involved in policymaking will also contribute
positively to the public good.

The CSM posits that ad hoc advisory groups are a principal mode for
making policy. Actors will be invited to take part in these advisory groups

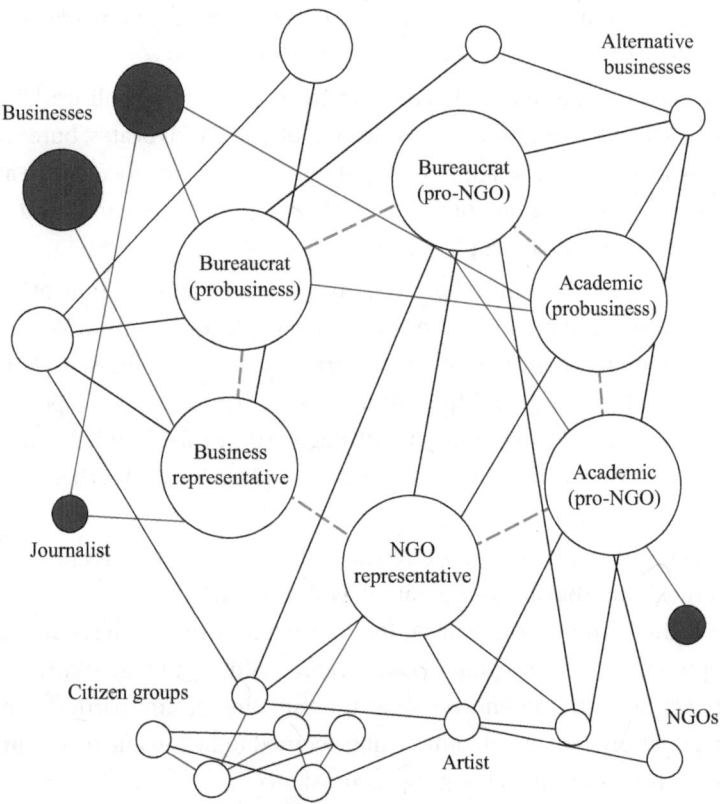

Figure 3.3
The Connected Stakeholder Model (CSM) (network-based, multi-interest policy-making). The white dots represent people connected to the advocate through their networks. The dark dots represent people not connected to the advocate's networks.

based primarily on their connections to relevant networks. Policymakers will seek out advisers who have numerous connections to multiple stakeholder communities in order to maximize their ability to understand the potential repercussions of policy on affected groups and improve their ability to craft creative solutions to public policy problems. Any individual who occupies a "seat at the table" will bring to the discussions diverse interests that cannot be hierarchically organized because they stem from multiple institutional affiliations, complex social relations, and varied life experience. Many of the actors are likely to know each other personally and have long-standing network-based connections to one another.

This model is fundamentally based on diverse networks of people who interact with one another, and others in their network, over long periods of time. People are valued for their access to and influence within a network, and growing and maintain their network is a common goal for all actors. To use Kingdon's language, these networks are involved in all stages of policymaking—setting the agenda, specifying policy alternatives, deciding among alternatives, and implementing decisions—although some parts of the network may be more involved during some stages than during others. Different parts of any individual's network may be activated at any given time, and it is likely that several networks will be active simultaneously as the actor tries to influence multiple policies at different stages in different policy subsystems.

CSM and Advocacy Strategies

How does this model relate to the strategies discussed in this book? The core questions of this book are the following: What advocacy strategies are most effective in creating behavioral change among governments, corporations, and individuals? And why are those strategies effective? The CSM offers a framework for understanding how very different types of strategies can combine to influence policymaking. In some cases the strategies are designed to influence policymakers directly—"make friends on the inside" is the most obvious case of this. When you cultivate a relationship with someone who is in a position of influence (or who you expect will or might be in a position of influence), then you gain policy access, enabling your ideas to be included in the policymaking process.

In other cases, the strategies are aimed at people who are closely connected to decision makers. These indirect advocacy efforts anticipate the likelihood that the people affected by the efforts will talk to not just one policymaker but several, so the advocate's ideas will already be known and understood by several policymakers when they sit down to discuss the policy. "Make it work locally" and "make it work for business" both fall into this category of influence. While it is unlikely that any given policymaker will have direct experience of any particular local pilot project or any particular green business product, information about local success stories will make its way through policymakers' networks to influence their decision-making.

Finally, some strategies are aimed at long-term, broad-based effects on the public, which in turn influence policymakers' networks. In some cases, such as art performances, short-term exhibits, documentaries published online, and public protests, they are intended to have immediate, intense influence on public opinion. For example, the Chinese documentary about air pollution *Under the Dome*[36] was published online on February 28, 2015, just ahead of the National People's Congress and the Chinese People's Political Consultative Conference meetings in Beijing that year. It had more than three hundred million views before it was censored in China. Responses to the film in the public were immediate and intense, with thousands of people pledging to change their personal behavior—for example, walk more, drive less—in response to their new awareness. Although it is difficult to connect policy changes directly to the film, many new regulations to address air pollution went into effect in 2016, a year after the film's release.[37]

In other cases, the advocacy efforts and cultural change occur over a longer period of time. Simon Avenell's work has demonstrated that environmental lawsuits in Japan, most of which failed in the courts, ultimately contributed to the development of the concept of "environmental rights," which entered not only Japanese jurisprudence but the rhetoric of activists and lawmakers alike.[38] Chapter 9 will show how the South Korean start-up tech company Tree Planet has promoted the idea that trees are companions, living creatures with individual characteristics who should be valued and cared for, rather than merely a decoration or a source of energy. Similarly, Japan's Cool Biz campaign, also discussed in chapter 9, has promoted cultural shifts in Japanese and Asian fashion, reframing a proenvironmental lifestyle from a sacrifice that individuals must make for the planet to a lifestyle that is creative, fun, healthy, and liberating. These "game changer" strategies have shifted cultural ideas in ways that promote better environmental behavior on the part of individuals, businesses, and the government.

This book explains how individuals and groups can utilize specific advocacy strategies to influence the people in their networks, which can ultimately influence policymakers. The CSM theorizes that policy-relevant networks that advocates create and utilize are not merely the sum of individual actors' efforts. I argue here that the networks themselves exert a somewhat independent, or, more precisely, interdependent, effect on the policymaking process. In particular, formal and informal networks help advocates and policymakers work around institutional collective action problems. As a result, the networks can have a catalytic effect, amplifying the effectiveness of advocacy efforts.

We can think of the networks affecting policymaking in a variety of ways, and I will focus on three here: information exchange, ally empowerment, and citizen engagement.[39] In all cases, the central feature of a network's value in influencing policy is the ways that it is able to reduce the collective action problems found in environmental policymaking and help network members identify and facilitate the development of win-win collaborations and effective policy.

Perhaps the most important role of networks in policymaking is the way they expedite the development and dissemination of policy-relevant knowledge. The environmental issue area is exceptionally complex, involving the full range of stakeholders from individual citizens to global corporations and everything in between. Given the scale and diversity of stakeholders, it is very difficult to identify the most policy-relevant knowledge and disseminate it to those who need and can use it. Additionally, institutional barriers frequently hinder information sharing both across sectoral lines (e.g., from the NGO sector to the corporate sector and vice versa) and inside organizations and governments; local governments with innovative solutions frequently have difficulty getting the attention of their central governments, and officials in one ministry may have difficulty collaborating with colleagues in a different ministry because of bureaucratic divisions of responsibility.

Networks—both formal and informal—enable people to share information across these institutional barriers. Networks help policy-relevant actors direct their resources in ways that can generate new policy-relevant knowledge, and once the new knowledge is generated, they can facilitate its dissemination to others. Many examples in the subsequent chapters illustrate how this can work. One of the best illustrations can be found in chapter 6, which describes how the KitaQ composting network helped develop, pilot, scale, and disseminate the innovative Takakura composting method, which has dramatically reduced municipal solid waste in dozens of cities in Southeast Asia and beyond. What started as the effort of the Japanese city of Kitakyushu to help its sister city of Surabaya in Indonesia deal with its trash grew into an elaborate effort that brought together local grassroots volunteer organizations, NGOs, and local and national governments in multiple countries.

A second, vital role that networks play in catalyzing advocacy is the way they can help advocates empower their allies. Cultivating and empowering allies is covered as a specific advocacy strategy in chapter 5, but it should be emphasized that the process of empowering allies is nearly always carried out through networks, and it is through networks that the other strategies can be combined with the strategy of empowering allies. Chapter 5 has several good examples that highlight the ways that advocates can use their networks to help their allies gain policy-relevant information in a timely manner, facilitate the development of allies' networks, and assist allies in overcoming bureaucratic and institutional barriers.

Examples of networks being used to empower allies can also be seen in chapters that are primarily focused on other strategies. Chapter 9, on game changers, discusses how then–environmental minister Yuriko Koike used her diplomatic connections to hold a Cool Biz fashion show of Asian political leaders. The success of the event helped raise the profile of her colleagues in other countries, enhancing Koike's *guanxi*, and contributed to the proenvironmental cultural shift toward wearing more comfortable clothes in the office during the summer, which leads to a reduced use of air conditioning.

Chapter 8, which focuses on art as an advocacy strategy, relates the story of the Daylily Art Circus, an event that brought joy and relief to victims of the 2011 triple disaster in Japan. Through its mobile circus, the artists used pop-up exhibits in unaffected areas in southern Japan to raise funds and spread awareness, compassion, and connection, which they then brought up north to the communities that were devastated by the environmental and manmade catastrophe. The art circus created formal and informal networks among individuals, artists, local governments, schools, and NGOs that facilitated compassionate connection and thus empowered not only their allies but also the victims of the disaster.

Finally, networks play an important role in citizen engagement. They do so in two ways: they can be used to engage citizens, and they are a way that engaged citizens can have their concerns heard by policymakers. As with the information catalyst and empower ally functions, the citizen engagement function can be combined with all of the advocacy strategies covered in this book. Chapter 7, about "make it work for business" strategies, and the "game changer" example of Ma Jun featured in chapter 9 both illustrate how making environmental performance more transparent can activate business supply chain networks, which can then engage and motivate suppliers, banks, and consumers to engage in more environmentally friendly behaviors.

All of the examples featured in chapter 6, focused on the "make it work locally" strategy, illustrate the vital role of networks in scaling citizen concerns. Whether it is blocking the building of new petrochemical plants or promoting policies to combat air pollution, grassroots activists without many direct connections to policymakers are able to use a combination of public campaigns, successful local pilot projects, and activation of their networks to pressure policy makers to enact proenvironmental policy change. Networks also help governments engage citizens in the implementation of proenvironmental policies and the dissemination of proenvironmental practices across a broader population. The Citizens' Green Seoul Committee demonstrates how networks can be used to engage and energize thousands of citizens, inspiring them to participate in green space-beautification campaigns, as well as recycling and energy-saving efforts.

Perhaps the most creative and fun examples of how networks can be used to facilitate citizen engagement are found in chapter 8, which focuses on the role of art. Whether they are the social media networks of individuals that helped *Under the Dome* go viral, or the community-based networks created by Ichi Ikeda's Moving Water project in Kagoshima, art can both create and utilize networks that have catalytic effects on citizen engagement. Citizens who had not thought about environmental issues before are exposed to the art through one of their networks and subsequently find themselves moved and inspired to act in proenvironmental ways.

In sum, the CSM offers a framework for understanding how advocates are able to persuade individuals, businesses, and governments to change their behaviors in proenvironmental ways. Individuals utilize a variety of strategies in conjunction with others in their networks to generate change. The advocacy strategies featured in this book—making friends on the inside, making it work locally, making it matter (with art), and being a game changer—are effective because they activate networks connected to policymakers. These networks function to develop and disseminate policy-relevant knowledge, empower allies, and engage citizens.

Benefits and Limitations of the Model

There are several benefits that the CSM of policymaking has over other models. First, it more accurately represents reality. Multistakeholder models assume that each stakeholder has a single stake for which he or she is fighting and that actors have little connection to one another except as competitors, which ignores the reality that policymakers and those who advise them are usually connected to one another and frequently have many stakes for which they are advocating. Similarly, while policy subsystems are a useful analytic tool, advocates and government officials are frequently engaging multiple subsystems at once, such that limiting analytic focus to a single subsystem is likely to miss much of the negotiation that may be occurring outside any given policy subsystem.

Second, conceptualizing policy actors as nodes in a set of interconnected networks containing multiple perspectives rather than institutional actors with defined interests dramatically broadens the possible policy outcomes and changes the scholar's conceptualization of the policymaking process itself. Stakeholders who are not "sitting at the table" can be heard. Actors with institutional affiliations that would ordinarily suggest hostility can easily become key allies. Policy outcomes have the potential to be much more creative and collaborative. Certainly, policy negotiations can still be expected to be contentious, but the process is not a game to be won. Conceptually framing policymaking as a collaborative process where individuals with different backgrounds, interests, and resources collectively craft policy opens up more possibilities for outside-the-room collaborations, informal resolutions, and flexible understandings among key actors.

Third, new actors become visible and relevant. While considerable research has investigated the role of scientists and other experts in the policymaking process, these experts are not usually conceptualized as political actors in their own right. Instead, at worst they are viewed as mere pawns of business, government, and NGOs, and at best they are seen as being required to work through those actors to affect policy.[40] Frequently, academics (and others) serve critical "network node" functions, maintaining connections to multiple civil society organizations and other actors. The CSM recognizes that when academics take part in policymaking, it is not just because of their technical expertise or as a pawn of another actor, but precisely because their multiple connections to diverse actors give them broad perspectives from which they may approach policy questions. Although they may not have the financial resources of business actors or the social or political resources of an NGO, academics and other experts have the potential to play powerfully important roles because of their ability to synthesize the views of so many and craft creative solutions that benefit multiple stakeholders simultaneously.[41]

Similarly, because the CSM allows for anyone connected to a network to influence policy, even those who might not be at the table can be recognized as playing an important role in the process. These "outside the room" actors might be politicians, retired government officials, powerful celebrities, journalists, or other powerbrokers who pull strings from outside to manipulate events "inside the room." In the CSM, these external manipulations, and the potential for them to affect policy, are not seen as exogenous to the policymaking process but rather are incorporated into it.[42]

Fourth, the role of institutions is reconceptualized. Rather than acting primarily as constraints on the policymaking process, institutions serve an important role in creating opportunities where multiple actors can meet and make connections with one another, facilitating the expansion of actor networks. In this way, institutions form a kind of structural framework for the creation of networks that influence the policymaking process. These institutional structures help give shape to the actors' networks and can act as focal points for action. However, since the networks reach beyond the institutions themselves and crisscross other relationships, it becomes fairly easy for actors to use alternative networks to work around any institutional barriers they might face. Indeed, as many of the stories in this book will illustrate, it is often the case that networks form primarily for the purpose of creating an informal mechanism to get around institutional barriers.

To reiterate, formal institutions are important and create important and identifiable opportunities and constraints for actors working to develop policy. The CSM expands this common view of institutions by calling attention to the ways that they can nurture the creation of personal networks that enable the development of informal mechanisms through which actors can overcome institutional barriers.[43] Additionally, the CSM helps identify how networks enable feedback gained from policy implementation to loop back into the policymaking process. Networks can link policymakers to policy implementers, even when the two groups of people are located in different institutions, when they have different interests or beliefs, or when they are in disparate geographic locations. Indeed, victims, clients, and those affected by the policy are also connected to networks linked to the policymaking process, making it easier to adjust and amend the policy even after it has been established.[44]

Finally, perhaps the largest intellectual benefit of the CSM is that it is not limited to democratic countries and does not assume democratic political processes. Actors create networks with each other and use those networks to inform themselves and influence policy in nearly every political system. They do so at all levels of policymaking from small villages to international organizations. The model does not assume that the networks are fully transparent, and it incorporates asymmetries of power. As with many other policymaking models, it contains a normative preference for more voices and perspectives to be heard during policymaking—it assumes that decision makers with the broadest understandings of the ramifications of policy for diverse constituencies will make the best decisions. It does not, however, require that those diverse perspectives come from people who are independent of one another or that they be interacting with one another in a democratic political context.

Although it has many benefits, the CSM also has several important limitations. While a benefit of the model is that it endogenizes more actors into it, it also complicates the identification of relevant actors, since they may include those who are not sitting at the table. Similarly, while the model assumes that the actors are acting rationally to grow and strengthen their networks and create policy that improves the conditions of those people and organizations to whom they are connected, it is not possible to derive an actor's interest directly from his or her institutional affiliation, nor is it possible to rank those interests hierarchically.

Thus, two core concepts in many policymaking models become more difficult to identify: the actors and their interests. It should be noted that the CSM places greater theoretical emphasis on the networks—their density, strength, and reach—such that precise identification of all of the actors and their interests becomes less important. What matters is identification of the networks that are connected to the policymaking discussions and the interests that are included in that network matrix. However, by deemphasizing the actors themselves, questions of accountability and identification of precisely who is responsible for which policy become even more difficult to answer.

A second limitation of the model is one that it shares with many other policymaking models—the model assumes a functioning state bureaucracy and a competent civil society. The CSM assumes that decisions made by policymakers will be implemented. It assumes that actors strive to act in the best interests of the multiple people and organizations to which they are connected. Further, it assumes that once policies are developed, government and private actors tasked with implementing the policy as intended will strive to do so.

However, many places in the world do not have sufficient capacity in either the state or society to implement policies that are made. When that happens, the networks serving to bring constituent interests to the attention of policymakers will devolve into pure patronage networks in which money and power rule. There may be a way to reconceptualize the model to account for the transformation of a symbiotic, public good–generating policymaking network into a parasitic, private good–generating network. Indeed, at a further stage of theoretical development, the model may offer great insight into how policy processes can be corrupted. At this point, however, the model is not applicable to low-capacity governments and societies.

A final limitation of the model, which is also found in other policymaking models but is exacerbated in the CSM, is the difficulty of identifying the beginning and end of a policy process. The CSM offers greater dynamism and flexibility in the policy inputs, since networks can change and multiply over time. Furthermore, since the networks continue to give feedback to the actors even while policy is being implementing, it is assumed that actors engage in post hoc negotiations and make adjustments to the policy. Thus, it becomes extremely difficult to determine when policymaking ends. The starting and end points of any scholar's inquiry into a particular policy thus become somewhat arbitrary.

4 Environmental Politics in East Asia

Citizen advocacy about environmental issues in East Asia emerged directly in response to the negative human health and economic consequences of rapid industrialization. Across the region, environmental activism began when people living near polluting facilities began to feel the effects of industrial pollution. Local residents organized to try to stop harmful pollution. Initially, governments and corporations were predictably resistant to change and were often violent in their attempts to coerce communities into accepting the costs of pollution in exchange for a variety of economic and other benefits. When residents refused to be bought off, government and corporate actors engaged in a wide spectrum of responses, ranging from violent suppression, coercion, and co-optation to compromise and even innovation.[1]

While initial movements in the region took the form of classic NIMBY (not in my backyard) struggles that dissipated once the pollution issues were resolved, over time they became more sophisticated. Activists found ways to work with local and national governmental officials, as well as corporations, to gain better environmental outcomes. They broadened the scope of the issues they focused on from strictly pollution to include recycling, conservation, climate change, and others. Now, activists across the region, whether they are private citizens concerned about the environment, nongovernmental organization (NGO) professionals, business entrepreneurs, or civil servants, are able to take advantage of the internet and social networking technologies to connect with others in their own country and abroad who are seeking positive environmental change.

Although regime type affects the resources available to activists, as the previous chapters demonstrated, it has not significantly changed the advocacy methods that advocates have employed. The biggest national-level differences in East Asia can be explained by the differences in the timing of when environmental issues became politically salient in each country. The timing of environmental movements in East Asia mattered in three ways. First, it mattered with respect to the political opportunities available in domestic politics—for example, was the ruling party vulnerable and willing to address environmental concerns? Second, timing mattered with respect to the political opportunities available in global politics—for example, was the environment on the global political agenda? Finally, it mattered in terms of sequence, because countries that experienced environmental movements later were able to learn from those who had gone before. The next section describes, in the order in which they occurred, the environmental movements of the four countries that serve as this book's focus: Japan, followed by Taiwan and South Korea, and most recently, China. This will be followed by a profile of environmental activism in the region today.

Environmental Movements in East Asia

Japan
Japan's environmental activism began at the turn of the twentieth century as a consequence of the dramatic industrial expansion initiated by Japan's Meiji Restoration. During the 1890s and early 1900s, village protests erupted first against copper mines: the Ashio mine in Tochigi Prefecture, Sumitomo's mine in Ehime Prefecture, and Hitachi's mine in Ibaraki Prefecture. In all three cases, sulfur gas pollution emitted from smokestacks, combined with heavy metals and acid released as wastewater into local streams, decimated the livelihoods of nearby farmers and fishers, causing serious health and livelihood problems for the local residents. Although the government and corporations initially tried to suppress and buy off the protesters, the plants eventually made significant investments to reduce pollution and compensate victims. Once the pollution was reduced and the victims' families were compensated, the protests died off.[2]

Environmental activism resurfaced again during the early postwar period. High-growth policies favored economic growth over environmental protection, resulting in toxic outcomes that threatened lives and livelihoods across the Japanese archipelago.[3] By the end of the 1960s, environmental pollution had become a major, national political issue. Once again the protests began as residents living near the industrial plants complained to the companies and local governments, and then they began taking more aggressive political action to pressure companies to clean up their plants.

Unlike in the prewar cases, which were eventually resolved through a combination of government pressure and corporate measures to placate residents, in the postwar cases residents were unsatisfied with government and company responses. Local residents also now had access to democratic processes—they could elect proenvironmental political leaders into public office and take corporate offenders to court. By the early 1970s, 40 percent of Japanese were living in areas with progressive chief executives, and by 1975 many of Japan's major cities, including Tokyo, Kyoto, and Osaka, had progressive mayors.[4] Joining counterparts in the United States and western Europe, pollution victims took to the streets to draw the attention of the public to their concerns. They also took the polluting corporations to court. To the surprise of many, environmental advocates won their lawsuits in what came to be known as the Big Four pollution cases, requiring significant and immediate action on the part of the companies, and giving elected officials a wake-up call about their responsibilities to keep people safe and healthy.[5]

The government reacted relatively quickly to the widespread and growing concerns about the environmental costs of its growth-first economic policies. Unlike its undemocratic, prewar predecessors, the postwar ruling party, the conservative Liberal Democratic Party, was sensitive to electoral pressure from opposition parties, and it reacted quickly to pass a sweeping array of environmental legislation in 1970 in what has come to be known as the Pollution Diet. Companies complied, and from that time forward, Japanese corporations, especially large ones, have tried to stay ahead of pollution issues to avoid the commercial consequences of a negative corporate image, as well as prevent additional intrusive government regulation.[6]

The Japanese case demonstrates that while the timing of an environmental crisis can put political pressure on ruling parties, it does not always mean the departure of those parties from power. In both the early 1900s and 1970, Japan's ruling party and major corporations were able to mollify the public, finding policies that enabled them to address environmental health issues without having to give up political power.

After a few decades off the global political agenda, the environment has once again risen to the top of international political discussions. A combination of the end of the Cold War, political instability in the Middle East, rising energy needs, growing concern about energy security, and rising scientific evidence about the multitude of threats related to climate change has brought the environment to international and national political agendas around the world. Citizens have also developed very sophisticated methods of networking with each other within and across borders, creating additional pressure on political and corporate decision makers.

The contemporary form of environmental politics in Japan can be traced to the events leading up to the Third Conference of the Parties of the United Nations Framework Convention on Climate Change, held in Kyoto in November 1997. Before the conference, Japanese environmental activists created the Kiko Forum [Environmental Forum] in Kyoto to connect local environmental organizations to each other and to groups from around the world. The successful creation of the Kyoto Protocol by participating governments was mirrored in the successful solidification of the global NGO community through the actions of the Kiko Forum and others.[7]

These networks grew dramatically as internet technology spread. In 2000 there were fewer than fifty thousand internet users in Japan (about one third of the population), and that number had more than doubled by 2010.[8] By 2016, internet usage exceeded 90 percent for Japanese between 13 and 69 years of age. Examples of environmental networks in the country include the Kiko Network (the new name for the Kiko Forum), the Japan River Restoration Network, and the Climate Action Network Japan, to name a few. Commercial networks related to the environment also proliferated—for example, the Japan Green Purchasing Network, Eco-Networks, and GreenBizJapan. These groups usually have connections to international NGOs, but large international environmental organizations, such as Greenpeace, the Nature Conservancy, and World Wildlife Fund, do not generally have offices in Japan, or their offices are a fraction of the size of those found in other countries.[9]

In the aftermath of the triple disaster of 2011, which included a devastating earthquake, a tsunami, and a nuclear disaster that took the lives of nearly twenty thousand people in the northeast region of Japan, these networks of organizations were critical in providing immediate assistance to those in need in disaster areas. They also helped focus the Japanese public on the importance of a wide variety of environmental concerns, especially those related to energy, conservation, and community resilience.[10]

Environmental politics reached a fever pitch following the disaster, with all segments of society getting involved. Japanese government officials have been working with global organizations and governments to improve international nuclear safety protocols and standards, disaster management, and eco-city development.[11] The corporate sector has also responded with innovative plans. For example, within weeks of the disaster, billionaire Masayoshi Son gained the cooperation of nearly all of Japan's governors for a bold new plan that would dramatically enhance Japan's renewable power infrastructure and expand renewable energy across the region.[12]

Private citizens and civic groups have also formed innovative groups. Safecast established an interactive map just days after the disaster that serves as a crowdsourced platform where individuals from around the world upload and view radiation readings from individual Geiger counters in an effort to offer a nongovernmental source of radiation information to the public. Networks of antinuclear groups such Sayonara Nukes have taken advantage of public outrage and new social networking technology to organize regular, simultaneous protests all over Japan, with the numbers of participants often reaching into the tens of thousands.[13]

It is now nearly ten years after the disaster, and environmental advocacy efforts have become more focused. Public protests about nuclear power plants have died down, No Nukes Japan stopped mapping protest events in 2012, and nine of Japan's forty-two operational reactors have been restarted.[14] Japan liberalized its electricity market in 2016 and is moving rapidly toward the expansion of solar, wind, and hydrogen as sources of energy for the country.[15] Many cities and NGOs have been active in efforts to curb greenhouse gas emissions—the city of Tokyo reduced its energy consumption more than 20 percent between 2000 and 2015, with the industrial and transportation sectors making astounding 41 percent and 42 percent reductions, respectively.[16]

Contemporary environmental politics in Japan is sophisticated and complex. Millions of ordinary citizens are working through local organizations to improve their local environments. Hundreds of small organizations help link Japanese citizens with local policymakers and with international environmental organizations around the world. Dozens of groups work closely with counterparts around the world and with high-level government officials in Japan. However, for the most part, environmental politics in Japan, as in the rest of East Asia, remains somewhat bifurcated. At an elite level, there are a small group of technocrats from the government, civil society, and business working on high-level issues of national and international

environmental policy. At the grassroots level, millions of Japanese are working in local groups to improve the environment in their communities. Both
groups are making significant headway, but many challenges remain, and
local efforts are often disconnected from national ones.

Taiwan and South Korea

As was the case in Japan, South Korean and Taiwanese environmental
organizations began as community protests against instances of local environmental pollution. Rapid industrialization under a developmental state
prioritizing economic growth had the expected result of high levels of pollution. When farmers and fishers found their livelihoods threatened by contaminated soil and water, and when residents found themselves and their
loved ones getting sick at unusually high rates, they began to protest.

The beginning of South Korean and Taiwanese environmental movements strongly resembled the early stages of environmental organizing in
Japan. Disgruntled and largely disempowered residents appealed to their
local government officials and directly to the factories themselves. Usually
they found a sympathetic ear, but their complaints were largely met with a
variety of appeasement measures that were intended to make the political
problem go away for the local elites but not take significant steps to address
the pollution issues.

Of critical difference in the South Korean and Taiwanese cases as compared with the Japan one was their timing. The timing of their movements
mattered in all three dimensions mentioned earlier. The domestic political
context at the time of the emergence of environmental movements meant
that they became inexorably entangled with democratization movements
and strongly linked with liberal political parties. The international context
at the time meant that groups could take advantage of the rising global concern with environmental issues, the political openings brought about by
the end of the Cold War, and the increasing international support for both
environmental activism and third-wave democratization. Finally, because
South Korean and Taiwanese environmental movements followed those in
the United States, Europe, and Japan, they were able to learn tactical and
strategic lessons about grassroots organizing and political advocacy from
those who had gone before.

Taiwan   Taiwan's first environmental protesters were victims of industrial
pollution: farmers, fishers, and local residents who had their health and

livelihoods threatened by nearby industrial production. From 1980 to 1987, 97 percent of environmental protests were reactive—victims seeking redress against damage that had already occurred. Examples include a lawsuit in 1981, when villagers in Hua-t'an Village demanded compensation from local brick manufacturers for damage to their nearby rice paddies. They eventually won NT$1.5 million (US$375,000).[17]

In a context where newspapers were increasingly reporting on local pollution issues, Taiwanese citizens were also affected by international events. Although not as geographically widespread as the Chernobyl explosion that would follow two years later, in 1984 a disastrous gas leak from a Union Carbide factory in Bhopal, India, killed two thousand people and made major headlines across the globe. Taiwanese were already feeling sensitive to the damaging effects of chemical pollution in their own communities, and they immediately recognized that they were similarly vulnerable to such a disaster. The very next year, protests and threats of violence against pesticide companies in Hsin-chu and T'ai-chung forced the closing of both factories for cleanup.[18]

The largest turning point for Taiwan's environmental movement occurred in 1986, when residents in the port town of Lukang began a protest against a planned titanium dioxide plant. The protest represented a shift away from a strategy of reactive protests against damage already done to proactive political action aimed at preventing damage that had not yet occurred. According to James Reardon-Anderson (1997), if the protest had happened earlier, it would have been crushed; if it had happened later, it would not have been noticed. "But it came just at the time when environmental consciousness in Taiwan had reached a critical mass and as the government was introducing political reforms that gave unprecedented scope to new forms of civic action." In that context, a small group of determined local activists "focused the attention of the entire island on this sleepy provincial town, raised the national consciousness about threats to the natural environment, and challenged the rules that government officials and industrial leaders in Taiwan had come to take for granted."[19]

All of these activities coincided with the rise of the middle class and the beginning of the third wave of democratization that would sweep East Asia and the world in the next decade. Taiwan ended martial law in 1987, and a number of groups that would lead Taiwan's environmental and democratization movement, such as the New Environment Foundation,

Taiwan Greenpeace, the Taiwan Environmental Protection Union, and the Homemakers' Union Environmental Protection Foundation, were founded shortly thereafter.

As was the case in Japan, in Taiwan, the fact that early environmental movements had emerged to fight local pollution problems made it difficult for them to grow beyond NIMBY protests. Groups tended to focus primarily on local pollution issues and would disband once a particular battle was over.[20] Throughout the 1990s the Taiwanese movement continued to grow, and political reform spread such that by the end of the decade, citizens were well placed to hold their governments accountable for poor decisions in the wake of the Asian (and global) economic crisis of 1997. In 2000 Taiwanese ousted the Kuomintang, which had ruled the island for nearly forty years, electing native-born Chen Shui-bian of the Democratic Progressive Party (DPP). Because environmental organizations and their leaders were closely linked with liberal political parties, activists enjoyed considerable access to policymaking during the eight years the DPP was in power.

Even though they had access, however, many activists felt as if their interests were pushed aside in favor of big business once the parties they had supported gained power. As one Taiwanese activist phrased it to me in 2010, "The DPP changed when it took power. When it got into power it didn't like the environment anymore." Another stated, "They want the votes, but they don't want to hear the voices."[21] However, even their limited access was severely curtailed when voters returned conservatives to power, electing Ma Ying-jeou of the Kuomintang in 2008. When the global financial crisis hit at the end of the year, Ma was quick to put "green growth" at the top of his strategies for economic recovery, making significant public investments to promote green technology and green industry related to the information technology industry, renewable energy, eco-tourism, and the like.

Although the conservative governments and businesses were promoting environmentally friendly economic development, environmental activists often felt completely shut out. Several suggested that the election brought back the very same people who were in charge under the military government. One Taiwanese activist was very blunt in assessing the situation for me in 2010: "[In Taiwan] corporations are a shadow government. Our government is their puppet."[22]

Responding to what they perceive as a lack of effectiveness and access, Taiwanese activists are beginning to shift their tactics away from protests and partisan politics. Environmental leaders talked to me about how the public and the policymakers have become anesthetized to public protests, such that they are no longer effective as a mode of advocacy. Instead, scientific and policy reports that give policymakers new information about an environmental problem and create an opportunity for dialogue about solutions appeared to be more effective.[23] Even after the DPP's return to power in 2016, Taiwan's environmental politics may be moving closer to the model found on the Chinese mainland, where advocacy has been aimed at working with the government rather than against it. It also exhibits the same bifurcation of environmental advocacy found in Japan: many small, local grassroots groups working on a volunteer basis to improve their communities and a few elite organizations working to influence policy, but not much in between.

South Korea   Of the four countries under examination in this book, South Korea is an outlier because unlike the other three countries, which have either no professional environmental advocacy organizations or only very small ones, South Korea boasts the largest in the region: the Korean Federation for Environmental Movements (KFEM) has more than fifty local branches and nearly one hundred thousand members, and it is deeply involved in local and national politics.[24] The highly politicized and well-organized nature of environmental politics in South Korea makes it somewhat distinct from the environmental politics found in the other three countries.

South Korea's environmental movement has grown out of and has contributed to a long history of protest politics. The country's tradition of mass protests can be traced back more than a century, beginning with the 1894 Donghak peasant movement protesting government corruption. Other famous mass movements in South Korea include the March First Movement protesting Japanese rule in 1919; the April Revolution, when student and labor organizations successfully ended the autocratic rule of Syngman Rhee in 1960; the failed student-led prodemocracy protests in May 1980; and the successful democracy movement of June 1987. This long history helped normalize protest, even violent protest, as a regular method through which civil society organizations would engage the state. Although considerably less violent than their predecessors, contemporary South Korean civic organizations, of which environmental organizations are one group, continue to favor confrontational modes of political engagement.[25]

Scholar and environmental activist See-Jae Lee argues that South Korea's environmental movement has passed through four stages: negation (1960s and 1970s), where neither the state nor civil society viewed the other side as legitimate; resistance (1980s), where the state acknowledged the existence of civic actors but did not view them as partners for dialogue and sought to suppress them; negotiation (1990s), where both sides recognized each other and struggled against each other for policy influence and public support; and participation (2000–present), where environmental organizations are incorporated into the state's decision-making process and participate in jointly developed policy projects.[26]

Early environmental organizations in South Korea were generally located in churches and universities and were focused primarily on raising environmental consciousness among the population.[27] However, as was the case for Japan, Taiwan, and now also China, high-speed growth policies led to rapidly deteriorating environmental conditions that threatened human health and livelihoods. Early environmental protest movements in South Korea, like their counterparts elsewhere in the region, began with residents' demands for compensation for damages caused by industrial pollution. The first major case to generate national attention was in the Ulsan and Onsan areas of Gyeongsangnam-do Province. Construction of the government-approved Ulsan Industrial Complex began in 1962, and in 1967 farmers began to demand compensation for agricultural losses. In 1971 they formed a pollution countermeasures committee, and 1978 they increased their demands to include financial aid for relocation, as well as compensation for residents experiencing health damages. Urban residents in large cities such as Seoul and Inch'on also began to protest against noise and air pollution.[28]

Responding to growing public concern, a small number of environmentalists, religious leaders, and prodemocracy activists formed the Pollution Research Institute (PRI) in 1982 in order to conduct pollution-related research independent of the government.[29] In contrast to the local NIMBY protesters, who were usually farmers, fishers, and local residents living near factories, the PRI had connections to national church leaders, leading academics, and student groups. It was the first professional environmental organization in South Korea with dedicated staff and office space.[30] Because of its close connection with student groups involved in the democracy movement, the antipollution movement was also seen as an antigovernment movement, so the government tried to prevent connections between local grassroots NIMBY groups and the professional environmental organization.[31]

In 1983 heavy metal pollution made the water near the Ulsan complex so toxic that the government suspended fishing rights. Immediately the people took to the streets to demand financial compensation for lost revenue, and the PRI began an independent investigation into the situation. In 1985 it released a very detailed report that claimed that more than five hundred people in Onsan suffered from cadmium contamination. The press offered considerable public exposure to the findings, and the issue became one of national interest. Soon after the PRI report, the Environment Administration conducted its own tests and reported that the illness spreading among the Onsan population was not a pollution-related disease. Residents and the PRI refuted the official test results and engaged in a series of public protests. Eventually, the government was forced to concede to the growing pressure from environmental groups and the public and resettle about forty thousand residents to new areas.[32]

Although martial law was lifted in 1981, it was not until the creation of a new constitution that guaranteed basic civil rights in 1987 that the legal protections needed for the creation of environmental organizations were established. Soon afterward key organizations, such as the Citizens' Alliance for Economic Justice, Green Korea United, and the KFEM were established and became important organizations for the combined environmental and democratization movements.

Throughout the 2000s, South Korean environmental organizations expanded their membership and broadened the scope of their issues, moving beyond merely responding to situations of environmental degradation to initiating preventive campaigns. One such campaign, led by the KFEM, was the anti-Doggang Dam campaign, which aimed to prevent the construction of a new dam and protect the natural ecosystem of the river. The campaign had a sophisticated strategy that included promoting tourism to the river. It also utilized cultural symbols, such as the annual Jeongseon Arirang festival, and ecological symbols, displaying rare species of fish and otters that live in and along the river. In 1999 President Kim Dae-jung's government formed a citizen-government joint investigation panel to research the dam, and a year later the committee recommended that construction be canceled. On Environment Day, June 5, 2000, Kim announced the New Millennium Vision for the Environment and pledged to repeal construction plans.[33]

In the 2000s, after receiving a considerable political boost with the victory of Kim in the 1997 presidential elections, South Korea's environmental groups grew increasingly politically involved and enjoyed closer connections to the government, serving on joint panels and being appointed to key government positions. They joined several other NGOs in forming the Citizens' Alliance for the 2000 General Elections, which was a protransparency, anticorruption campaign that blacklisted candidates with records of political corruption or illegal activities and succeeded in defeating fifty-six of the eighty-nine candidates it targeted. President Roh Moo-hyun of the Millennium Democratic Party pledged to create a "participatory democracy" in South Korea, appointing many civil society activists to government positions and expanding NGO participation in policymaking.[34]

Just as had been the case in Taiwan, in South Korea, the liberal renaissance had ended by the end of the decade, bringing the environmental movement's access to policymaking to a wrenching end. In late 2007 South Korean voters elected Lee Myung-bak of the conservative Grand National Party, and he assumed his post as president in February 2008. Once again paralleling the events in Taiwan, when the global financial crisis hit at the end of the year, Lee promoted "green growth" as a key component of a strategy for economic recovery. The Four Rivers Project, a massive plan to reengineer the country's major rivers, became the core of South Korea's Green New Deal, an economic stimulus package that pledged $40 billion (equivalent to 4 percent of total gross domestic product [GDP]) for four years to promote sustainable economic growth.[35]

Environmental activists in South Korea faced the same problem as their counterparts in Taiwan. Although the conservative government and businesses were publicly promoting environmentally friendly economic development, activists felt completely shut out. As one South Korean activist explained it to me in an interview in 2010, "The ex-government valued governance and wanted to hear civil society. It didn't decide everything on its own but always had channels with civil society. This government doesn't."

However, in contrast to the Taiwanese activists who are shifting strategies and trying to find more ways to work with the government, South Korean activists are returning to the protest repertoires of the past. In April 2008, fewer than two months after the new conservative government assumed office, it reversed the decision of the previous liberal administration, re-allowing the import of US beef, which had been banned since 2003 after evidence of bovine spongiform encephalopathy (BSE, or mad cow disease) had been identified in the US. Playing on food safety fears spurred by BSE and the concurrent tainted milk scandal in China, building on rising anti-US nationalist sentiment among the youth, and capitalizing on national discontent with the new conservative leadership, the KFEM joined other environmental and social groups to support public protests against the national government. The protests, which came to be called the Candlelight Protests because they were usually held at night and protesters brought candles, grew to be national in scope and attracted hundreds of thousands of protesters from late May through August. Interestingly, although Taiwan experienced a similar reversal in its ban on US beef, its protests remained isolated and small, failing to grow as they did in South Korea.[36]

The Candlelight Protests in 2016–2017, combined with the impeachment of President Park Geun-hye on corruption charges (she was ultimately sentenced to twenty-four years in prison),[37] gave new energy to liberal opposition parties' and environmental organizations' efforts to reclaim political power. They were successful, and Democratic Party leader Moon Jae-in became South Korea's twelfth president on May 10, 2017. With his election, environmental activists once again had exceptional policy access, as he named activists to key policy positions in his administration.

South Korea's environmental movement shares with those of China, Japan, and Taiwan its origins as NIMBY protests against pollution that was harming human health and livelihood. It shares with Taiwan's movement its close links to the democracy movement. However, it remains unique in the region for its highly political organizational structure. Just as South Korean activists were able to learn from Japanese and Taiwanese, Chinese government officials and environmental advocates have been paying very close attention to the South Korean and Taiwanese experiences and are working to ensure that the movement in their country takes a different path.

China

Environmental policy did not feature prominently in China's politics until Mao Zedong, and then it was largely as antienvironmental policy. Judith Shapiro argues compellingly that Mao's policies were not just proindustrialization; collectively they amounted to a "war against nature."[38] It wasn't until Deng Xiaoping's efforts to introduce some market reforms in the late 1970s that China's economy started to take off. Reforms that began as experiments in special economic zones were expanded to include much of the country by the late 1980s, spreading wealth and pollution in their wake.

By the 1990s, with the end of the Cold War hostilities and the rise of the other economies in East Asia, foreign investment was flowing into China and new policies made it easier for Chinese to travel and study abroad. By the time that Beijing hosted the United Nations' Fourth World Conference on Women in 1995, China had begun the groundwork to enter the World Trade Organization, which it joined in 2001, and was becoming more economically, politically, and socially integrated into the rest of the world. According to many activists, the World Conference on Women was an eye-opening experience for the Chinese political leadership. In addition to the official representatives from governments around the world, more than five thousand members of a variety of NGOs poured into the city for the event. Their large numbers and the productive role that they played at the conference helped make the Chinese leadership aware of the rising importance and usefulness of the nonprofit sector.[39] This impression was solidified for the Chinese Communist Party (CCP) in 2008 after a devastating earthquake in Sichuan killed nearly seventy thousand people. Just as they had in Japan after the 1995 and 2011 earthquakes, international and domestic NGOs rushed to the scene to assist with the rescue and reconstruction efforts. Their numerous successes helped to demonstrate the usefulness of the NGO sector to the Chinese leadership, which moved quickly to support the sector and integrated it into the CCP-directed political system.[40]

As was the case for the other countries in the region, environmental activism began primarily at the local level with citizens protesting pollution by particular factories in their villages. Because of the size of the country and the scope of the problems, local protests are much more widespread in China than elsewhere, with official statistics recording tens of thousands of separate environment-related protests every year.[41] Policy responses to public protests are generally not local because the central government is the primary crafter of environmental policy. It does this in close consultation, both formal and informal, with a wide variety of stakeholders, including academic experts, who often also represent NGOs, and the business community.[42] One of the main institutional mechanisms for this consultation in all four countries is the use of government-organized NGOs. In China as elsewhere, these organizations provide an institutional location where environmental activists, technical experts, and policymakers work together to address environmental issues.[43]

Unlike the environmental movements in other countries in the region, China's movement came of age when it could take advantage of the experiences in other countries and draw on the enormous resources of the international NGO sector.[44] These international environmental NGOs see themselves as utilizing a wide array of delicately balanced political tools that provide negative and positive pressure, with many NGOs in all fields tending to favor the latter type of activity over the former. As one activist I interviewed phrased it, "Big confrontational actions don't work in China. We try to push the boundaries of what civil society can do slowly. ... We have and need to establish ourselves as a constructive partner."[45]

Starting from the late 1990s and accelerating through the 2000s, the number of international NGOs in China proliferated, and they have also expanded their staffs and offices. The primary reasons for their involvement are the scale and scope of the environmental problems in the country. China is now the largest emitter of greenhouse gasses in the world,[46] and there is no doubt that the fate of the global environment is heavily influenced by what happens in China.

While environmental activists in China benefited from the timing of their environmental movement by being able to draw on the experiences and resources of environmental movements in other countries, the Chinese government also learned valuable lessons from the Japanese, South Korean, and Taiwanese experiences. From Taiwan and South Korea, the CCP leadership learned that local environmental protests can transform into national democracy movements if allowed to grow. From the Japanese experience it learned that if the party in power can accommodate or even exceed the public's demands for environmental action, the party can remain in power and even enhance its legitimacy. As a result, the Chinese government has been working at multiple levels to prevent the formation of national environmental movements that might transform into democratization movements and has been striving to develop aggressive environmental policies to address its citizens' rising concerns.

In response to spreading protests, the central government is ramping up its efforts to improve the environment in the country. At the same time, it has diversified and strengthened the legal and political mechanisms through which it can monitor and control the activities of the NGO sector.[47] Thus, although China supports many proenvironmental activities and will listen to environmental activists, it is working very hard to ensure that advocates remain cooperative partners with the party-state and do not engage in any activities that might challenge the CCP.

Patterns of Environmental Activism in East Asia

Like their counterparts from around the world, environmental activists in East Asia are working in a number of different issue areas. This section will provide an overview of the activities of East Asian environmental organizations, examining especially the issue areas in which they engage and the types of strategies they employ. The data are from an original database built from information gathered from environmental groups in China, Japan, South Korea, and Taiwan, about one hundred groups in each country. More details about the methodology for collecting the data can be found in appendix B.

Figure 4.1 demonstrates that there is remarkable similarity across all four countries in terms of the issues about which they are most active. Pollution, conservation, and biodiversity got the most attention—a majority of groups in most of the countries had activities that engaged with these three issue areas. Energy, climate, and recycling issues were the next most popular, with environmental justice and transportation attracting less attention.

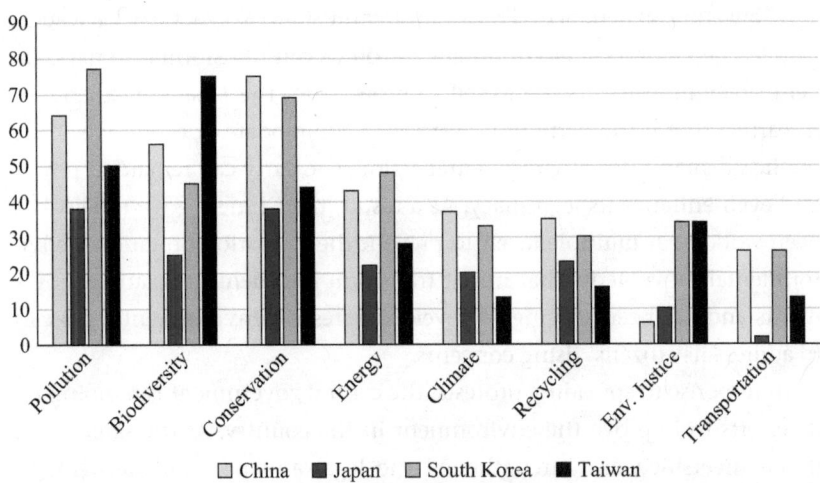

Figure 4.1
Environmental issues in East Asia (percent of each country's organizations engaged in each issue area).

Overall, China and South Korea are more engaged in more issues than Taiwan or Japan—they led engagement for six of the eight issue areas, but the patterns are not consistent. Japan generally had lower levels of engagement on the issues than the other countries. It may be that Japan's lower figures are a result of the fact that Japan's environmental movement is the oldest, so its environmental groups tend to specialize a bit more, resulting in a smaller proportion of the groups engaging on any particular issue.

Figure 4.2 shows that environmental organizations in East Asia are not just working on similar issues, they are employing similar strategies. The two most common strategies are public education aimed at the grassroots

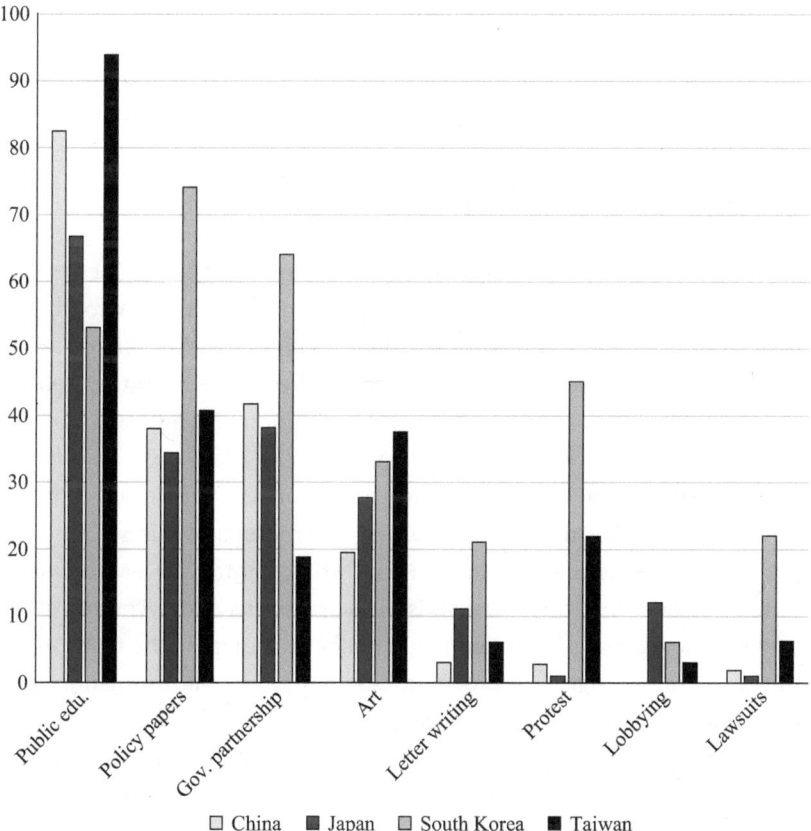

Figure 4.2
Advocacy strategies of environmental organizations in East Asia (percent of each country's organizations that employ each strategy).

level and policy papers aimed at elites. Forming partnerships with government is also very common across the region, with more than half of the South Korean organizations engaged in government partnerships, and more than a third of Chinese and Japanese organizations doing the same. Art was moderately common everywhere, with between 20 and 40 percent of organizations supporting or engaging some form of environmental art in conjunction with their advocacy efforts.

Letter writing, protest, lobbying, and lawsuits were much less common everywhere, with the notable exception of South Korea. Although we see very few patterns of engagement that follow regime type, the two countries whose history has linked environmental politics with partisan politics—South Korea and Taiwan—are also the two countries that engage in the most protests and have the most lawsuits.[48] For South Korea especially, it is likely that its history of contentious protest and partisan political culture has contributed to the prevalence of more contentious forms of advocacy in the country.[49]

Conclusion

East Asia is well known for its strong governments, global businesses, and weak advocacy sector, so it is somewhat surprising that environmental policy in the region has progressed so far. This chapter has offered a brief overview of the history of environmental politics in the region, as well as a sketch of the kinds of issues and strategies common among environmental organizations there.

Across the region, high-speed industrialization led to toxic pollution and intense citizen pressure to address local environmental crises. East Asia's pro-growth developmental states then evolved into eco-developmental states,[50] responding to the demands of their citizens and becoming global leaders in a number of environmental policy areas. Japan, the first country in the region to engage in proenvironmental policy transformation, now enjoys some of the best air and water quality among developed countries, and it has become a global leader in a number of green technologies, including electric, hybrid, and hydrogen cars; recycling and waste management; and green construction.[51] Like Japan, South Korea and Taiwan have largely cleaned their air and water and are now expanding their national forests, as well as devoting resources to green-growth technologies.[52]

Finally, while China still has a long way to go to improve the toxic environment faced by its citizens, it has set ambitious goals and has been exceeding them, especially in the energy sector. It pledged to cut emissions per unit of GDP by 45 percent by 2020, and by the end of 2017 it had cut its carbon dioxide emissions per unit of GDP by 46 percent from 2005 levels. Similarly, by 2018 it already surpassed its 2020 goal for solar deployment and now accounts for more than half of global photovoltaic demand. By the end of 2016, China's wind power generation exceeded that produced by all of Europe and was more than twice the wind power produced in the United States.[53]

Although all four countries have made considerable progress improving their environmental policies in some areas, they continue to struggle in many areas of environmental policymaking that negatively affect business. Japan remains one of the few countries in the world to defy global norms against whaling;[54] the illegal wildlife trade continues to thrive in China;[55] indigenous groups struggle to gain environmental justice in Taiwan;[56] and South Korea's industrial air and water pollution problems continue to worsen.[57] While this book focuses on advocacy success, it is important to remember that environmental challenges remain large in the region, success is by no means guaranteed, and there are still many, many areas where environmental policies lag far behind where they need to be to ensure a healthy future.

The next five chapters will take a closer look at specific strategies that have proved to be effective for advocates in the region and across the world. These chapters are intended to offer more details about exactly how these strategies work in different contexts, with the hope of providing insight into how advocates can be effective in generating behavioral change among individuals, corporations, and governments, even under hostile political conditions.

5 Make Friends on the Inside: Cultivate and Empower Allies

Again and again when I talked with environmental advocates, whether they were located in grassroots organizations, international nongovernmental organizations (NGOs), businesses, or government, one strategy was consistently discussed as the most important: make and empower allies with decision-making power. This is not surprising. For advocacy to be effective, advocates must convince people with power to change policies for the better. Accessing and cultivating allies with power is perhaps the most important of all advocacy strategies presented in this book.

Public policy shapes the rules that govern the decision-making processes for citizens, governments, businesses, and nonprofit organizations. If policy makes antienvironmental behavior illegal (and that policy is enforced), then there will be less of the targeted antienvironmental behavior. If policy helps proenvironmental behavior become more profitable and easier, then more businesses and consumers will engage in the promoted proenvironmental behavior. Policy is a powerful tool to shape the behavior of everyone within its jurisdiction[1] and sometimes even those beyond it.[2] Therefore, to the extent that advocates are able to convince policymakers to craft policies that promote the behavior that they wish to see—in this case proenvironmental behavior—their advocacy will be successful.

At the core of the "make friends on the inside" strategy is the recognition that policymakers need friends on the "outside" because they cannot know or do everything themselves. Across the world, and especially in East Asia, governments are short-staffed. Here are some very basic numbers about the four countries at the heart of this study to illustrate that fact: China nearly doubled the number of its core environmental policymakers when it transitioned the Ministry of Environmental Protection into the Ministry of Environment and Ecology in 2018, but the new ministry is still estimated to have only 500 full-time staff in Beijing and just over 1,000 distributed across seven provincial offices to conduct inspections.[3] Japan's Ministry of the Environment has about 2,000 staff,[4] South Korea's Ministry of the Environment has about 3,500 staff,[5] and Taiwan's Environmental Protection Agency has about 1,600 staff.[6] In contrast, the Environmental Protection Agency in the US has more than 14,000 employees.[7] While that may seem like a lot of people compared with the other countries, in fiscal year 2017 the agency was responsible for cleaning up more than four hundred million cubic yards of water, 245 million pounds of hazardous waste, and eight billion gallons of untreated discharge, as well as handling its nonenforcement responsibilities, such as research and education.[8]

East Asia's staff numbers are not so small when compared with other governments—for example, Germany's Ministry of the Environment has about 1,500 staff,[9] the UK's Department for Environment, Food, and Rural Affairs has 3,500,[10] and the European Commission's Directorate-General for the Environment has 500 staff.[11] These small numbers worldwide mean that no matter where you are, there are not very many government policymakers to oversee a very large and diverse set of policy issues. These policymakers need friends outside government to help them identify upcoming policy issues, discover policy options, and assist in formulating policy that will not only be acceptable to stakeholders but also be effective in achieving the outcomes that they are seeking.[12] If an advocate can become one of the people whom policymakers turn to for consultation and advice, it makes it much more likely that the advocate's perspective will be reflected in final policy outcomes.

The Connected Stakeholder Model introduced in chapter 3 illustrates how advocates are connected to the policymaking process through multiple personal and professional networks. The model also discusses the importance of network "nodes," those people who are linked to multiple networks simultaneously and therefore have particularly powerful effects on the whole matrix. These network node individuals are in particularly good positions to have a disproportionate effect on the shaping of policy. "Making friends on the inside" is an advocacy strategy that involves targeting these network nodes in particular, cultivating them as allies, and seeking to enhance their power and influence within their various networks.

This chapter will begin with a discussion of which people are most likely to serve these network node functions, whom advocates should be cultivating.

It emphasizes the importance of bureaucrats, politicians, policy-connected academics, and other policy professionals. It will then describe several ways that advocates can gain access to target individuals, such as by utilizing personal connections, attending conferences and other events frequented by policymakers, and serving together on policy-relevant committees. The chapter will focus on specific examples to highlight four particularly effective methods that advocates use to cultivate allies: be useful, hire retired bureaucrats, keep in touch, and share the credit.

Finally, the chapter will use the example of the China-US Energy Efficiency Alliance to illustrate how advocates can scale the "make friends on the inside" strategy by empowering their "friends on the inside" and expanding the network of friends involved in policymaking. The chapter will conclude by linking back to the core questions of this book, explaining why making friends on the inside is such an effective advocacy strategy.

Which Friends on the Inside Should Advocates Cultivate?

Who are the most important policymakers and policy influencers that advocates should target? My findings support other research suggesting that there are four types of policymakers who are particularly useful for advocates to target when they are seeking to influence policy: bureaucrats, politicians, academics, and other policy professionals. All of these people are likely to be consulted as part of the policymaking process and are also likely to be connected to multiple stakeholder networks at once. In other words, these individuals will frequently occupy the role of network node according to the Connected Stakeholder Model and be disproportionately influential in the policymaking process.

Bureaucrats working in the relevant ministries are frequently primarily responsible for drafting government policy. Generally, they do this in close consultation with some form of a policy advisory committee.[13] Therefore, the most influential policymakers for most forms of policy are the bureaucrats who are responsible for designing the policy.[14]

Politicians can also be important friends for advocates. In countries with green parties, green party politicians can help raise environmental issues on the political agenda, increasing the pressure on bureaucrats and other policymakers to develop proenvironmental policies.[15] Similarly, politicians from nongreen political parties can serve as important allies for advocates, especially if they seek to change not just policy but also legislation.[16] While national-level politicians can have broader influence if they change law, they also always have many other policy issues competing for their attention, and environmental issues regularly fall down in their list of priorities.[17] In contrast, environmental issues are frequently quite pressing at the local level, so local-level politicians, especially mayors and governors, are often more willing to prioritize environmental policy. Frequently, these local-level politicians can be more important allies than national-government politicians, since cultivating a local-level political ally can enable advocates to combine a "make friends on the inside" strategy with a "make it work locally" strategy, a particularly effective combination, as demonstrated in chapter 2.

As the next chapter will discuss in greater detail, mayors and governors can be vital allies for advocates seeking to demonstrate the effectiveness of specific policy prescriptions and promote the dissemination of successful policy models.[18] Furthermore, local government leaders can sometimes go on to become national-level leaders, and they bring their local-level experience and focus on environment with them. South Korea's president Lee Myung-bak served as mayor of Seoul before becoming president. While mayor, he restored the Cheonggyecheon River, which flows through the center of Seoul, and as president he reoriented South Korea's economic strategy to focus on "green growth." Similarly, Xi Jinping introduced an "ecological province" strategy to Fujian Province while he was governor, and then when he became president of China he initiated policies intended to turn China into an "ecological civilization."

Occasionally, the process works in reverse, with national-level politicians bringing their national-level environmental policy experience to the local level. For example, Koike Yuriko served as Japan's minister of the environment before being elected governor of Tokyo in 2016, and Hau Lung-pin led Taiwan's Environmental Protection Administration before serving as mayor of Taipei from 2006 to 2014. The environmental focus of these leaders remains evident in both cities, which have extensive environmental action plans and are frequently ranked among Asia's greenest cities.[19] In all of these cases, political actors grow more powerful over time as they become connected to more networks. To the extent that advocates were helpful early in the politicians' careers, the advocates have increasingly powerful allies in important policymaking roles.

Academics are another set of policy influencers that advocates can target. Academics are generally considered to be apolitical participants in the policy process, merely offering technical advice to policymakers.[20] However, in contrast to the conventional view of the apolitical academic, this and subsequent chapters will argue that academics commonly act as network nodes, are often important allies for advocates, and are political actors in their own right. They are frequently instrumental in shaping policy, sometimes becoming influential voices in national environmental movements and even cabinet ministers. Academics are able to use their technical expertise and perceived nonpartisanship to shore up the legitimacy of their proposals among both policymakers and the public, sometimes even becoming elected political leaders or heads of NGOs.[21] Because of their perceived neutral status, they are able to belong to diverse networks in the nonprofit, for-profit, and governmental sectors, giving them access to many different people and perspectives. Furthermore, in contrast to civil servants and politicians, who are often very difficult to reach without a prior personal connection, academics tend to be more accessible, since they are more likely to answer email, attend conferences, and host open office hours.

A final group of "friends on the inside" that advocates should cultivate is policy professionals. This is a catch-all term that I use to refer to individuals who are neither members of the government nor academics but who are active in the policy area of concern and are frequently involved in policymaking. Most commonly, policy professionals are full-time employees of nonprofit organizations who are active in the policy area and are connected to multiple networks of stakeholders. In policy advisory committees, these individuals are often asked to represent the "nonprofit stakeholder" position in policy dialogues, bringing not just their technical expertise to bear on the subject but also their political perspective on the policy.[22] Because of their acknowledged policy expertise, these individuals also often serve as network nodes, connected to multiple, diverse networks of stakeholders with an interest in the policy area.

Also included in the policy professional category are those who work in for-profit organizations who have developed an expertise in the relevant policy area. Similar to their counterparts in the nonprofit sector, these policy professionals have multiple years of experience working on relevant policy issues, have developed a professional expertise, are active in related policy communities, and are considered by their peers to have relevant knowledge for policy development. Like their nonprofit counterparts, these professionals are commonly asked to represent the for-profit, business stakeholder perspective on policy, reflecting the economic and other interests of the business community while seeking better environmental outcomes.[23]

While policy professionals may be asked to participate in policy advisory committees because of their technical expertise and their insights into the substantive policy issue of concern, experts who are highly sought after as advisers are generally the people who are connected to the most networks and have the most diverse set of connections. Often individual people will build their own capacity to serve as network nodes by moving among the nonprofit world, the for-profit world, and the government over the course of their careers. The career trajectory of these people serves to enhance their capacity to connect to more networks and gives them greater insight into the perspectives of diverse stakeholders.

Since there are usually thousands of people who might fall into any one of these categories, how can advocates tell if they are targeting the network nodes, the people who have the most influence over policy? The easiest way to determine who is most influential is if the relevant policy advisory committee has a public listing of its members—advocates should try to make friends with the people who regularly serve on those committees. Frequently, however, committee membership is not public. In those cases, advocates should seek out the people who regularly appear at policy-related events—not just in the audience but also as keynote speakers and on panels.

Advocates can target individuals from their own countries and communities who appear on the programs for large international conferences related to environmental policy (e.g., the Intergovernmental Panel on Climate Change, the Conference of Parties [COP], the C40 Cities Climate Leadership Group, ICLEI), as well as local gatherings—events and panels on climate change, sustainability, waste management, transportation, clean energy, and other such topics held in capital cities and local communities. It is usually the case that for any given national or local community, there will be a relatively small, identifiable set of individuals who regularly appear at these events. People who are frequently asked to serve as panelists and discussants at policy-related conferences and other events are also likely to be asked by government to serve on policy advisory committees.

These are the influential people who are connected to multiple networks with whom advocates should seek to build relationships.

When seeking to identify individuals to cultivate, advocates should not just gravitate toward those with whom they have a natural affinity; they should try to get to know any and all people who may be in a position to influence policy. As was discussed in chapter 3, the more policy-connected people that you know, the better. Cultivating allies does not just mean finding policymakers who already agree with you; it is best accomplished when you can persuade those who thought they were your enemies that you are in fact on the same side—that your policy preferences are something they should support as well.[24]

Relatedly, the people who exert the most influence over policy are not always the same; they change over time. During one administration, one group of individuals may have a lot of influence, but when another administration takes power, a whole new set of people will be placed in decision-making roles. A few years later, some of the first folks might be back again. Or people who you thought were your friends might turn away once they have power. As one activist from the Taiwan Environmental Action Network put it to me during a 2010 interview, "We no longer have any friends on the inside. In the previous DPP [Democratic Progressive Party] administration, there were some friends in the Executive Yuan, but now [after the Kuomintang won power in 2008] there aren't any." Another activist, from the Taiwan Environmental Protection Union, expressed a frustration common to many activists around the world—their political allies were very friendly and supportive when running for office and in their early days of governing, but that changed after they were in power for a while. "The DPP changed when it took power. Once they were in power, they didn't like the environment anymore."[25] Who makes policy shifts over time, so advocates should seek to cultivate allies from among all potential policymakers.

How to Access Policymakers

Although people are able to gain access to policymakers in a multitude of ways, three avenues appear to be the most common: forming personal connections, attending the same events, and serving together on committees. Personal connections are usually the easiest, and often the most powerful, way to connect with a policymaker—alumni of the same school, people who grew up in the same town, family members, and members of the same church or social club all make it possible to form a personal connection with a policymaker. Furthermore, relationships that begin with a social connection are inherently less instrumental, so they make it easier for the advocate to build trust with the policymaker. As the extensive literature on the role of *guanxi* can attest for the case of Asia, these personal connections can be vital in gaining access to policymakers and facilitating communication between advocate and policymaker.[26]

Usually, however, advocates do not have a personal connection with any of the policymakers they wish to target, so they must seek other ways to meet them. Perhaps the most common and productive way to do this is to attend a policy-related conference, forum, workshop, roundtable, lecture, or other event that policymakers will likely be attending. Smaller conferences with fewer people and more focused conversations will often make it easier to meet and get to know the policymakers and potential policymakers whom you want to access. Conferences with titles like "Workshop on the Joint Crediting Mechanism and Low-Carbon Technologies in the Philippines"[27] will allow for greater interactions with people who are more directly involved in your target policy area. Larger conferences, such as the COP meetings or ICLEI's World Congress, may offer an opportunity to meet high-profile policymakers, but longer conversations during such events are often impossible, and finding someone who is directly linked to your policy area of interest can be difficult. Nonetheless, as was discussed in chapter 3, these venues serve as important gathering points for interested parties to connect with one another, helping to strengthen and expand advocates' policy networks both within and beyond their own national setting.[28]

Finally, perhaps the best way to gain access to policymakers is to be directly involved with them through common service on some kind of policy-related committee or working group. When advocates are able to work alongside policymakers directly in a policymaking context, a personal connection, as well as the development of professional respect, is a natural outgrowth. Gaining a reputation as a productive committee member will commonly result in more invitations to participate in policymaking opportunities, either because policymakers who are participating seek to work with productive advocates again or because they recommend those advocates to their colleagues when the latter seek members for other groups.[29]

How to Make and Empower Friends on the Inside

Once an advocate has gained access to a policy influencer, how can the advocate cultivate that person to become his or her ally? While there are several ways to encourage policymakers to become allies, this section will highlight four that were discussed as particularly effective by interlocutors in all the countries where I conducted research: be useful, hire retired bureaucrats, stay in touch, and give credit to allies.

Be Useful

As the last chapter indicated, East Asian governments historically have been highly suspicious of environmental advocates, viewing them as hostile, disruptive actors akin to terrorists.[30] However, as their societies developed and diversified, and as global civil society grew around the world, East Asian governments came to recognize that it was impossible for them to meet all of society's needs and that civil society and the nonprofit sector could be leveraged to provide services that the government found it difficult to offer. Fundamental to this attitudinal transformation has been the ability of NGOs and other community organizations to demonstrate to local and national governments that they can be useful. Whether it was helping out in times of disaster[31] or providing necessary social services,[32] East Asian governments reluctantly recognized that an active civil society was a necessary and desired component of advanced, developed societies.

The challenge for advocates has been to shift their relationship with the government from one where their organizations passively executed and supported government policy to one where advocates are able to influence the development and setting of policy priorities, as well as the construction of policy measures to address those priorities. This transformation in the relationship between policymakers and advocates can take decades.[33] One of the best ways for advocates to increase their influence over the policymaking process is to become useful partners.

Being useful requires that the advocate and the advocate's organization provide the policymakers with things that they need—for example, technical expertise, new data, volunteers, political access, publicity, or innovative policy ideas. As advocates demonstrate to policymakers that they can offer something useful, the advocates become more respected as partners in policymaking and can then gain greater access to policymakers during the policymaking process, not just after the policy has already been developed.

The story of Greenpeace in China is instructive on how advocates can dramatically improve government attitudes toward their advocacy efforts. In this case, the Chinese government went from being suspicious that Greenpeace was trying to cause political trouble for the government to valuing its participation in environmental policymaking. In 1995, as part of its global action in support of the Comprehensive Nuclear-Test-Ban Treaty, Greenpeace activists went into Tiananmen Square and held up an anti-nuclear banner. Following their common protocols for this kind of event, they contacted the press beforehand.[34] Although they were successful in obtaining significant international press coverage of their event, they were quickly removed from the square, and the incident had devastating consequences for their activities in China. Greenpeace's activists were kicked out of the country, and it was nearly a decade before the organization was able to open an office to the mainland, well after counterparts like World Wildlife Fund and the Nature Conservancy had established offices in Beijing.

Greenpeace's publicity stunt gained the organization a reputation as a troublemaker from the perspective of the Chinese government. As one Beijing-based journalist who covered the event at the time put it to me in a 2010 interview, "That [incident] made them [Greenpeace] a pretty dirty word around here for a while. The sad thing is that they had been working very effectively behind the scenes with one of the Ministries about how to get the Chinese to use refrigerators with lower/no CFCs [chlorofluorocarbons] to cut greenhouse gas emissions. That was a really important issue, but suddenly, after that incident, everything stopped."

Greenpeace was still interested in working in China, so it changed its approach, deciding to be less confrontational and more strategic. In early 1997, it established an office in Hong Kong while it was still under British rule, and for the next five years, all of its activities in China were coordinated out of that office. Its first action was to offer Kelon (a large Chinese manufacturing company) free transfer of Greenfreeze, a new technology that enabled refrigeration without the use of CFCs, allowing Kelon to become the first Chinese company to produce ozone-friendly refrigerators. In the following years Greenpeace actively supported local and governmental efforts to curb toxic waste dumping and improve water quality.[35]

In 2004 Greenpeace East Asia started a renewable energy program, and the following year it released a policy brief highlighting the benefits of renewable energy, focusing on the potential for China to harness more than one thousand gigawatts of wind energy.[36] The report was shared with industry partners (European and Chinese wind power associations), as well as the Ministry of Environmental Protection and the National Development and Reform Committee. The report provided new data that helped support existing government efforts to develop the wind power industry. China became the world's largest producer of wind energy less than a decade later and has positioned itself to become a "renewable energy superpower."[37]

In 2005 Greenpeace East Asia partnered with local scientists to conduct a detailed assessment of the Yellow River, Asia's second-longest river. Although the river is one of China's most polluted,[38] Greenpeace focused its report, *Yellow River at Risk*,[39] on the negative effects of climate change on the river rather than pollution. The focus on climate change enabled Greenpeace to call attention to many of the most pressing problems related to the river—flooding, threats to fresh water supplies, desertification, soil deterioration, and biodiversity reduction—without pointing fingers at the government or polluting companies. The detailed data, graphs, images, and local stories offered compelling evidence that government officials at the local and national levels could use to support policies and programs designed to improve the water quality of the river. The report also prompted additional media coverage of problems related to the river, and the media was not as circumspect as Greenpeace in limiting itself to climate issues. Although it is a state-run company, CCTV's coverage of the Yellow River that year focused not only on climate-related challenges but also on pollution.[40]

Simultaneous with their work at the local level, Greenpeace and other similarly situated international environmental organizations, such as World Wildlife Fund and the Nature Conservancy, assisted national-level Chinese officials as they began to enter the world of global politics in a more significant way. At the time of the 1997 COP meeting in Kyoto, which generated the Kyoto Protocol, China's gross domestic product (GDP) was not quite $1 trillion, placing it at number seven in the world, below Italy and slightly higher than Brazil.[41] In 2001 China joined the World Trade Organization, and by the 2005 COP meeting in Montreal, its GDP had more than doubled, to $2.2 trillion, putting it in the number four slot ahead of the UK and just behind Germany. By 2010 China had become the second-largest economy in the world after the US, with an annual GDP of nearly $6 trillion.[42]

China's meteoric economic growth resulted in a corresponding expansion in its role in international diplomacy, including climate and other environment-related negotiations. Previously content to attend these meetings and remain largely in the background, Chinese diplomats started to find themselves both in great demand and with more at stake in international forums.[43] As explained in chapter 3, Chinese civil servants, like those in much of East Asia, are generally rotated to new positions every two or three years in order to give them broad exposure inside their ministries. This means that the people attending these high-profile global conferences were generally new at their jobs, making it difficult for them to know whom they should contact and whom they should avoid. Building off their growing track record of helpful collaboration at the local and national level, international conferences offered Greenpeace and similar organizations an invaluable opportunity to be helpful to Chinese officials. International climate conferences are a core part of the mission of these global NGOs, so they were exceptionally well placed to offer useful advice about protocol and sessions and could help set up desired meetings for Chinese officials who were attending for the first time.[44]

By 2010 Greenpeace's office in Beijing had been open for several years, and it had begun to rebuild its reputation with Chinese officials by providing useful data related to local environmental issues and acting as a helpful ally in international forums. Furthermore, the Chinese government had now recognized that it was facing an environmental crisis and had embarked on a number of ambitious policy initiatives, including Hu Jintao's Green GDP project announced in 2007 and the elevation of the State Environmental Protection Agency to ministry status with the establishment of the Ministry of Environmental Protection in 2008. As a result of these developments, Greenpeace was able to be a bit more direct with its advocacy efforts. Taking on the issue of pollution in China's largest river, *Swimming in Poison* (2010)[45] documented the harm caused by toxic chemicals in the Yangtze and recommended that China increase the range of chemicals that it regulated.

Like Greenpeace's earlier report about the Yellow River, the Yangtze report relied on original research collected in collaboration with local scientists and was filled with compelling new data, charts, pictures, and stories of local communities. While the report documented hazardous levels of specific pollutants found at locations along the river, as well as the toxicity of different fish species, it was careful not to blame any specific industry, government, or company for the problem. It is impossible to draw a direct link to the publication of Greenpeace's report, but soon after *Swimming in Poison* was released, the Ministry of Environmental Protection set up a committee to investigate toxins, and by the end of the year it had issued MEP Order 7, which updated and augmented the regulation of the use and release of chemical substances in China.[46]

Greenpeace's experience illustrates how advocates can shift from being enemies of governmental policymakers to becoming valuable friends by being useful, providing new data, conducting practical analyses, and facilitating political access for potential allies in government. While some advocates may argue that working in this collaborative way with policymakers and making strategic decisions against confrontation is a form of "selling out" or being "co-opted," the Greenpeace story illustrates how an organization can remain true to its mission and still cultivate allies within the government. Once allies have been nurtured among influential policymakers, they are more likely to read the advocates' reports and be persuaded by the evidence and arguments presented. Sometimes, they may even respond by changing policies in the desired direction because they have been persuaded by the advocates' arguments.

Finally, it must be noted that in the Chinese context in particular, being useful is not just a strategy to gain access and influence; it is a strategy for survival. For advocates in Japan, South Korea, and Taiwan, being useful to policymakers can be a helpful way to build trust and gain influence with policymakers, but those who are not useful are still able to pursue their own agendas and activities. In China, however, the situation is much more cut-and-dried. As one advocate put it succinctly to me in a 2011 interview, "People write about the paradox of the government's handling of NGOs, but it isn't really a paradox—those that are useful are allowed to exist, and those that cause trouble are shut down."

Hire Retired Bureaucrats

One of the most effective ways to gain access to policymakers is to invite one to be on your board of directors or hire one to work at the senior levels of your organization. Retired bureaucrats not only have a wealth of information about the decision-making processes of their ministry, they have direct personal contacts among the current decision makers. Their contacts inside and outside the government mean that they are connected to multiple networks, enabling them to access diverse perspectives and reach out to many stakeholders, as well as policymakers. The combination of their insider knowledge about the decision-making process of the relevant bureaucracy and their personal connections to decision makers, as well as stakeholders, means that these retired bureaucrats can be powerful allies for advocates.

This practice of bringing retired bureaucrats into an organization is found around the world and is particularly common in East Asia. In his 1982 book, *MITI and the Japanese Miracle*, Chalmers Johnson explains how this practice, called *amakudari* (descent from heaven) in Japanese, helped contribute to the close coordination of Japan's industrial policy as the country recovered from the war, facilitating rapid economic development, especially in its large manufacturing sector. Margarita Estévez-Abe's *Welfare and Capitalism in Postwar Japan* (2008) illustrates how the same phenomenon operates in the social welfare policy area.

Although much smaller in terms of scale when compared with for-profit industrial entities, a similar practice occurs in the nonprofit sector, where bureaucrats who are forced to retire early by government policy (e.g., by age fifty-five or sixty) seek to remain active in their fields of expertise. They retire from a ministry to work for a government-organized NGO (GONGO) or NGO, or sometimes start a new NGO, relishing the opportunity to remain active in policymaking without the constraints of being a government employee. Their decades-long careers in public service mean that they are often able to help their new organizations find issue areas that are in line with the government's preexisting priorities. They also have numerous personal connections inside and outside the government that may enable them to bring their organization's priorities to the attention of policymakers who are more likely to be sympathetic, or even reshape the policy ideas of those who have remained in government such that they are more in line with the preferences of the NGO or GONGO.[47] Their network connections make it possible for the retirees, and the organizations with which they are affiliated, to be highly attuned to the needs and interests of policymakers. This enables them to be particularly savvy at seizing political opportunities, framing their issues and policies in ways attractive to policymakers, and engaging in all parts of the policymaking process, from problem definition to policy formation and through the politics of decision-making.[48]

This kind of close connection with the government does not come without a cost. The closer the connection with the government in terms of people or funding, the more cautious and less controversial the NGO will be in its activities. GONGOs, which generally receive most of their funding directly from the government and often have their top leadership dominated by former bureaucrats, will find it very difficult to stray much beyond the priorities articulated by the government. They cannot risk their funding by engaging in controversial activities. Even NGOs that do not receive significant funding from the government will frequently avoid controversy in order to preserve a good relationship with the policy actors whom they seek to influence.[49]

In the Chinese context, Peter Ho and Richard Edmonds have called this kind of highly government-connected advocacy "embedded activism."[50] In China and other authoritarian contexts, it is often difficult for activists to engage in any form of advocacy outside some kind of embedded framework. By embedding themselves in the state's policy and implementation structure, advocates are frequently able to engage in activities that improve the environment and society in line with their mission, sometimes even using the very agents and agencies that are seeking to constrain them as channels for policy change.[51] Unfortunately, even embedded activists can find the space in which they are able to act squeezed as state priorities shift.[52]

One example of an organization that has been exceptionally adept at affecting both policy and outcomes by utilizing its friends on the inside has been the Japan-based Institute for Global Environmental Strategies (IGES). Of all of the organizations that I visited in researching this book, IGES was one of the most effective. Established in 1998 as part of Japan's commitment to environmental leadership at the COP 3 meetings in Kyoto, IGES has been working for more than twenty years as a resource for local and national governments in the Asia-Pacific region as they grapple with increasingly intense environmental challenges.[53]

IGES has a classic GONGO structure: its founding was a governmental initiative, its top leaders are former government officials, and its funding comes largely from government sources. Since its inception, IGES has engaged effectively in "embedded activism,"[54] utilizing its close connections with government to develop and promote proenvironmental policies at the local, national, and international levels across the Asia-Pacific region and the world.

By incorporating retired bureaucrats into its organization, it has been able to ensure that its policy proposals are framed in ways that are appealing to policymakers, making it more likely that they will be implemented. For example, immediately after the 3/11 disaster, IGES was able to rapidly expand its activities related to energy conservation, renewable energy, smart cities, green finance and markets, and other issues to take advantage of—and promote—national government initiatives in those policy areas. By organizing events such as the "IGES-YCU Joint Seminar on Low-Carbon and Smart Cities: Seeking Local Energy Solutions after the Nuclear Crisis,"[55] held just four months after the Fukushima disaster, IGES helped policymakers learn what they needed to know and connect decision makers with people they needed to meet during policy-critical time periods. IGES could then present its independent research to policymakers, who had already been primed to be receptive to IGES's recommendations related to smart city and green market development.[56]

Over the past twenty years, IGES has been developing high-level connections through its international partners as well. For example, its 2017 annual report claimed that the previous year had "25 cases of high level influence." Three examples discussed in the report were a pilot project in participatory watershed management in the Philippines that won the Dubai International Award, local and national waste management plans that were official adopted by Myanmar's Ministry of Natural Resources and Environmental Conservation and the Mandalay City Development Committee, and the development of an online Sustainable Development Goals analysis and visualization tool that has been accessed by users from 115 countries.[57]

Hiring former bureaucrats can be an effective way of gaining policy access for advocates. Perhaps even more importantly, it enables advocates to judge correctly what is of interest to policymakers at the moment and frame their proposals in ways that will be more attractive. In response to a question about how advocates can make themselves heard by policymakers, one of my interview subjects put it clearly in an interview with me in Beijing in 2010: "You need to understand where there are openings, what the internal dynamics of the government are. How can you come in and create a win-win cycle? You do this by aligning your ideas with what the government is trying to do anyway." Former bureaucrats have direct lines of communication with current policymakers, making it easier for them both to discover what the current interests of policymakers are and facilitate access that might shift those interests to be more in line with those of the advocates.

Keep in Touch

Time and time again in my interviews with advocates of all kinds—such as NGO professionals, local and national officials, business people, and academics—they talked about how important personal relationships were in making policy. When individuals get to know each other as people with families, hobbies, and personality quirks, it becomes easier to trust and work with them in a professional way. When working relationships become iterated over time and proliferate across diverse contexts, trust develops. Once a relationship is formed, reconnection and collaboration become much easier, even if years pass without any close collaborations. Because it is impossible to know what issues will arise in the future, and which of your acquaintances and friends might be in a position of influence for any given policy issue, keeping in touch is a vital component of cultivating friends on the inside. Indeed, as time goes by, advocates who were once on the outside may find that they become the friend on the inside whom others seek to cultivate.

A good example of this comes from South Korea. Myung Rae Cho was a professor of urban planning at Dankook University when I first met him in 2010. Like many faculty members whom I met during my research, he wore many hats—scholar, teacher, technical expert, environmental advocate, local volunteer, government consultant, and others. He mingled freely with many different types of people, from neighbors who were engaging with local pollution at the grassroots level to global intellectual elites who were wrestling with how to address the world's looming environmental crisis and its social ramifications.

One of his earliest engagements with the advocacy field was through the Citizens' Coalition for Economic Justice (CCEJ), which was formed as a social movement in the late 1980s to question the high-growth model of development that was generating significant pollution, as well as economic and social inequality. CCEJ claims to be South Korea's oldest NGO,[58] and it played an important role in South Korea's democratization process.[59] Cho's work with this organization led him to become involved with other groups, including the Citizen Movement for Environmental Justice, which broke off from CCEJ in order to focus its efforts on the environmental components

of economic justice. Through his involvement with these grassroots organizations, Cho was able to understand how environmental issues related to
pollution and overdevelopment were negatively affecting people's everyday lives, even as his professional training in environmental planning and
urban studies helped him develop solutions to those problems.

While researching the history of South Korea's environmental movement,
Seoul's urbanization, and South Korea's green growth strategies,[60] Cho worked
as an adviser to local governments to help them navigate the complex environmental and social issues they were facing as a result of rapid industrialization and urbanization. He served as Suwon City's international adviser, as
well as chairing Seoul City's environmental impact assessment commission
for several years. He was also an expert adviser for the Presidential Committee for Balanced National Development under President Kim Dae-jung.
All of these committees offered opportunities to broaden his connections
to people on all sides of diverse local and national environmental issues.

As a member of the Citizens' Green Seoul Committee (an organization
discussed in greater length in the next chapter), he headed up a subcom-
mittee to study the restoration of the Cheonggyecheon River. The large
river runs through the center of Seoul but was covered as part of the city's
urbanization process in the 1960s, with a large section being covered by a
major highway in the 1970s. By the late 1990s the cement was starting to
crumble and significant maintenance was needed. Rather than repair the
highway, newly elected mayor Lee Myung-bak wanted the project to be an
example of "green growth" that would create a large green space in the center of Seoul while being supportive of the construction, leisure, and service
industries. They broke ground in 2003, demolition and landscaping took
two years, and the park opened to the public in fall 2005.[61]

The project was wildly heralded as a success, and it was claimed that it
increased pedestrian activity by 76 percent, reduced vehicle volume by 45
percent, reduced air pollution by 10 percent, and decreased the urban heat
island effect by 4.5 percent.[62] It is frequently used as a case study of success
to demonstrate green urban design and redevelopment.[63] The popularity
of the project is also commonly credited with helping Mayor Lee become
President Lee in 2008.[64]

However, the political process was not as smooth as many of the lauda-
tory case studies suggest, as Cho documents in his own study, which reveals
the competing understandings of nature held by the city government and

the local NGO community.[65] As part of the Citizens' Green Seoul Committee, Cho was responsible for bringing the NGO's concerns to the attention of the city government and negotiating solutions. As he explained it in a 2010 interview with me, the committee was very active in collaborating with the city government to improve Seoul's environment, but as he began to reveal problems with the restoration plans, relations turned sour. "For a while there I was public enemy number 1 of the former mayor, now president." Expressing sentiments similar to those of his counterparts in Taiwan when the administration switched from a liberal party back to a conservative one, he said, "Now, we're completely excluded [from policymaking]."

Although Cho may have felt excluded from policymaking while the conservatives controlled the government, he kept in touch with his colleagues both inside and outside the government. As an example of how he managed, he described to me his interactions with Professor Shim Myung Pil, who was an old activist friend and professional colleague who had been asked to head up Lee's Four Rivers Restoration Project.[66] Intended as a way to scale the success of the Cheonggyecheon River restoration, the new, very large-scale project targeted South Korea's four largest rivers (Han, Geum, Nakdong, and Yeongsan) for development, and it was almost universally criticized by environmentalists.[67] Cho also opposed the project, but he still reached out to Shim. Over dinner they were able to share memories of past collaborations, as well as airing disagreements over the current project. They were able to reinforce a personal connection even as their politics no longer aligned.

A key component of the "cultivate friends on the inside" strategy is that you stay in touch with your friend, even though you might not feel so friendly when he or she is in charge of a policy that you do not like. Furthermore, in politics things often shift—people who are in power move out and vice versa. Indeed, this is what happened with Cho. While the conservatives were occupying the Blue House, Cho remained active, serving as copresident of Environmental Justice (an NGO) and chairman of the Sustainable Development Committee of the Seoul Metropolitan Government. Eventually, the liberals gained power again, and he moved from outside to inside, heading up the Korean Environment Institute and the Council of the Heads of Environmental Research Institutes. In 2018 Cho was appointed by President Moon Jae-in to be South Korea's environmental minister, making the final shift from being an advocate seeking to cultivate friends on the inside to becoming the ultimate friend on the inside whom other advocates sought to cultivate.

Give Credit to Allies

A final method that is highly effective in cultivating and empowering friends on the inside is to publicize success and give credit to allies. As mentioned earlier and will be discussed in greater detail in the next chapter, policy-makers frequently encounter enormous internal bureaucratic hurdles to developing and disseminating effective policies. Advocates can be of great assistance to allies when they are able to mobilize media attention to policy successes, drawing central governmental attention to good policy enacted locally and raising the reputation and profile of allies and their depart-ments with the public.[68]

I will use the example of Hong Kong's Fair Winds Charter to illustrate how advocates can give credit to allies as part of a successful advocacy strat-egy. The Fair Winds Charter was developed through a collaboration between Civic Exchange, a Hong Kong–based think tank, and the shipping industry, which resulted in changes in local and, eventually, national regulations. As a clear indication of credit sharing, although it was the brainchild of Civic Exchange, the charter is described on Civic Exchange's website as "the first voluntary scheme initiated by Hong Kong's shipping industry to reduce ship emissions by requiring ocean-going vessels to switch to a low-sulphur fuel while at berth."[69] Ultimately, the charter influenced the development of new shipping regulations not just for Hong Kong's ports but also for the Pearl River Delta, Yangtze River Delta, and Bai Bohai Rim regions as well.[70]

The process began in the early 2000s when researchers from Civic Exchange collaborated with scholars from the Hong Kong University of Science and Technology to investigate the contribution that the shipping industry made to Hong Kong's air pollution problems; they found that it was significant and growing.[71] They shared their report with Hong Kong's Environmental Protec-tion Department, which commissioned its own study of marine emissions in 2008. That same year and for the next several years, Civic Exchange held a series of workshops designed to engage a wide variety of stakeholders in the issue, including port and local craft operators, the trucking sector, and fuel suppliers, as well as government officials and academics, in order to inform everyone about their findings on the importance of the shipping industry's contribution to Hong Kong's air quality and to hear from stakeholders about concerns and potential opportunities related to emissions control. Civic Exchange ensured that the relevant government officials from both the Environmental Protection Department and the Marine Department were invited to the workshops and kept in the loop as discussions progressed.

Having support from key business leaders was critically important to gaining access and trust from industry. Arthur Bowring of the Hong Kong Shipowners Association; Peter Ng, Tim Smith, and Roberto Giannetta of the Hong Kong Liner Shipping Association; and other key industry leaders were interested in the reports' findings about the development of green ports in Los Angeles and Long Beach in the US, and saw the opportunity for Hong Kong, one of the world's largest ports, to become an industry leader. These business leaders helped connect Civic Exchange researchers to the technical and operational people in the industry. As Civic Exchange's chief research officer at the time, Simon Ng, explained to me in a 2019 interview, "If an NGO had just come in and talked about the environment, they [the shipping industry people] would have been hostile. When they could see that we were coming in with a different angle, that we were trying to help the industry become greener, which would benefit the whole industry, they became more receptive. [Introductions from their own leaders] really helped them to be more receptive to listening to us."

By 2010, although government officials remained reluctant, the industry was ready to move forward. At a workshop hosted by Civic Exchange, industry leaders began to develop a voluntary agreement in which Hong Kong's major shipping lines would agree to require their ships to switch from highly polluting marine bunker fuel to low-sulfur distillate fuel while at berth in Hong Kong. In October 2010 a total of seventeen companies signed the Fair Winds Charter, which was the world's first voluntary at-berth fuel-switching scheme. After making this strong public commitment, the industry then requested government regulation to level the playing field and retain Hong Kong's position as a leading international port.[72]

Simon Ng talked about how the signing of the charter was a turning point for Civic Exchange's relationship in the policymaking process. In the beginning, when no one is talking about the problem, the NGO sector can be out front to bring attention to the issue, but once business is engaged, the advocates do their best to support their allies in business and government:

We want to give them the face, give them the credit in our public relations and promotion of the initiative. We don't mind giving them more credit. The same is true for government. Once they [the industry participants] sign the charter, we know the effort will move forward, so we invite the government to step in and take over, and we [Civic Exchange] can step way back to help work out the details.

[Can you elaborate on how these credit-giving and public relations efforts work to empower your allies?]

The shipping industry to a large extent deserved the credit, as they embraced the idea, paid for the cleaner fuel, and stuck with their commitment. With public endorsement, the industry could no longer turn back, but they were also pleased to carry on with their reputation enhanced.

It is not just giving credit but also sharing what is going on elsewhere. We have a sharing culture. By sharing, you offer technical and policy design and support to others. ... I'm still close with those people today. That trust is important. They also empower us. They bring their new information to us too. Sharing and learning at the same time.

[You said you are still close with these people. In what way are you close?]

We are still friends today and will get together from time to time.

[Why does your personal relationship matter?]

Because of these connections and our success working together on that project, we think we can do more. ... You gain experience with one project, and then we can do it with another project. The process of NGO-business collaboration first and then handing over to the government. We can do that same process in other cases too.

Christine Loh, the founder and CEO of Civic Exchange, also emphasized the importance of collaboration when she talked with me in 2019 about the Fair Winds Charter and its ultimate influence on China's maritime policy.

You need to engage the stakeholders very early and start to talk with them, so you can understand their concerns and perspectives. [If] you bring people together to do something productive, the thinking shifts, and people begin to understand each other. ... It is a co-learning experience. You have to ask the shipping sector how ships work and really understand that before you can start to talk about how that links to air pollution. ... It is the power of evidence to convince people, not just passion. That is why we partnered with HKUST [Hong Kong University of Science and Technology] to commission research on both the policy side [what other large ports around the world are doing] and on science side [how shipping emissions affect public health].

In this case, Loh's collaborative process and trust-building led her not just to make friends on the inside but also to become a friend on the inside. She was appointed as undersecretary of the environment for Hong Kong in

September 2012 and served until 2017. From her position in the government, she was able first to encourage the Hong Kong government to issue new regulations and then to bring Hong Kong's experience to the attention of national-level policymakers.

Loh could bring the success of the industry's charter and the local government's new regulations and their scientifically demonstrable positive effect on air quality and health in Hong Kong to the attention of regional and national policymakers.[73] In December 2015 China's Ministry of Transport announced new domestic emissions control areas that would require all ships entering China's main ports to use low-sulfur fuel by 2019.

Making friends on the inside is one of the most effective strategies for achieving positive policy change. Publicizing your success to others and giving allies credit for making the success possible not only builds trust with current allies, it can enhance the career prospects of those allies, which may put them in more powerful positions in the future. Raising the profile of allies is also a method to draw the attention of national policymakers to local successes—national-level policymakers are much more likely to listen to their own local government officials and industry leaders than they are to a local NGO. By amplifying the voice of allies in government and industry, advocates are also able to enhance the impact of their message. Ultimately, this improves not only the chances of success of one particular advocacy effort but also the capacity for advocates and their allies to be effective in the future.

How to Scale: Empower and Support Your Friends

Once an advocate has identified potential allies among influential policymakers and cultivated those individuals so they become friends on the inside, how can the advocate scale his or her influence such that it grows beyond access to one person or influence on one policy? Advocates should not just make friends on the inside; they should also empower and support those friends. Friends on the inside who are successful are able to advance their careers, allowing them to become more influential and expand the scope of their networks and their activities. To the extent that advocates can support policymakers to be successful, advocates and policymakers will be able to work together to find innovative ways to diversify and expand collaborative activities.[74]

This section will use the case of the China-US Energy Efficiency Alliance to illustrate how advocates can scale local success to become nationally and even internationally effective. The case highlights the process of how advocates can make new policymaking friends by being useful, staying in touch, and giving credit to allies. They can support those allies by offering technical assistance, political connections, and positive public relations to make small projects successful and noticed by top officials. By supporting a lower-level policymaker in his efforts, advocates were able to support his career advancement. As his career advanced, he gained greater scope and authority for collaborative projects, and the advocates' good track record in working with him made them valuable partners in those larger and more ambitious projects, while at the same time enabling both advocates and officials to diversify and increase the number of allies they had with policymaking authority. In this way, cultivating a single friend on the inside by working on a single, local project can be scaled to include multiple projects over a much larger geographic area. This is the process through which an advocate can expand his or her network, create more nodes in the policy-relevant network matrix, and germinate the formation of new networks that reach new stakeholders and new policymakers.

I first learned about the China-US Energy Efficiency Alliance during a 2015 interview with Barbara Finamore, senior attorney and Asia director of the China Program of the Natural Resources Defense Council (NRDC). Her narrative of how early energy efficiency collaborations between the NRDC and local officials in Jiangsu Province grew into the creation of the China-US Energy Efficiency Alliance[75] highlights the importance of personal connections, sustained long-term commitment to allies, and the numerous ways that advocates can support and empower their allies in government, resulting in better policy that benefits everyone.[76]

The first really big project that we [NRDC] did [in China] was related to energy efficiency in Jiangsu. We went to a conference in Chong Qing hosted by the Demand-Side Management Center set up by the Asian Development Bank. The head of the Jiangsu power company was the head of that, and I kept up with him for years. First we did the demand-side management project. For that project the utility companies pay customers to be more efficient. It was very successful and got the attention of the central government. It took ten years, but eventually the central government made the rules nationwide.

So, we kept in touch. We [NRDC] brought people down to Jiangsu. We brought people from California. We brought California officials to China. We brought Jiangsu officials to California. We brought central government officials to California. We brought Governor Schwarzenegger to China because California was a leader in demand-side management because of their energy crisis. There can be a gap in the connection, but it is still there, and now he [my Chinese contact] is very important.

There are people who sat through all those meetings, who were very quiet, but who sat in all the meetings. Eventually they move up the administrative ladder, and now those people are running the regulation companies. They're not quiet anymore.

[It seems like you're not just empowering allies by giving them information, but you're also empowering them by helping them to make political connections. Can you expand on that?]

We brokered a memorandum of understanding [MOU] between the California public utility commission and the Jiangsu Utility to cooperate on energy efficiency. We brought the California officials over to Jiangsu—they're sister provinces or something. The MOU had two parts—the first was government to government, and the second included the NRDC as implementers. I helped found the China-US Energy Efficiency Alliance 10 years ago—that alliance is now helping other communities form these kinds of agreements.

What started as a small collaboration to assist a local utility company in a single province in China generated not only better energy efficiency outcomes in the province but better energy policy for the entire country and a framework for developing international public-private collaborations on energy efficiency and innovation. At the core of the entire enterprise were a small number of individuals who supported each other, kept in touch, and were able to work collaboratively in more diverse and influential ways over the course of many years. Over time, small, local collaborations grew into a larger network of individuals, organizations, corporations, and governments that were ultimately able to institutionalize productive relationships that facilitated proenvironmental changes at the local, regional, and national levels through technology transfer and policy innovations.

Conclusion

This chapter has discussed what is perhaps the most important advocacy strategy—make friends on the inside by cultivating and empowering allies who are or will be in positions to influence policymaking. Once advocates have identified which individuals they wish to get to know, they can gain access to these policymakers and potential policy influencers by utilizing personal connections, attending events, and serving together on committees with them. Finally, once advocates have gained some access to policymakers, they can cultivate their friends on the inside by being useful, hiring retired bureaucrats, keeping in touch, and giving credit for success, all of which help empower their allies.

Why is making friends on the inside such an effective advocacy strategy? Ultimately, advocates are seeking to convince governments, corporations, and individuals to change their practices. Advocates gain access to policy decision makers and diverse stakeholders through their personal and professional networks. Their network connections enable them to discover how best to frame their policy initiatives so they will be well received by policymakers. Advocates do not do this alone—they work with and through other people and organizations to generate policy change. Making friends with the people in charge of crafting the policies advocates seek to change is an effective way of gaining access to and, perhaps, influence over those policies.

As the foregoing examples have illustrated, advocates gain influence by finding ways to work with policymakers productively over a long period of time, starting with small projects and working toward larger ones. At its core, utilizing a "make friends on the inside" strategy is ultimately about creating win-win policy innovations—policies that are wins for the advocate because they move policy in the direction sought by the advocate, and policies that are wins for the government (or corporations) because they advance their goals as well. When it works well, it creates a win-win cycle where advocacy success leads to policy success, which leads to career advancement and greater prestige for government and business allies, which leads to more policy innovation and collaboration with the advocates, which generates more policy success, and so on.

When I asked one savvy advocate in Beijing in a 2011 interview how advocates can create these win-win cycles, she emphasized that advocates need to work for many years in the background, keeping an issue alive, helping to shape how policymakers are thinking about the issue and policy solutions, waiting until the political opportunity arises to put it forward. In the end, "[advocates] can create a win-win cycle by helping align [NGO goals] with what the government is trying to do anyway." This idea of keeping in close contact with policymakers and cultivating them so you can be part of all stages of policymaking illustrates John W. Kingdon's classic theory of policymaking "streams"—problems, policies, and politics. Effective advocates should be engaging with all three streams simultaneously, so they can help shape the problems that policymakers are examining, offer concrete policy solutions to those problems, and influence the politics around the process by supporting their allies in the process.[77]

This chapter has shown how making friends with those in a position to influence policy, and those who are likely to be able to influence it in the future, is an effective way of generating win-win policy outcomes for advocates. By cultivating friends on the inside and participating in productive collaborations, even if they start small, advocates build a network of allies who can be activated when opportunities arise to make progress on issues of interest.

As these networks become stronger, larger, and more diverse, advocates increase their capacity to influence more policymakers. Indeed, as was the case for two of the individuals highlighted in this chapter, Myung Rae Cho and Christine Loh, if an advocate is particularly adept at making friends on the inside, he or she might wake up one day no longer an advocate but rather a policymaker, appointed to a position of authority by one of those same friends who had been cultivated.

6 Make It Work Locally: Local Models, Global Solutions

One of the biggest challenges for advocates everywhere is how to scale their advocacy—how to replicate and expand success achieved in a single local situation into something that can work in other contexts. In East Asia in particular, scholars have regularly asserted that advocates tend to focus on only their local issues and fail to transform local advocacy success into national or international advocacy success, resulting in a stay-local phenomenon that has been crippling to the expansion and development of civil society and advocacy in the region.[1]

This chapter takes a different view, asserting that East Asian environmental advocates are regularly able to scale their success. While it may be rare to see some of the types of advocacy scaling found in North America or Europe, such as the transformation of a local advocacy group into a multimillion-dollar, professional nongovernmental organization (NGO), or the growth of a single advocacy event into a nationwide movement that involves hundreds of protests and thousands of participants, we do see local success scaled in East Asia. Advocates who are successful at "making it work locally" are often able to leverage a single local success into something that has a longer-lasting, broader impact. In fact, the research for this volume suggests that "make it work locally" is one of the most effective strategies available to advocates seeking large-scale, long-term change.

The Connected Stakeholder Model helps us understand why and how a "make it work locally" strategy is particularly effective for advocates seeking policy change. In all cases, advocates are able to create and utilize their networks to facilitate the development and dissemination of good policy. This chapter will use one example each from South Korea, Taiwan, China, and Japan to illustrate four different ways that local advocacy success can be scaled.

An examination of the creation and work of the Citizens' Green Seoul Committee will show how community-based organizations focused on neighborhood environmental concerns can institutionalize citizen voices into a city's policymaking processes, and that decision- and policymaking process can then serve as a model for broader policymaking throughout the country. This is an example of success scaled through institutionalization. Advocates took what had been relatively ad hoc arrangements and regularized them, ensuring that their voices would be incorporated into policymaking for years to come and enabling other branches of the government to adapt similar procedures for their own policy areas. To use the language of the Connected Stakeholder Model, formal networks were intentionally created to connect diverse stakeholders, facilitating both the development and the implementation of good environmental policy.

A discussion of the development of Taiwan's anti–naphtha cracker protest movements will explain how a NIMBY (not in my backyard) protest that focused on halting industrial development of a specific site can generate lessons that can be productively utilized by later activists in different localities—success was scaled through incremental innovation. By learning from their predecessors, advocates were able to innovate and grow more sophisticated in their capacity to resist unwanted industrial development. These advocates diversified the kinds of stakeholders who were engaged in their advocacy effort and extended their networks from one town to another until they eventually covered the entire country.

The 26 Degree Campaign in Beijing illustrates how a small coalition of local NGOs was able to leverage its local success into new national policy standards—success was scaled through local-national policy connections. Although the movement was initially intended to address a specific problem in a particular locality—air pollution in Beijing—its success in that one locality served as a kind of pilot for new regulations. When the new regulations were successful locally, it became possible to develop and apply the same rules nationally. Seizing on political opportunities made possible by the upcoming Olympics, activists were able to identify and recruit influential people who served as critical nodes in policymaking networks, facilitating the rapid scaling of the city's successful policy to the national level.

Finally, the KitaQ network shows how a Japan-supported pilot composting project was able to expand not just nationally but internationally—scaling through transnational networking. Japanese environmental advocates identified a good pilot project in which they could test a new composting technology. When they were able to demonstrate significant municipal solid waste reduction and positive social and health benefits in that first site location, the composting system was copied and implemented elsewhere. The Institute for Global Environmental Strategies (IGES), a Japan-based NGO, facilitated the development of an international network of municipalities that could implement the system, spreading the local success internationally.

Since the particulars of what makes an advocacy effort successful are often highly contingent on local factors (e.g., Was the mayor supportive? Was it sunny on the day of the protest?), the emphasis in the discussion of these cases will be not so much on how the advocates achieved local success but rather on how they were able to scale that success. Of particular interest here is how a "make it work locally" advocacy strategy can achieve not just a one-time solution to a local problem but also longer-term solutions in geographically larger and more diverse areas.

Citizens' Green Seoul Committee: Scaling Success through Institutionalization

The Citizens' Green Seoul Committee[2] was established by municipal ordinance in 1996 so "Seoul citizens and enterprises [could] together participate in city administration to create a pleasant environment and to contribute to the conservation of global environment."[3] The committee was an example of a local implementation of Agenda 21, a nonbinding resolution adopted at the 1992 Earth Summit in Rio that that commits countries and localities around the world to adopting local regulations, policies, and governance structures to address the impending climate crisis.[4] Set in the context of South Korea's recent democratization—its democratic constitution had been adopted less than five years earlier, and Seoul's mayor was elected for the first time just one year earlier—the committee, which included many prominent environmental, feminist, and democracy advocates among its early members, was also viewed as an early attempt to put the democracy and environmental movements' vision of "participatory governance" into practice.[5]

From its very early days, the committee was established as a collaborative effort that would bring members of the government, business community, and civil society together to co-develop policy. It is co-led by three chairpeople—the mayor, an elected member representing civil society, and an elected member representing enterprise—and one hundred additional members who consist of ex-officio members (relevant city officials such as the chief officer of the Climate and Environment Headquarters and the director-general of the Water Circulation Safety Bureau) and members commissioned by the mayor for their knowledge and expertise related to the environment (these members are often academics). The one hundred members then serve on one or more of four subcommittees (climate and energy, ecosystems, resource circulation, and environmental health). Meetings of the full committee are called at least twice a year, with agendas and project plans submitted ahead of time for deliberation. Commissioned members serve two-year terms, can serve a maximum of two terms, and are required to recuse themselves if conflicts of interest emerge in the course of their duties.[6] The design of the committee helps ensure that multiple perspectives are included in the policymaking process and requires that those voices change over time, remaining relevant.

One of the committee's earliest projects was the Nam Mountain Restoration Project, which sought to remove a number of apartment buildings built during the Japanese occupation and replace them with a botanical garden and a traditional *hanok* village. This was quickly followed by a series of projects and programs designed to increase the livability of Seoul and engage local residents in their neighborhoods—for example, the Revitalize Our [Neighborhood] Hill campaign, the Great Streets to Walk program, the No-Wall Movement, and the Create a Beautiful City to Walk program.[7]

By 1998 the Citizens' Green Seoul Committee had established branches in each of Seoul's twenty-five administrative districts (these district committees also contained one hundred members each), and each of these district committees was active in gathering citizens, community businesses, and representatives of civil society to design, plan, and (eventually) manage the new community green spaces. As green spaces in neighborhoods were restored, residents as well as city officials began to realize the benefits of co-developing urban policy, and Seoul's participatory governance system as it was worked out in the Citizens' Green Seoul Committee gained legitimacy.[8] Seoul's fundamental governance model began to shift from "government-oriented projects to public-private, partner-oriented projects," a shift that was compatible with new ideas about urban governance and was also fiscally attractive after the Asian financial crisis of 1997 significantly constrained all government budgets in the region.[9]

Mayor Goh Kun, who served as mayor of Seoul from 1988 to 1990 (appointed) and 1998 to 2002 (elected), played an important role in developing and expanding the governance model of the Citizens' Green Seoul Committee. During his early days as mayor, he worked closely with the committee on the Revitalize Our Hill campaign, and over the course of his administration he worked to replicate the committee's successful governance model in other areas of city administration beyond those related to the environment.[10] For example, his Committee for Administrative Reform was formed largely of citizens and was designed to offer guidance to the mayor about city priorities and planning; the Joint-Inspection Team was a public-private group that inspected public works projects to reduce corruption; and the Regulation and Reform Committee offered guidance on which government regulations should be eliminated or revised.[11] These efforts to engage citizens in governance were highly regarded, and Goh served as prime minister under three presidents (Roh Moo-hyun, Kim Dae-jung, and Kim Young-sam) and briefly as interim president (March–May 2004) when Roh was unable to serve.

The current mayor of Seoul, Park Won-soon, has expanded the model even further as he sought to replace an "economy-centered development agenda with a people-centered living welfare agenda."[12] Furthermore, because of Seoul's large size (twenty-six million people in the greater metropolitan area)[13] and its dominant position in South Korea's government and business, policies made by Seoul mayors affect the whole country. Park's One Less Nuclear Power Plant campaign, launched in 2012, has been actively engaging citizens to participate in a number of energy conservation, energy sharing, and renewable-energy-generation efforts to decrease Seoul's total energy use. The results have been impressive, not only affecting Seoul's energy and climate but also shifting the energy profile of the whole country. The campaign exceeded its energy reduction goals (2.04 tonnes of oil equivalent by June 2014) and reduced its municipal solid waste and air pollution even while generating new jobs and creating more green space.[14] Seoul's success has contributed to the ability of the Ministry of Trade, Industry and Energy to be even more ambitious with its renewable energy goals for the country—it aims for the newest planes to be composed of 35 percent renewables by 2040.[15]

The Citizens' Green Seoul Committee began as many local environmental committees do—as a local effort of the mayor's office to engage the citizens more in the care of their community. It has been enormously effective in working with a succession of mayors to transform Seoul into one of the world's most sustainable cities.[16] As the current secretary general, Seong Hwan Min, reflected to me in a 2019 interview, "Making something better for twenty-five years is pretty unusual. This kind of collaborative environmental governance isn't common, but it works."

The committee's success did not remain isolated in the policy area of the environment or even in the city of Seoul—it has been able to scale its success through institutionalization. It created and has updated a set of institutionalized governance practices that have outlived the charismatic visionary leaders who originally founded the group. The committee's success in shaping successful environmental policy in ways that have benefited the city and its residents has spread to other policy areas within the city administration, and Seoul's mayor has been disseminating this model internationally as well. In an interview with me in 2019, Park expressed his sense of responsibility for using his experience as Seoul's mayor to disseminate best practices around the world.

> We have a great responsibility for the next generation and for the future. We have a responsibility to help the cities around the world because climate change isn't confined to one city, but crosses [national] boundaries. ...
>
> The challenge of climate change is about average citizens because the citizens are feeling the issue directly. It is our mission to solve it for future generations. We [mayors] are closer to the citizens than the central government and have more diverse solutions at our disposal.
>
> We can't do it without the citizens. We asked the citizens to be part of the eco-mileage program, to use public transportation rather than cars. Now, more than two million citizens have participated in eco-mileage. This wonderful outcome can be done [only] by the power of the citizens, not by me.

Park has been active in sharing Seoul's experience with other mayors around the world. In 2015, Seoul hosted ICLEI's World Congress, and Park was elected to serve as the organization's president. He also serves as a leader in the Global Covenant of Mayors for Climate and Energy. The collaborative governance model of the Citizens' Green Seoul Committee is one of the core features of urban policymaking that Park seeks to disseminate to cities around the world that seek to emulate Seoul's success.

The Citizens' Green Seoul Committee has been able to scale its success by institutionalizing its multistakeholder governance model. It connects diverse stakeholders in a large network of citizens who are working together to improve the livability of the city. The term limits of committee members mean that people regularly cycle off—thereby extending the network itself, as those rotating off retain personal and professional connections with those remaining on the committee even as they build new connections with others. When some of the former members of the committee gain power and seniority—in civil society, business, and government—they gain access to additional networks, as well as more influence over policymaking in national and even international settings. These connected stakeholders are then able to make and implement policies that improve the environment for everyone.

Antipetrochemical Protests in Taiwan: Scaling Success through Incremental Innovation

As chapter 4 explained, NIMBY protests were the origin of nearly all environmental politics in East Asia. The environment generally did not become a political issue until it had negative consequences for human health and livelihoods—people getting sick and dying, fish and crop yields dropping off. Globally, NIMBY protests are usually short-lived. The cause of a protest is highly local—residents don't want the nuclear power plant, petrochemical company, dam, or other construction—in their own backyard. Eventually, the problem is resolved one way or the other; the power plant, factory, or dam is either built or not built, and with either outcome the protests and advocacy stop.[17] Occasionally, NIMBY protests can generate positive long-term policy changes[18] or grow into larger social movements.[19]

This section describes how a local advocacy effort in Taiwan to halt the construction of a new naphtha-cracker industrial plant (a large, industrial petroleum-processing facility) was able to scale its success through incremental innovation. Advocates who had been successful in one community were able to form an increasingly large and diverse network of advocates who could become reactivated when a new community was threatened with similar development projects. Taiwanese advocates have been able to scale one success into several by connecting newly threatened communities with those that had successfully blocked development, enabling experienced advocates to pass on successful advocacy strategies and empower new advocates to innovate on old strategies to take full advantage of new resources and technologies. As stakeholders in one NIMBY fight are able to connect with stakeholders involved in similar fights in different parts of the country, the connected stakeholders are able to activate their numerous personal and professional networks, increasing the political pressure on policymakers to make proenvironmental decisions.

Taiwan's petrochemical industry began during the period of Japanese occupation and was targeted by the postwar government as an area for investment to spur rapid industrialization and economic growth.[20] The state-owned China Petroleum Corporation built its first naphtha cracker in 1968, and it constructed three more between 1975 and 1984. The country at the time was under military rule, the population was focused on rapid economic development, and the environmental and health costs of naphtha crackers were not yet widely understood. As a result, these first four industrial complexes faced very little local opposition.[21] By the late 1980s, however, the politics around these large-scale industrial complexes had shifted dramatically. Inspired by protests that came to be called the Lukang Rebellion (discussed in chapter 4), in which local activists successfully blocked DuPont's proposed construction of a titanium dioxide plant near the city of Lukang,[22] and legally empowered when martial law was lifted in 1987, Taiwanese began to resist new naphtha cracker installations whenever they were proposed.

The first sets of anti–naphtha cracker protests occurred in 1987, immediately following the lifting of martial law. That year, Taiwan hit peak economic growth, with its gross domestic product growing more than 15 percent.[23] Formosa Plastics, one of Taiwan's largest petrochemical companies, first applied for government approval for a naphtha cracker project in 1973, eventually winning approval in 1986.[24] Its first choice for a site for the plant was in Yilan County, located on Taiwan's northeastern cost about fifty kilometers from the capital city of Taipei.

While convenient from an operational perspective, the location was hostile politically. The nonpartisan local magistrate (the chief county executive, a position similar to a governor) had formerly worked for Formosa Plastics and was vocal in promoting a proenvironmental political agenda.[25] Local Democratic Progressive Party (DPP) activists linked protection of Yilan's environment with their broader prodemocracy, anticorruption political agenda, supporting local activist groups and working with the magistrate to stage protests in Yilan and also in Taipei.[26] Eventually, Formosa Plastics gave up its efforts to build its sixth naphtha cracker in Yilan, and the victory of the environmentalists and the opposition party over the large corporation and the conservative Kuomintang (KMT) government contributed to the creation of Yilan's moniker as the "holy land of Taiwan's democracy."[27]

Simultaneous with Yilan's successful effort to block the construction of a naphtha cracker in its backyard, Kaohsiung residents were fighting China Petroleum's efforts to build the fifth naphtha cracker. Kaohsiung City had long been heavily industrialized. The Japanese had built a navy fuel plant there, and it was already home to Taiwan's first and second naphtha crackers, as well as the center of Taiwan's steel and shipbuilding industries. While all of this heavy industry meant that Kaohsiung residents were used to pollution, it also meant that they were paying a heavy and disproportionate cost for Taiwan's economic miracle.

When martial law was lifted in 1987, residents began to protest against the proposed new complex. They blockaded construction sites, had violent clashes with police, allied with the DPP opposition party, and encouraged local Buddhist temples to stage martial arts demonstrations in conjunction with their antiplant environmental protests. In the end, the KMT-ruled government and China Petroleum were able to offer locals a package of community-based benefits such as subsidized gas, free school lunches, and a promise to relocate the new cracker within twenty-five years. These, combined with intense police crackdowns against protesters, weakened the opposition; construction began in 1990, and the plant went into operation in 1994.[28]

Although the protests against the fifth and sixth naphtha crackers occurred simultaneously with protests against DuPont's plans for a new chemical plant in Lukang, these efforts were not well connected. All three anti-chemical-plant protests took on NIMBY protest characteristics[29] and were largely focused on local personalities, impacts, and dynamics. By 1991 the ruling KMT and industry allies had figured out how to target communities more likely to accept large-scale projects. They proposed gigantic complexes that benefited multiple industries, built on reclaimed land that would not require the sale of farms or other private property, and preemptively offered locals attractive packages of financial and community benefits.[30] Although the Taiwan Environmental Protection Union, Taiwan's leading environmental group, tried to counter these efforts and organize locally, its efforts were too weak and too late.[31] Formosa Plastics was able to site Taiwan's sixth naphtha cracker as part of a gigantic industrial facility in Mailiao, Yunlin County, which now occupies thirty-two square kilometers and includes fifty-four industrial plants, producing everything from PVC piping to spandex, as well as an oil refinery, a power plant, and a port.[32]

By the late 1990s, however, the dynamics had shifted. Politically, the DPP gained power in 2000 and held it until 2008. As a governing party, the DPP became more probusiness, so advocates began to seek out a broader coalition of supporters and move away from partisan politics.[33] As discussed in chapter 4 the dramatic expansion of the global environmental movement, as well as the development of the internet and social media, greatly enhanced the resources that local environmentalists could enlist in their efforts and made it significantly easier for advocates to find one another, coordinate activities, and activate networks of supporters when new threats arose.

The protests against the eighth naphtha cracker, which began amid the 1999–2000 presidential campaign, illustrate how sophisticated the advocates had become. Academics in particular played important roles as network facilitators, technical experts, and organizational focal points. What looked like a made-to-order opportunity for the KMT vice-presidential candidate to bring a big industrial plant, and its accompanying jobs and income, to his hometown of Chiayi unraveled when the KMT lost the presidential race and control of the Legislative Yuan at the national level, and a DPP candidate won the local magistrate race in 2002.[34] Efforts to site the eighth cracker were shelved for a bit but were revived in 2008 when the KMT regained control of the national government and sought to make a new industrial park in Kuokuang a national investment project.[35]

Activists were ready. Academic advocates, many of whom had been involved in the protests related to earlier crackers, mobilized. Science, technology, and public policy faculty created panels and forums at the annual meetings of their professional associations, presenting papers and other research on the hazards of the plants in which they highlighted the negative health and economic outcomes in communities like Mailiao, which had accepted the sixth naphtha cracker, and juxtaposed those negative outcomes to the vibrant and diverse economy and society in Yilan, which had blocked plant construction.[36] Subsequently, many of these conference papers and presentations were turned into authoritative peer-reviewed research publications, offering strong scientific evidence of harm caused by the crackers. These scientific findings were then further publicized by the press, increasing the pressure on public officials to disallow the project.[37]

Local stakeholders such as farmers and fishers joined with students, medical experts, lawyers, NGO advocates, and religious leaders to oppose construction of the plant. Celebrities, youths, and members of the cultural community joined the advocacy effort as well, creating online memes that went viral, publicizing images of the natural wetlands that would be destroyed by the project, and composing songs like "The Song of the White Dolphin" whose popularity enabled the activists to spread concern about the issue far beyond those living near the proposed development site.[38] The advocates spread a message about the importance of economic diversity and local identity. Tu Wen-ling, a professor of environmental planning at National Chengchi University and longtime environmental advocate, explained to me during a 2019 interview how she talks to residents living in threatened communities:

> People make a choice at the beginning. You think it is a choice about economic development, but you will lose your identity and your autonomy for economic development, which will become dominated by the big corporation. Then, you'll be at their mercy. Instead, if you have a more modest economy, you can promote more diverse forms of economic development. Yilan is a good example. They might not be as rich as [they] may have been with the naphtha cracker, but everyone has their small thing. Their economy and lives aren't dominated by one industry.

Although Taiwan's petrochemical industry has an enormous influence on the country's economy—it accounted for nearly a third of the country's gross domestic product in 2010[39]—it has not been able to site a new petrochemical complex in more than twenty years, and significant opposition arises even when the industry seeks only to update an existing plant. What began as a small set of NIMBY protests by small groups of residents in dispersed communities has, through incremental innovation, become a large, diverse, geographically extensive network of allies who can mobilize rapidly when threats arise. Furthermore, these advocates are not just limiting themselves to antipetrochemical development; many cross-pollinate with the antinuclear community, which is borrowing and sharing its diversified resistance tactics.[40] Taiwan's environmental advocates were able to "make it work locally" and then, through incremental adjustments, scale their success not only to other communities on the island but also to other issues where advocates face similar political, commercial, social, and health dynamics.

26 Degree Campaign in Beijing: Scaling Success by Seizing a Political Opportunity

In 2000, the then-small community of environmental advocates based in Beijing began a monthly salon to gather informally and discuss environmental issues. During a few of their gatherings, they started to discuss what they could do to raise awareness and improve energy conservation in the city. Many of the advocates had experience living abroad, so they drew on their collective ideas, eventually focusing on the problem of overcooling.[41] As Sheri Liao, founder of Global Village, explained to me in a 2015 interview in Beijing,

> I think I got the idea when I was in the US. I would feel very confused because I would go into the supermarket in the summer, and I would have to wear a sweater [because it was so cold]. "This is ridiculous!"
> I thought at the time that China would never do that kind of thing, but later I found that China was following the same path. So, I discussed it with some NGO people, and we came up with the idea of twenty-six degrees in summer and twenty in winter.

In the summer of 2004, six NGOs (Global Village of Beijing, World Wildlife Fund in China, the China Association for NGO Cooperation, Friends of Nature, the Institute of Environment and Development, and Green Earth Volunteers) launched the 26 Degree Campaign to urge hotels, office buildings, malls, government offices, embassies, and other public spaces to set their air conditioners to twenty-six degrees Celsius or higher during the summer months to reduce energy consumption.[42] The NGOs partnered with the media to raise awareness of the twenty-six degree goal and pressure large buildings, especially hotels, to raise their indoor summer temperatures to save money, improve health, and lower Beijing's CO_2 and air pollution.[43]

NGO volunteers went around Beijing with thermometers and measured the indoor temperatures of a wide range of large buildings—including hotels, malls, and government buildings—and worked with journalists to expose how far below twenty-six degrees these large and highly populated spaces were, emphasizing both how much CO_2 and polluting emissions would be saved if the air conditioners were set to a higher base temperature and the negative effect that the frigid air had on the health and comfort of the public. Beijing's mayor at the time, Wang Qishan (he is now serving as the vice president of the People's Republic of China), got involved in promoting the

26 Degree Campaign. He sent his deputy mayor, Zhang Mao, to hotels for surprise temperature inspections, drawing attention to the importance of energy conservation and air conditioner settings in the summer as a way to reduce Beijing's terrible air pollution and to mitigate climate change.[44]

The 26 Degree Campaign advocates were quite savvy with their choice of advocacy issue, and they got lucky with their timing. Just a month after they launched the campaign in June, there was a massive heat wave in Beijing that caused brownouts across the city as electricity supplies ran short,[45] further heightening the public's awareness of the connection between air-conditioning use and electricity supplies. Additionally, since the government had made a commitment to hold a "green Olympics," local government officials, as well as the hospitality industry, were particularly attuned to the importance of addressing environmental issues ahead of the 2008 Olympics.

Following up on the summer's 26 Degree Campaign, in November volunteers began distributing "26 degree commitment cards" to office buildings, shopping malls, restaurants, and hotels. Beijing Global Village targeted the eighty Olympic Games service hotels in particular and succeeded in convincing ten of them to commit to keeping the temperature setting of their public spaces to twenty-six degrees or higher.[46]

The next summer, in 2005, the 26 Degree Campaign got a large boost when Premier Wen Jiabao gave a speech entitled "Spurring the Development of a Conservation-Minded Society" in which he declared that all government offices would have their air conditioners set to twenty-six degrees or higher and promoted more casual business attire during the summer, emphasizing that jackets and ties were not required for most government employees during normal business functions. He gave his speech the day before Prime Minister Koizumi Junichiro appeared at a press conference wearing a short-sleeved Okinawan shirt to launch Japan's Cool Biz campaign, discussed in chapter 9. Wen would have known about Koizumi's planned announcement, as well as the Cool Biz fashion show planned for the world expo in Tokyo on June 5.

Wen's speech was followed by a notice by the Departmental Affairs Management Bureau of the State Council and the Central Committee of the Chinese Communist Party that all of their departments must set air conditioners at twenty-six degrees or above. Beijing's municipal government then committed its own office buildings and those of the party to following the same standard, the mayor sent an open letter to all the corporations in the city urging them to do the same, and several national embassies, government and academic institutes, and other entities also agreed to commit to this standard.[47]

After the official announcements, NGO volunteers entered the public buildings that had committed to the new standard, as well as some that had not, and reported their temperatures, applying pressure to those that were not meeting their commitments. They also urged NGOs outside Beijing to engage in similar actions in their own communities, and fifty-one new non-Beijing NGOs joined the effort.[48]

In the years that followed, the system was expanded and institutionalized, with measurements taken on the same days (July 7 and 23)—volunteers received training on which buildings counted as public spaces that could be accessed and measured, as well as how to take and record the temperature—and much of the effort was coordinated using the social media platform Weibo. By 2012 it involved fifty volunteers in Beijing and had expanded to nine cities. By 2015 volunteers in thirty cities were coordinating over WeChat, and the effort was funded by HSBC as part of a larger project related to low-carbon households.[49]

The 26 Degree Campaign is a particularly successful example of how a "make it work locally" strategy was scaled utilizing local-national policy connections. The fact that the locality where the campaign took place was the capital city and that the capital city was hosting the Olympics were both significant. These two factors meant that the central government was particularly attuned to the events occurring at the municipal level. Additionally, the local advocates were able to leverage good connections with the press to help engage the local government, the national government, and the corporate sector to promote the effort. As will be explored in greater depth in the next chapter, it was also helpful that following the goals of the campaign would also generate positive economic value for companies.

Beijing's NGO advocates, its proenvironmental government officials, the press, and supportive members of the business community all worked together to achieve local success and enable the campaign's local recommendation to become national policy. Advocates at all levels and in all sectors were persistent—renewing and expanding the campaign every year until it was nationally successful. Furthermore, they continue to this day in a more organized, institutionalized, and well-funded fashion. Winning a policy victory does not necessarily ensure long-term behavior change, so advocates frequently need to commit to long-term support for their policy gains to be enduring.

KitaQ Composting Network: Scaling Success through Transnational Networking

A final example of how a "make it work locally" strategy can scale comes from an international collaboration initiated in Japan and spread globally. IGES was established in 1998, with support from the Japanese government, to promote sustainable development around the world. Its mission has been focused on "conducting strategic research on policies and practical measures ... based on a foundation of natural and social scientific research as well as technological research on global environmental issues."[50] From the beginning, the organization has focused on the Asia-Pacific, "as the region holds more than half of the world's population and is experiencing rapid economic growth, and thus plays a critical role in the protection of the global environment."[51]

In 2001 IGES identified solid waste management as a critical issue for improving sustainability in the Asia Pacific[52] and began researching ways that Japan's success in managing municipal solid waste could be adapted for other Asian contexts.[53] Soon afterward, an opportunity to move forward with a pilot project in Indonesia arose through IGES's connections with the Kitakyushu city government via IGES's Kitakyushu office.

In 2001 Surabaya, Indonesia's fourth-largest city with a population of three million people, faced a solid waste crisis when local resistance forced the closure of one of its largest landfill sites for the city.[54] In order to address this problem, in 2002 the city invited a team of experts from its sister city, Kitakyushu, to come to Surabaya to investigate its municipal solid waste challenges and develop a solution. The investigation revealed that while waste collection in the middle- and high-income areas was sufficient, the poorer areas suffered from inadequate collection.[55] Together the Kitakyushu International Techno-cooperative Association, Pusdakota (a local Surabayan NGO), and IGES worked with Takakura Koji, a Japanese scientist, to develop an easy household composting method using fermenting microorganisms that would be fast, clean, and efficient in the Indonesian context. They also developed a collection and distribution system that would engage households while also cleaning and beautifying their neighborhoods.[56]

Through this multiorganizational and multicity collaboration, IGES helped Pusdakota establish a model composting facility as a pilot demonstration project, using organic material from vegetable markets and street maintenance activities. In the first year, Pusdakota worked with first ten and then ninety households to distribute the special composting baskets designed to enable households to compost their kitchen scraps using the Takakura method. Participating residents would receive the compost basket for free and learn how to use it to generate compost. Additionally, Pusdakota would buy finished compost (US$0.07 per kilogram), allowing participants to supplement their income while improving the hygiene of their household, reducing their solid waste, and generating rich compost for use in their own kitchen gardens.[57]

When the pilot project in Pusdakota proved successful, the City of Surabaya expanded the pilot to include more composting centers across the city. The city would buy the composting baskets, helping to support Pusdakota's activities, and then Pusdakota would give the baskets away to residents who were willing to undergo training and be part of the program. Pusdakota and PKK, a local women's group, also trained environmental cadres to work with other NGOs and community groups to distribute the household compost baskets, explain how to use them, and promote not just composting but also the separation of waste and recycling within the household. The environmental cadres also follow up with participants and troubleshoot common problems, so the dropout rate is low.[58]

In the first five years of the program, Surabaya reduced its municipal solid waste by 30 percent (from 1,819 tons per day in 2005 to 1,241 tons per day in 2010), its sixteen composting centers had created seventy-five new jobs for low-income residents, seven thousand tons of compost were produced for use in city parks and roadside stands of trees, and green space in the city had increased by 10 percent.[59] Surabaya's success generated national policy change—community-based composting was added to the Indonesian government's National Solid Waste Management Law in 2008. Surabaya also won many awards for its efforts, including the Adipura Award (Indonesia's clean city award), the 2005 Energy Globe Award (from Austria), the UN-Habitat award for best practices in 2007 and 2008, and Association of Southeast Asian Nations' Environmentally Sustainable City Award in 2011.[60]

With the success of its pilot project in Surabaya, IGES sought to scale its success in several different ways. First, IGES and Kitakyushu expanded and diversified their collaborations with Surabaya City. What began as a joint project to solve a landfill-overflow problem and develop a better way to manage municipal solid waste grew into joint projects related to water quality management, water purification, clean energy, and public health. Over time the diversity and complexity of their cooperative efforts keep growing; in 2018 IGES signed an academic partnership agreement with PT Sarana Multi Infrastruktur (an Indonesian finance corporation) to work with the Institute Technology Surabaya, among other collaborators, to develop new research on green finance, research related to Sustainable Development Goals, and sustainable infrastructure.[61]

Additionally, IGES facilitated the dissemination of community and household composting as a method to improve urban sustainability in other cities across Southeast Asia. It did this by convening a series of workshops in 2008–2009 that gathered municipal leaders together to explain how the Surabaya model worked. These workshops were held first in Surabaya, and then in Bangkok, Thailand, Bago, the Philippines, and Sibu, Malaysia. At each workshop, municipal officials and NGO representatives would come and learn about Surabaya's model and were invited to apply to the Japan International Cooperation Agency for grant assistance to develop a program in their own city.[62] By 2011 the fifteen cities that were using the method developed in Kitakyushu had formed the KitaQ System Composting Network to share information and facilitate its adaptation in other cities.[63] By 2018, more than thirty cities across the globe had adapted the Takakura Composting Method for their own communities.[64]

Finally, IGES's experience in nurturing the Kitakyushu-Surabaya collaborations has enabled it to promote other, highly sophisticated and diverse collaborations that bring together municipal-level governments, local and international NGOs, international development funding agencies, and local and international corporations to promote a wide variety of sustainability initiatives. For example, in 2012 officials from Hai Phong attended one of the IGES-hosted meetings to disseminate the KitaQ composting system in Siem Reap, Cambodia. At that meeting, which included a diversity of representatives from municipal governments in the region, including Hai Phong, and NGOs such as Clean Air Asia, IGES staff presented information about the Surabaya composting pilot. Hai Phong was already involved in technical cooperation with Kitakyushu, but after 2012 the relationship between the two cities deepened.

Facilitated by IGES, Kitakyushu and Hai Phong expanded their collaborations around water quality and waste water management, as well as capacity
development for plant management. In 2014 the two cities signed a sister
city agreement and worked with Nippon Steel, Sumikin Engineering, Amita,
and NTT Data Institute Management Consulting to develop a low-carbon
development plan for Hai Phong City, modeled on the Surabaya example,
to turn the city into a green port city.[65] Hai Phong had also become one
of seven pilot cities in Clean Air Asia's Cities Clean Air Partnership program,
which helped the city develop a wide range of activities related to improving air quality.[66] By 2018 Kitakyushu and Hai Phong were collaborating
in a range of policy areas, not just composting but also municipal water
and sewage treatment, small-business development, transportation, renewable energy, sustainable agriculture, and cultural exchanges. The people
and entities involved in the partnerships included private global companies, local companies, city officials, national government officials, funding
agency officials, NGO specialists, and academics in both countries.[67]

What started as a small pilot program in a single neighborhood in a single
Indonesian city has grown into a new waste management plan for the fourth
most populous country in the world, an international network of cities that
are reforming their municipal waste management systems, and deepened and
expanded city-city collaborations that involve multiple levels of governance,
as well as the nonprofit and private sectors, pursuing a diverse range of proenvironmental projects. IGES was able to utilize a "make it work locally"
strategy through careful piloting, clear documentation of the successful
pilot, facilitation of dissemination opportunities, networking of interested
cities and organizations, and nurturing of innovation that resulted in many
new spin-off projects and relationships that grew from the initial project.

Conclusion

Around the world, most environmental advocacy efforts begin when residents mobilize to solve a pressing issue threatening their local community.
While this origin story is common, what is much less common is for advocacy that emerged out of a need to solve a particular community's problem
to move beyond that particular community. Whether it is a NIMBY protest seeking to halt a development project, citizens seeking redress from a

polluting company, or residents advocating for cleaner and greener public spaces, local advocacy efforts typically stay local and do not scale.

There are many reasons for the parochial nature of most environmental advocacy. Two of the most common and obvious are the following: (1) residents have only so much time and energy, so once their problem is solved, they don't have much interest in trying to reach out to others who are facing similar problems. (2) The factors that led to success in their own context are so particular (e.g., a friendly mayor, a charismatic community leader, timing during an election year) that it becomes difficult to find ways to translate the success found in one community to another.

This chapter has highlighted an example of a "make it work locally" advocacy effort located in each of the four countries of interest. As the cases demonstrated, there is no single strategy that can help advocates turn a community-level success into one that transforms multiple communities. The advocates and projects profiled here used institutionalization, incremental innovation, local-central policy connections, and international networking to scale their success to other communities.

Although the cases do not point toward a single method for scaling "make it work locally" strategies, they do highlight the benefits of this strategy and help explain why it is so effective for advocacy. Across all four cases, we see three aspects of a "make it work locally" strategy that help advocates become successful: providing a proof of concept that inspires others, learning from others, and networking.

Perhaps the single most important element of a "make it work locally" strategy is that it enables advocates to prove that an idea for environmental innovation can work. If the innovation can work in one place, it can inspire others to implement the same policy or project in their own communities. The Citizens' Green Seoul Committee showed how participatory governance can be a productive way to develop urban policy. The successful protests against Taiwan's sixth naphtha cracker demonstrated that citizens can mobilize successfully against large corporations and their political allies. The 26 Degree Campaign in Beijing highlighted the ways that small-scale community activists can come up with a good idea (limit air conditioner use) and convince public officials to turn that idea into policy. Finally, Surabaya's success—technical, political, social, and environmental—with its pilot composting project inspired others to try it in their own cities.

Without examples of success, it becomes difficult to convince others that improvement is possible.

Second, once a concept has been successful in one place, it makes it feasible for others to learn and innovate. Adapting an existing model is much easier than trying to generate a new solution from scratch. Communities that have been inspired to follow the example of those who have succeeded can learn not only from their success but also from their failures. They can avoid pitfalls that the first community fell into. They can modify policies that worked to fit their own circumstances, and the experience of the community that performs the second implementation can then be used to form even more knowledge for subsequent communities. In theory, the accumulated experience of experimentations with similar policies or projects should make it easier and easier for later-implementing communities to replicate the success of their predecessors. All four cases show how learning among advocates and policymakers helped shape subsequent iterations of the strategy, enabling the initial success to be scaled beyond a single community.

Finally, all four examples highlight the importance of networks and networking to advocacy success. As was discussed in chapter 3 fundamental to scaling the success of advocacy is the scaling of advocacy networks. As advocates are able to build and diversify the network of individuals and organizations with which they collaborate, they are able to expand their policy influence. In Seoul, what began as an ad hoc group of environmental and democracy activists became a formal committee integrated into the city government's environmental policy process. This governance model was then deepened when branches were developed in each of Seoul's twenty-five districts. The network was also broadened as the city copied and adapted the success of the Citizens' Green Seoul Committee to other areas of city administration and helped shape the evolution of South Korean citizens' ideas about democracy and how it is practiced in their communities.

In Taiwan, local activists worked to fight industrial development in their community and engaged with academics and other policy professionals to help legitimize their claims. This network of academics and policy professionals was then reengaged as new communities were threatened, and they learned something more about the techniques that worked each time they fought a new NIMBY battle. Their network, which now includes not just academics but also musicians, celebrities, and technology experts, is diverse and coordinated enough to engage and innovate the moment any new battleground is identified.

Beijing activists were able to leverage the assets of a small number of activists in a small number of largely all-volunteer organizations into a citywide effort by connecting with one another. None of the organizations would have been able to carry on the campaign by themselves, and they certainly could not have facilitated the transformation of their effort's success at the municipal level into national-level policy change without connecting with sympathetic allies in the corporate and governmental sectors.

The most obvious example of how building a network is critical to scaling "make it work locally" success is the KitaQ endeavor, which was able to build a diverse network of local and international actors to disseminate Surabaya's successful composting program in numerous communities across Southeast Asia and even generate new policy collaborations in the development of a green port city in Hai Phong, Vietnam.

In sum, "make it work locally" is one of the most important foundational strategies for advocates seeking change. By demonstrating that their ideas can generate positive outcomes in a specific place, they can inspire others to copy their example of success. Advocates learn from one another, avoiding the mistakes that others have made and offering new innovations related to their own implementation efforts. The networks that advocates create can further enable them to scale success.

While this chapter has highlighted the techniques that advocates can use to scale local success, perhaps the most important lesson of this entire book is that individuals seeking change should start now and start local. The cases in this chapter, and those presented throughout this book, demonstrate that there is no need for advocates to have a lot of money, professional expertise, political connections, or, really, much of anything except themselves, a bit of energy, a sense of what the problem is, and an idea of how to solve it.

By starting small and starting now, individuals and local groups can begin the process of bringing positive change to their own community. They can talk with their neighbors about the problem, build support for trying a solution, and then work with others to implement that solution. If they are lucky, they'll be successful and solve the problem. If they are motivated, they might be able to use some of the strategies suggested in this book to scale that success, but even if they don't do anything to scale their success, that success still matters—their community will be better off than it was before. Even if their initial attempt fails, advocates will have learned more about the problem they face and the pitfalls of finding and implementing a solution. As a result of their efforts, their community will likely be better educated about the problem, and everyone will be in a better position to develop a new solution that has a better chance at success in the future.

7 Make It Work for Business: Crafting Win-Win-Win-Win Outcomes

Of all the advocacy strategies presented in this book, perhaps the one that comes the closest to guaranteeing success is "make it work for business." For this strategy, advocates find the sweet spot where proenvironmental outcomes also generate economic profit. In doing so, they create win-win-win-win outcomes that are good for business, good for governments, good for society, and good for the planet. Furthermore, through the networks they create with diverse stakeholders, proenvironmental advocates are better able to scale their success from one business to many, sometimes shifting entire markets and policies in ways that encourage better environmental behavior and outcomes.

This chapter will begin with a discussion of why probusiness advocacy is such a powerful tool, and why it is especially prevalent in East Asia. The chapter will then use two corporate examples—Walmart and Toyota—to explain how integrating environmental considerations into core business can generate large profits by reducing costs or generating new value. Additionally, since a critical challenge for advocates is not just identifying a handful of vanguard companies but rather convincing all companies (and citizens and governments) to behave in more environmentally responsible ways, the largest section of the chapter will examine how ecolabeling, green financing, and government policies can transform entire markets by generating positive profit incentives that reward firms for environmentally beneficial behavior and punish them for actions that harm the environment. The chapter will conclude with some examples of how nongovernmental organization (NGO) advocates can support allies who are employing a "make it work for business" strategy to encourage proenvironmental changes on the part of businesses, consumers, and the government.

Business Takes the Lead

For many years it was thought that business was the natural enemy of the environment, seeking maximum profit without a care for how the pollution it created harmed people and planet. As described in chapter 4, environmental concerns in East Asia, as in most parts of the world, have been driven by human health crises—environmental pollution was poisoning people, and citizens demanded that governments and businesses fix the problem. Early corporate responses were often to deny, shift blame, and make only cosmetic adjustments, as businesses viewed environmental issues primarily as public relations problems rather than issues that required a systemic, corporation-wide solution.

Since companies were focused on solving a public relations problem rather than addressing their environmental irresponsibility, citizens, journalists, and scholars soon observed the unequal distribution of pollution and environment-related violence, watching it fall disproportionately on the poor and politically vulnerable.[1] As economic globalization expanded, many feared a "race to the bottom" would doom us all to a world where businesses seeking additional profit would avoid environmental regulations as much as possible and relocate their polluting factories to areas with the lowest regulatory standards and lax enforcement. Then, localities seeking economic growth would reduce their regulations and loosen their enforcement to attract investment and jobs to their communities and countries. The result would be a rapid spread of environmental degradation into the remotest corners of the planet.[2]

As it turns out, although a number of industries, such as textiles, waste, and natural resources, have experienced a race to the bottom,[3] the pattern has not been universal and in several areas has been shifting in a more positive direction.[4] There is considerable evidence that the global trend now is closer to a race to the top than a race to the bottom. Rather than migrating to areas with the lowest environmental standards, globally competitive corporations frequently seek locations with higher environmental standards, more uniform government enforcement, and better living conditions for their workers.[5] Perhaps surprisingly, it is often the largest, most polluting firms that are the ones pushing governments to increase their environmental regulatory standards and enhance local enforcement.[6]

What can account for this puzzling behavior? Why would large companies seek higher environmental standards? Why is this dynamic especially common in East Asia? When discussing this topic, my interview subjects gave four main reasons for East Asian firms' proenvironmental policy orientation: high prevalence of export-oriented businesses, business innovation, the cultural importance of personal relationships and reputation, and the pattern of financing.

As is well known, East Asian economies grew rapidly by relying on export-oriented development that focused on manufacturing goods that would ultimately be purchased by consumers in North America and Europe.[7] Since those export markets started setting high environmental standards for products aimed at their consumers, East Asian–based firms found it economically advantageous to develop products that equaled or exceeded the environmental requirements of those markets. As the East Asian firms grew in size and global reach, it made economic sense for them to produce all products using the highest standard, since complex supply chains made it costly to produce similar products using different product specifications.[8]

Additionally, with the development of communication technologies, journalists and NGOs increasingly targeted global brands for investigation. When they discovered that multinational corporations, their subcontractors, or their subsidiaries were abusing workers, selling harmful products, or polluting the environment, activists would hammer the companies across numerous global media platforms, costing the brands millions of dollars in lost sales and requiring millions more to restore consumer trust.[9]

As the region prospered, East Asia became not just a base of production but also a valuable market. Global multinational corporations soon discovered that they were at a commercial disadvantage compared with local firms, since the latter did not need to meet European or North American production standards. As a result, large multinational corporations increasingly found themselves working with NGOs to pressure East Asian governments to raise local environmental standards. Local standards that were on par with those in Europe and North America benefited these global firms, since their production facilities and methods already met the higher standard. These regulatory changes would even the competitive playing field within East Asian markets by disallowing local producers to take advantage of low standards.[10] The following section about Walmart in China illustrates how this phenomenon has worked for one multinational corporation operating in China.

Following a similar logic but operating in reverse, if local firms had a proenvironmental business innovation, they too would press for regulatory changes to expand the usage of their environmentally beneficial technology. A good example of this is Gammon Construction's successful advocacy for the expansion of B5 biodiesel use in Hong Kong. B5 is a blended fuel in which regular diesel is blended with 5 percent biodiesel fuel. In Hong Kong, the biodiesel is generally made from cooking oil and grease collected from commercial kitchens and blended for vehicle use.[11]

In 2013, seeking innovative ways to reduce its carbon footprint, Gammon encouraged its energy partner, Shell, to set up blending facilities and began using B5 fuel in some of its off-road plant and equipment working on its construction sites. At the time, B5 was a legally allowed fuel source for off-road use, but it was not permitted for use by vehicles traveling on regular roads. Although Hong Kong produced biodiesel locally and was expanding its production capacity, nearly all of the biofuel it produced was being exported to Europe and Southeast Asia, where biodiesel was mandatory for most blended diesel products.[12] Gammon recognized that increasing its use of B5 would help it meet its sustainability goals, so after all of its plant and equipment was using B5, it encouraged Shell to open a B5 retail gas station at the end of 2015. Gammon was then able to expand its use of B5 to its new concrete-mixing trucks, and Shell was able to sell B5 diesel on a retail basis to other corporate partners.[13]

In parallel, Gammon, along with Shell, ASB (Hong Kong's largest biofuel producer), and environmental groups like Friends of the Earth, lobbied the government to mandate biofuel use in diesel products.[14] Their efforts were partially successful. By 2016 the Hong Kong government was mandating that firms use B5 biodiesel when fulfilling public works construction contracts (although not in all diesel products), and by 2019, Shell was operating two additional B5 retail stations and its partnerships had extended to Alliance concrete mixer trucks and Maxim's Cakes, which was using B5 in all of its one hundred delivery vehicles.[15]

In 2019 I spoke with Emma Harvey in Hong Kong, group sustainability manager of Gammon Construction Limited, who was involved in the effort to expand B5 use at Gammon and throughout Hong Kong, about "make it work for business" advocacy strategies. She reflected on the role of industry leaders and the government, as well as on how laggard firms can be convinced to adopt more sustainable policies.

> For these kinds of efforts you need an industry leader, and we were that leader. We were committed to expanding our B5 use because of its lower carbon emissions. We proved that it was feasible, but we were disadvantaged for providing it in our contracts because our competitors weren't using it. Our competitors adopted it only in the places when they were required to.
>
> The challenge is that B5 and other sustainable actions often make business sense only after they are scaled, and the early adopters, like us, end up losing money while that scaling happens. For everyone [including the leaders] there is a short-term cost but long-term savings.

The ability to think long term and absorb the up-front costs of innovation is one of the reasons we've often seen larger, multinational corporations take the lead on developing and piloting proenvironmental innovations. Firms that operate in multiple locations, especially those that sell to markets like Europe, the US, Japan, and Singapore, with higher environmental regulations, often have both the financial capacity and the incentive to make investments in eco-friendly technologies. Because their early adoption puts them at a commercial disadvantage in local markets, they cooperate with other like-minded actors in the business and NGO communities, expanding the network of proenvironmental stakeholders, who then use their collective political influence to pressure governments to change regulations in ways that shift markets in environmentally friendly ways. The new regulations generate incentives for smaller and less environmentally minded firms to improve their environmental behavior because better behavior has become commercially advantageous.

Another factor that enables East Asian firms to place a higher value on environmental aspects of their business is related to the cultural importance of personal relationships in all aspects of business. As Yoshihito Iwama, director of the Environment Policy Bureau at Keidanren, phrased it to me in 2011, "In the Anglo-Saxon model there is a lot of emphasis on the shareholders. In Japan a lot of stakeholders are involved: the shareholders, the employees, the community, the suppliers, the customers, etc. The family idea is quite important. Companies need to take care of family and neighborhood." While many volumes have been written about how culture, especially Asian culture, affects business practices,[16] I will focus on three elements that are particularly important with respect to firm interest and ability to prioritize environmental performance when conducting business.

A high proportion of East Asian firms, especially the large ones, are family owned, so a firm's relationships with customers, suppliers, and employees are especially personal. When a business is family owned, the reputation of the firm is tied up with that of the family, so corporate leaders are likely to be more concerned with leaving a "legacy" than a regular, board-hired CEO. Additionally, family-owned firms may be more willing to invest in preserving, enhancing, and expanding their relationships. Finally, family-owned businesses are usually less dependent on short-term stock value to gain revenue, so they are better able to make environmental investments that may take decades to pay off.[17]

Whether they are family owned or not, East Asian firms tend to have much longer-term relationships with their employees, suppliers, and customers than those in Western countries. Although the model has faded in recent years, Japan's postwar political economy was based on a "lifetime employment system" where employees are hired immediately after they graduate from school and work their entire lives for the same firm.[18] While South Korea does not have the same lifetime employment system, the dominant position of the *chaebol* companies (which are family owned) has exerted a similar effect on its employment and social welfare systems.[19] Historically, China's corporate commitment to its employees, customers, and suppliers was even more stable than that of Japan or South Korea because state-owned enterprises, which guaranteed employment as well as housing and social services, dominated its commercial sector for decades.[20]

A final factor that has enabled East Asian firms to prioritize environmental issues in their businesses is their relative lack of reliance on the stock market for financing. Stock market financing forces firms to demonstrate quick, positive value to shareholders, whereas bank or other forms of institutional financing generally allow for longer time horizons to demonstrate a return on investments.[21] Japanese, South Korean, Taiwanese, and Chinese firms all have heavier reliance on bank financing and lower reliance on capital markets for their financing.[22] Since environmental investments usually require a longer time horizon to pay off, the configuration of East Asian corporate financing means that East Asian firms are able to wait longer for their investments to return a profit, making it easier to make environment-related investments.[23]

In East Asia, the high proportion of export-oriented businesses, environmental innovation, the prevalence of family ownership, the cultural importance placed on long-term relationships, and the lower reliance on stock market financing have all helped "make it work for business" become a common and successful advocacy strategy in the region. As chapter 2 has demonstrated, this strategy is not specific to East Asia; it is a common and effective strategy everywhere in the world. The next sections will provide more specific information about how it works within specific companies and how those successes can be scaled by transforming markets.

Proenvironmental Policies Reduce Costs: Walmart in China

Walmart sources most of its products from China,[24] and China now represents one of its largest markets with more than four hundred stores and a rapidly growing online retail business.[25] In the 1990s and early 2000s, Walmart's engagement with China, like that of other large retailers and manufacturers at the time, was seen as part of a global trend that moved manufacturing facilities away from countries with high environmental and labor standards (such as the US, Europe, and Japan) toward lower-cost and lower-standard countries such as China, Mexico, and Brazil.[26]

That approach changed in 2005 when CEO Lee Scott announced Walmart's new vision for the future in which "being a good steward of the environment and in our communities, and being an efficient and profitable business, are not mutually exclusive." He laid out Walmart's goals to get 100 percent of its energy from renewable sources, create zero waste, and sell products that sustain resources and the environment.[27] With so much of Walmart's supply chain and so much of its anticipated future market located in China, this meant a major change for its operations there. The rollout of Walmart's policies in this area have been slow, steady, and highly profitable.

As with many companies that embark on integrating sustainability goals into core business, Walmart began with the low-hanging fruit of immediate cost reductions related to its operations before moving on to requiring efficiencies and standards compliance from its suppliers, and then to bringing sustainability goals into the process of new product development.[28] Its first target was the energy efficiency of its stores, its trucking fleet, and its packaging. Because Walmart is so large, the cumulative impact of these changes has been enormous—huge monetary savings for the company, along with large carbon and pollution reduction. According to its 2018 sustainability report, by 2017 energy efficiency measures in stores and delivery centers had saved the company more than $100 million; it doubled the efficiency of its trucking fleet between 2005 and 2015, saving more than $1 billion; and 78 percent of its waste materials have been diverted globally from landfill and incineration.[29] Walmart's carbon emissions per unit of sales have dropped from 61 MT CO_2e/$M to 42.5 MT CO_2e/$M, which means that Walmart and its suppliers emitted 18 million fewer tons of CO_2 in fiscal year 2017 compared with fiscal year 2006. This drop occurred even though revenue had expanded from $300 billion to nearly $500 billion over the same period.[30]

Walmart's sustainability efforts took off after its 2008 Sustainability Summit in Beijing, which gathered one thousand of its suppliers, government officials, and NGO representatives together in order to roll out new sustainability goals and commitments in China. At that meeting Walmart executives underscored their commitment to pushing the new sustainability agenda down into the supply chain. Walmart would require all of its suppliers to demonstrate compliance with local environmental laws and regulations; it would work with its top two hundred suppliers to improve energy efficiency by 20 percent; and it aimed to eliminate defective products completely by 2012.[31] It also announced plans for a prototype store that would use 40 percent less energy, a plan to cut energy use in existing stores by 30 percent, and a commitment that all stores in China would reduce their water use by half.[32]

These early commitments were not just talk. Less than two months after the Sustainability Summit, Walmart signed a memorandum of understanding with China's Ministry of Environmental Protection to work together to develop new green supermarket standards.[33] Walmart was also one of the first global companies to work with the Institute of Public and Environmental Affairs (discussed more in chapter 9) and its Green Choice Alliance to identify polluting suppliers and work with them to improve their environmental performance.[34] Its extensive disclosure and internal management processes and its willingness to work with suppliers to engage in corrective action earned it the number three spot (out of forty-eight) in the Institute of Public and Environmental Affairs' 2012 *Cleaning Up the Fashion Industry* report.[35] By 2014 Walmart had surpassed its energy commitments—210 of its factories were saving 20 percent energy compared with

2008, which generated $279 million in savings. In 2011 it installed the first non-state-owned solar plant project as part of China's Golden Sun Solar Program, which was generating thirty-five thousand kilowatts of clean energy monthly.[36]

As Walmart expanded, so did its environmental initiatives in China. In 2012 it promoted the use of the Sustainability Index, a disclosure and management tool that would track product sustainability, committing to integrate its use throughout its supply chain such that 70 percent of its goods were from participating suppliers within five years.[37] In 2018 it announced Project Gigaton China, which aimed to reduce a gigaton (one billion metric tons) of emissions from Walmart's global value chain by 2030, with fifty million metric tons of that reduction coming from China.[38] The project included the creation of the Walmart Sustainability Hub,[39] which has data management tools, case studies, and webinars to help make it easier for suppliers to identify ways to reduce energy, waste, packaging, pesticides, and deforestation, as well as register their progress with Walmart.

Walmart's efforts to make money from integrating environmental concerns into its business have moved beyond the low-hanging fruit of production and distribution efficiencies into building new value through sustainable products. One of its most lucrative new business opportunities has been its significant investment in organic and sustainable agriculture. In 2008, the same year as Walmart's Sustainability Summit, China was rocked by a tainted milk scandal when contaminated infant formula killed six children and poisoned three hundred thousand more. Even ten years after the scandal, Chinese consumers still did not trust local companies because of food safety concerns.[40] The lack of trust in China's domestic food and drug safety has been compounded by additional scandals that seemed to follow year after year—for example, toxic wax on oranges in 2010, carcinogenetic cooking oil in 2011, and toxic fillers in infant formula in 2012.[41]

For Walmart, China's food safety concerns offered an enormous market opportunity. In 2016 it created its Beijing-based Walmart Food Safety Collaboration Center and committed $25 million to advance food safety in China.[42] By 2018, 100 percent of its seafood was sustainably sourced, thirty-four million acres of Walmart-destined produce were following a fertilizer optimization plan, and 100 percent of suppliers located in the Amazon and Cerrado regions were participating in Brazil's beef risk monitoring system for net zero deforestation.[43] Walmart is now the largest retailer of certified organic food in North America,[44] and with the expansion of its new e-commerce and grocery delivery capacities in China, it is clearly seeking to repeat that commercial success there.[45]

Not all of Walmart's Chinese environmental efforts have been smooth. In 2011 officials in Chongqing arrested two Walmart employees, detained thirty-seven, and forced the closure of thirteen stores in the city when it was discovered that regular pork had been mislabeled as "organic."[46] In 2014 Walmart was forced to recall donkey meat being sold in its Chinese stores when it was discovered that the DNA of other meat, including fox, had been included in the packages.[47] As of this writing, Walmart is involved in negotiations over a $300 million settlement deal with the US government over a global bribery scandal that includes accusations of bribery in China.[48]

Walmart's engagement in China is likely not the first thing to come to readers' minds when thinking about environmental advocacy. And yet, Walmart's path toward becoming an environmental advocate in China—one that is both pressuring and supporting tens of thousands of companies in the country to adopt better environmental practices and supporting the development of environmental consumer values—offers several examples of ways that companies can increase their profits by integrating environmental concerns into their core business development. I am not trying to argue that Walmart, however impressive its benchmarks may be, is fundamentally a sustainable business.[49] However, its experience does illustrate how global multinational corporations operating in East Asia are effecting large-scale, proenvironmental transformations in the region, and how they can use both market and political pressure to encourage others to improve environmental outcomes locally and globally.

Proenvironmental Policies Generate New Value: Toyota

Like Walmart, Toyota initially realized economic profit from incorporating environmental values into its business by focusing on the elimination of waste and the realization of higher business efficiency. The deprivation and austerity of wartime and the postwar period, combined with the 1952 Enterprise Rationalization Promotion Law, forced Toyota and other auto manufacturers to develop an almost obsessive preoccupation with efficiency and eliminating waste.[50] This focus on production efficiency, combined with rising public and governmental concern with growing environmental pollution, led to the development of highly fuel-efficient vehicles and a streamlined production system using a process that has come to be called Just-in-Time production.[51]

By the time of the oil shocks of the 1970s, which quadrupled gas prices in some places, Toyota was in an excellent position to expand internationally. Its affordable, reliable, and fuel-efficient vehicles were attractive to global consumers, and the Japanese government's industrial policies further supported the auto industry's global expansion.[52] By 1974 Japan had overtaken West Germany to become the largest automobile exporter in the world.[53] The following year Toyota became the largest foreign car company in the US, ousting Volkswagen and claiming the number four slot in terms of US market share.[54]

Although the elimination of waste in the production process and the creation of more fuel-efficient vehicles were both important proenvironmental actions, the Toyota corporation and consumers did not yet view Toyota as a "green" company. That shift in self-understanding and brand image came in the 1990s. According to a Toyota manager I spoke with in 2011, the process of developing Toyota into a green brand began just before the 1992 Earth Summit in Rio de Janeiro. In January of that year, Toyota established an environmental committee that was chaired by Toyota's president and created its own Earth Charter, which declared that "finding ways to preserve an abundant natural environment to pass on to future generations is the most pressing issue for people on earth today."[55]

In 1996 the International Standards Organization (ISO) began publishing the 14000 family of standards,[56] which established internationally recognized standards for environmental management. This meant that Toyota and other companies had a clear set of environmental management standards that they could follow and a method to certify compliance to consumers and financial organizations. In December of the following year, the first-generation Prius was launched. As the Toyota manager explained it to me during a 2011 interview in Tokyo, "The first-generation Prius didn't have much commercial success, but Toyota as an environmental brand image jumped a lot, way ahead of Honda or Nissan. That was a real turning point. Pollution problems may have started it, but we saw that the environment can improve the competitiveness of the company. It can enhance the brand

and the company. That aspect had not been commonly understood—the environment can be commercially useful."

In 1998 Toyota established its Environmental Affairs Division, and the following year it began including environmental specifications on all new or redesigned models. Subsequent years saw a continual expansion of the scope of environmental activities and the development of new products, including significant research and development investment in hydrogen fuel cell vehicles (the Mirai was launched in 2014).[57]

In 2015 the company announced a highly ambitious Environmental Challenge 2050, which aimed for "going beyond zero environmental impact and achieving a net positive impact." The 2050 plan has six challenges: zero CO_2 emissions for new vehicles, zero CO_2 emissions for the entire vehicle life cycle, zero plant emissions, minimal water usage, promotion of a recycling-based society through the deployment of end-of-life vehicle systems, and establishment of a future society in harmony with nature by expanding biodiversity conservation activities and expanding initiatives to "foster environmentally conscious persons."[58]

By the time of its 2018 sustainability report, Toyota was reporting its progress toward 2030 benchmarks: It had reduced new vehicle emissions by 13.7 percent compared with 2010 levels, had already achieved its 2020 goals for sale of next-generation vehicles (1.5 million units per year), and had sold its first fuel cell electric bus, Sora. It had reduced emissions due to transportation logistics and production activities by 35 percent and 45 percent, respectively, compared with 1991 levels, it had collected nearly one hundred thousand batteries from end-of-life vehicles for reuse and recycling, and it had reduced the waste volume per vehicle by 62 percent compared with 2002.[59]

Toyota's activities extended beyond its production facilities into the communities where its plants are located. In 2009 Toyota City, home to its headquarters, was selected by the Japanese government as an environmental model city in pursuit of a low-carbon society. The city has extended its electric vehicle charging station network, reduced overall emissions, expanded the green space in the city center, and promoted energy-efficient homes and buildings, including the instillation of rooftop solar panels. As a result of all these activities, the citizens in Toyota City are more aware of their city as a model eco-city, as well as of which kinds of behaviors can reduce their carbon footprint.[60]

All of Toyota's environmental initiatives are paying off. It won a number of international awards in 2017, including "A List" status for climate and water from the Carbon Disclosure Project. The Prius Prime was named the World Green Car of the Year (the Mirai won the same award in 2016), and Toyota's new North American headquarters in Plano, Texas, won LEED Platinum status.

Toyota's commitment to environmental measures also appears to be benefiting its bottom line—in 2018 Toyota had the largest global market share of any auto company.[61] Its net revenue and net income were both up compared with the previous year. The Prius recovered its initial investment within five years, and in spite of issues in 2009–2011 that forced the recall of several Prius models, by 2018 Toyota had sold approximately twelve million hybrid vehicles, for an effective CO_2 reduction of about ninety-four million tons.[62]

Toyota's story demonstrates how one of East Asia's largest corporations has successfully integrated environmental concerns into its core business and found that process to be not just good for the planet but also commercially successful. Toyota is now a well-recognized "green" company and serves as an inspiration for others.[63]

Scaling Probusiness Environmental Action

While individual companies can find commercial success by integrating environmental considerations into their core business development processes, to scale a "make it work for business" strategy, what is needed is to shift entire markets in ways that encourage the development of practices and products that are good for the environment and discourage those that are harmful. This section will discuss three methods that advocates in business, the NGO sector, and government can use to shift markets in proenvironmental ways: eco-labeling, green finance, and government policies.

With all three methods, proenvironmental advocates do not act alone. They build and expand networks of stakeholders, connecting to new types of businesses and different geographic regions and linking with NGO activists, academics, civil servants, and politicians. Unlike many areas of capitalist competition, these probusiness and proenvironmental initiatives are win-win: when more businesses, communities, and countries join, the commercial and environmental benefits expand for everyone.

Eco-labeling

An eco-label (ISO 14024) is a seal or logo that can be placed on a product based on a set of criteria that have been third-party certified. There are also other forms of environmental labeling, such as self-declared environmental claims (ISO 14021), which are claims made by companies about their products that are not third-party certified but are expected to be verifiable and accurate, and environmental declarations (ISO 14025), which are quantitative indicators of environmental performance that are primarily used for business-to-business or government procurement.[64] In line with Toyota's experience, there was a rapid growth in these schemes in the late 1990s, from fewer than one hundred labeling schemes before 1992 to more than five hundred by 2010. Eco-labels are now prevalent around the world and cover the full range of products, such as food, furniture, buildings, appliances, and tourist outings.[65]

How do eco-labels shift markets? The basic idea is that eco-labels give consumers environment-related information, so those who believe that the environment is important can then buy more environmentally friendly products. This shift in demand then leads producers to make more environmentally friendly products and fewer environmentally harmful ones. Additionally, the existence of eco-labels can raise the salience and awareness of environmental issues in society, which might not just affect consumption choices about particular labeled products but also generate broader proenvironmental behavior change.[66]

While the theory behind eco-labels offers considerable hope, the evidence concerning how environmental labeling affects environmental outcomes is mixed. There is fairly strong evidence that eco-labels (government-issued or third-party verified) have shifted markets in proenvironmental ways such that goods and services with higher environmental ratings increase their market share over time, although the extent of the positive results varies considerably by country and product.[67] For other forms of environmental labeling (e.g., private labeling schemes that make environmental claims but are not third-party verified), the evidence is much more mixed—some labeling schemes enhance environmental outcomes, others have no effect, and some may even have negative effects, including market distortions that harm small producers and those in developing countries.[68]

Since business-driven environmental initiatives are so common in East Asia, it is not surprising that the region's governments were early adopters of eco-labels. Japan's EcoMark, established in 1989, was one of the earliest in the world. South Korea's Eco-Label and Taiwan's Green Mark were both launched in 1992, and China's Environmental Labelling Program started the following year. In the following decades, all four programs have expanded rapidly and now cover thousands of registered products in each country.[69]

A downside of the success of environmental labeling programs has been their rapid proliferation, leading to confusion and even misrepresentation.[70] Consumers now have trouble figuring out what each of the different labels means and whether some are better than others. Many give up completely and begin to distrust all green labeling schemes. Governments are seeking to regulate labels and environmental claims, but it is proving to be difficult.[71]

Green Finance

Green finance is a general term used to describe financial tools (e.g., grants, loans, investments, and insurance) that increase financial flows to sustainable development priorities.[72] Nearly all forms of green finance have grown exponentially in the last few years. The 2018 UN *Sustainable Finance Progress Report* states, "The past 12 months has seen a surge in sustainable finance momentum. ... Sustainable finance policy has been characterized by strong growth, increased scope, and greater maturity."[73] While nearly all elements of sustainable finance have grown, I will discuss three components here: green funds, green investment measures, and green bonds.

Green funds are public or private funds that finance environmental projects through grants and loans. The oldest and most common are funds for conservation, which buy land in order to conserve it. For example, the Nature Conservancy, which was established in 1951 in the US, has protected more than 119 million acres of land worldwide and has total net assets of nearly $7 billion. Newer funds not only buy land for conservation but also support a variety of proenvironmental activities ranging from education programming to green infrastructure development. The scale of these funds is now getting quite large. The Green Climate Fund was established in 2010 by the 194 parties to the United Nations Framework Convention on Climate Change and began disbursing funds in 2016.[74] By 2019 it was spending $2 billion to support 102 projects, which will contribute to the resilience of 276 million people and avoid 1.5 billion tons of CO_2.[75] There has also been a large increase in the number and volume of funds available through national environmental funds, which are making significant contributions to biodiversity conservation worldwide.[76]

Some funds are quite specialized, such as the Japan Green Fund, which was established in 2003 to fund renewable energy projects in Japan.[77] The recent US-China Green Fund seeks to "greenergize" China with collaborative green development projects. It was launched in 2016, and by 2019 it had spent $420 million supporting thirteen companies and more than one hundred public projects. Examples of supported projects include a rural e-commerce network that sells green energy to village consumers and a hospital management company that provides green medical services in maternal and children's health.[78]

Green investing (and divesting) is a method through which individuals and institutions make investment decisions based on environmental criteria. The movement really took off after 2004, when former UN secretary general Kofi Annan invited top CEOs to participate in an initiative that would find ways to incorporate environmental, social, and governance (ESG) standards into capital markets.[79] The CEOs, banks, and many institutional investors responded, shifting their own investing patterns and developing new "green" and "sustainable" index funds. Additionally, the "divestment" movement, in which activists around the world seek to pressure institutional investors to shift investments away from fossil fuel and other carbon-intensive businesses, helped increase the demand for more investment instruments that could demonstrate that they were not harming the planet.[80]

The green finance movement gained a large boost in 2016 with the establishment of the Task Force on Climate-Related Financial Disclosures,[81] initiated by Michael Bloomberg and supported by the Financial Stability Board, an international body of financial regulators. Its final report and recommendations were released in late June 2017 and presented the following week at the G20 summit in Hamburg. Financial regulators around the world began to implement the recommendations in their local jurisdictions, significantly increasing the environmental disclosure requirements for firms listed on their local stock exchanges. One can see the influence of the task force's recommendations about ESG reporting in the frequent mention of them in the ESG reporting guidelines of major stock exchanges.[82] As a result, there has been a rapid expansion of green investing across financial markets. By the beginning of 2018, $12 trillion in the US was invested in sustainable, responsible, and impact investing, representing one-quarter of all assets under professional management.[83]

The area of green finance that has seen the most rapid growth recently has been the development of green bonds. In 2007 a Swedish pension fund issued the first green bond, seeking to reduce the risk to its investors by avoiding investments that contributed to climate change and investing in businesses and funds that promoted sustainability. Over the next few years the World Bank, the UN, and other organizations worked to develop disclosure and investment criteria that could help ensure that green bonds actually were green.[84] By the end of 2018, the total green bond issuance had grown to more than $500 billion.[85]

Green bond growth in Asia has been particularly large recently. China issued its first renminbi-denominated green bond in 2015, and by 2018 it had issued close to US$31 billion in green bonds.[86] In 2019 Japan launched the Green Finance Network to catalyze green bond issuance in the country, which rose to $6.7 billion, up 79 percent from 2018.[87] South Korea is the fifth-largest issuer of green bonds after China, India, Japan, and Australia.[88] It also hosts the secretariat or headquarters of several of the most influential green finance organizations, including the Global Green Growth Institute and the Green Climate Fund.

The rapid growth of green bonds and green investing was propelled not just by ethical considerations. Green bonds have outperformed conventional benchmarks in recent years, so they represent not just a lower-risk investment but also a more profitable one.[89] As a result of the new reporting requirements, as well as the expanded financial opportunities, companies are now scrambling to report their ESG information and are seeking out creative ways to improve their metrics as a means for lowering their cost of capital. Whereas environmental concerns used to be relegated inside companies to their public relations and compliance departments, now finance departments are taking an interest, and ESG performance has become a focal point for discussions among CEOs and board of trustees members. The lower cost of capital for companies with better ESG metrics is creating a strong market incentive for above-minimum compliance and continual improvement, even among firms without much interest in the environment. As Sungwoo Kim, a senior environment and energy consultant based in Seoul, who has been involved in advising companies and governments on these issues, explained to me in a 2019 interview, "The ESG reporting includes compliance, but it isn't just the degree of compliance, but also management systems and the board decisions, systematic detection and prevention measures. It is much more comprehensive than just compliance. This is the right way to approach environmental issues—not just minimum compliance."

The spread of green finance, which is making more and cheaper capital available to better-performing companies, is dramatically expanding the "make it work for business" strategy beyond just those firms with visionary leaders or those deploying innovative technologies. Capital market shifts affect all firms, so many companies that may not have considered their environmental impact before are now looking more closely at how they can change their operations to reduce their harm, generate more positive outcomes, and increase their access to capital.

Government Policies

A final way that markets can be shifted in proenvironmental directions is through government policies. Governments, both local and national, have the capacity to shift markets in a number of different ways, and I will discuss four here: subsidies, taxes, regulation, and procurement policies. As discussed in the introductory chapter and chapter 4, the developmental states of East Asia all employed highly sophisticated industrial policy to generate high-speed economic growth and have now adapted those policies to accommodate environmental concerns.[90] Therefore, it is not surprising that national and local governments have all been very active in using policy to reshape markets in ways that promote the better environmental outcomes sought by their governments.

Government subsidies to encourage firms to develop and expand green technology are the most obvious form of government intervention into markets, and East Asian governments have been quite generous in this regard. Two of the most common subsidies have been those for renewable energy and low-CO_2 cars (e.g., hybrid, electric, fuel-cell). China's subsidies and market development have been, not surprisingly, the largest. Its subsidies for the purchase of electric vehicles helped propel its market from essentially zero electric vehicle sales in 2010 to more than 1.2 million in 2018, more than half of all the plug-in sales in the world,[91] and its renewable energy subsidies have propelled it into the position of "renewable energy superpower"—China now is the largest producer of renewable energy and has the largest market for renewable energy in the world.[92] Since both the market for electric vehicles and that for renewable energy are now maturing, China is currently in the process of phasing out many of these subsidy programs.[93]

After the 2011 nuclear disaster in Fukushima, Japan implemented a feed-in-tariff program[94] to promote the expansion of the renewable energy market. The country now has the second-largest solar market in the world, and its wind, biomass, geothermal, and micro-hydro markets have also grown.[95] It also has a variety of green car subsidies that have boosted the adoption of hybrid, plug-in, and fuel-cell vehicles; and hybrid and electric cars were 40 percent of all new car sales in the first half of 2019.[96]

South Korea has continued to expand its subsidy program to promote the electric and fuel-cell vehicle market.[97] It had a feed-in-tariff program (which requires power companies to buy renewable energy from producers for a fixed price) from 2001 to 2011, it switched to a renewable portfolio standard (where power companies are required to increase the proportion of their electricity generated by renewables) in 2012,[98] and a feed-in-tariff program was reintroduced in the city of Seoul in 2013. South Korea's One Million Green Homes program, which went into effect in 2009, has subsidized the installation of solar panels on residential homes,[99] and in 2018 the government announced plans to build the world's largest solar park.[100]

Taiwan also has a feed-in-tariff program, which is encouraging rooftop solar as well as offshore wind,[101] and it subsidizes electric vehicles, with a focus on electric scooters.[102] It has special subsidies for foreign firms to encourage them to develop research and development centers and new technology in Taiwan.[103]

Another important policy tool commonly used by governments is to tax products and behavior that they would like the market to reduce. The last decade has seen the proliferation of emissions trading schemes, sometimes called a carbon market or cap-and-trade system. In this system the government sets a limit on the amount of emissions a company can emit, and then companies trade for carbon credits. This allows more efficient companies to earn money for their efficiency and forces less efficient companies to pay for their inefficiency. If the market works well, the price for emitting rises over time, incentivizing everyone in the market to emit less. South Korea opened its national carbon market in 2015, as did China. Taiwan authorized the development of an emissions trading scheme in 2017 but has not yet set a date for implementation.[104] Japan no longer has a national trading scheme in place, but the city of Tokyo introduced a local one focused on urban buildings in 2010. The schemes do appear to be reducing emissions in all cases, although they only cover a small fraction of the emissions in their respective countries.[105]

Another common tax, whether done at the national or local level, is on municipal solid waste. Taxes on waste have dramatically reduced the volume of waste sent to incinerators and landfills and helped boost markets in recycling and composting. One of the most successful examples is Taiwan, which introduced its producer responsibility scheme in 1998. In this system, producers must bear responsibility for dealing with the waste that their products generate during their entire life cycle and at the end of life. Recycling rates in Taiwan are now among the highest in the world. The waste fees collected through the scheme have supported the development of a new multibillion-dollar recycling industry, and curbside waste collection in many cities has become a kind of local ritual where bright yellow trucks play classical music as residents bring out their waste and recycling to the passing truck.[106]

The most powerful government policy tool is regulation—the government can make certain environmentally harmful behavior more costly or illegal, such that violators incur fines and criminal penalties for bad behavior. Sometimes the government does not just limit but will ban a whole class of activity—for example, cutting trees in a particular area, fishing in a specific area or at a certain time, or using a particular chemical or using a chemical above a certain concentration.

Advocates frequently concentrate their political efforts in this narrow area of governmental activity, since shifts in regulation can, if enforced, have very powerful proenvironmental effects. The best and most effective regulation shifts are ones that business has "bought into," such that business recognizes that limiting (or raising) certain behavior will benefit them in the long run. High emissions standards are one such example—as discussed earlier, Japanese auto manufacturers have been able to benefit commercially from their high-efficiency vehicles, which were induced by stringent government regulation on emissions.

A challenge with the more extreme forms of regulation is that they can have unintended consequences. For example, China instituted the National

Forest Protection Plan in 2000, following devastating floods in the upper Yangtze River. The plan banned all logging in some areas of the upper Yangtze and significantly restricted logging in others. The logging restrictions have been successful in protecting the forest in many areas and have expanded reforestation,[107] but deforestation worsened in areas with lax enforcement, such as the Tibetan sacred forests.[108] Additionally, more wood was imported from abroad, often from areas where illegal logging persists, essentially pushing the deforestation problem onto another country. The harm caused by those avoiding the ban has been so extensive that the ban may have actually had a net negative effect from a global ecological perspective, the opposite of its intent. [109]

A final form of government support for proenvironmental shifts in markets is through Green Public Procurement (GPP) policies, which require public authorities to purchase goods and services with reduced environmental impacts. GPP policies help governments achieve their environmental targets; expand the market for environmental products and services, which reduces their costs for everyone; offer an example for private companies to follow; help raise public awareness of environmental issues; and improve resource efficiency. Because governments are such large actors, GPP policies can significantly influence markets and spur innovation. The South Korean government estimates that its GPP policies were responsible for 643,000 tons of reduced greenhouse gas emissions and $382 million of economic benefits in South Korea in 2014.[110]

In sum, governments at both the local and national levels have policy instruments that they can use to shift markets in ways that encourage proenvironmental behavior by firms and discourage polluting behavior and investments. They can reward positive behavior with subsidies and by consuming from eco-friendly companies directly, and they can punish harmful behavior with taxes and laws that make certain behavior costly or illegal. By definition, all of these government activities are political, so there will be some firms that benefit and others that lose out, and it is often the case that the firms that lose are powerful (e.g., energy, transportation, chemical industries), which is why governments, especially in East Asia, tend to prefer the positive-supporting policies (subsidies, GPP policies) to the punitive policies (taxes, regulation), although combining the two usually has the strongest positive effect.

Conclusion

Perhaps the most important element to the success of the "make it work for business" strategy is transparency. Customers used to be satisfied to know that the products they used would not hurt them—they wanted reassurance that their car would not explode when they drove it and their tennis shoes wouldn't fall apart with a second volley at the net. Now, however, they want to know that their products are not causing harm to people or the environment all along a company's supply chain. Consumers don't want cars that pollute the air they or the factory workers breathe; they don't want shoes made by child labor or with dyes that are toxic. These added consumer expectations, and the technical capacity to verify and demonstrate supply-chain activities, are benefiting advocates who are seeking better environmental behavior because greater transparency can offer commercial benefits for businesses acting in an environmentally friendly way and can pressure those who lag behind in improving.[111] The vital role of transparency in transforming markets will be discussed at greater length in chapter 9.

As discussed earlier, businesses can use a focus on environmental factors to enhance the profitability of their business by reducing waste and by developing new products and services that benefit the environment. In East Asia, Walmart and Toyota are good examples of very large corporations that have used a focus on environmental sustainability to improve the profitability of their own companies, as well as influencing other companies in their industry and in the region to follow their example by making similar proenvironmental changes to their own businesses.

Beyond improving the profitability and environmental sustainability of their own businesses, advocates can help scale "make it work for business" success by developing trustworthy eco-labels, expanding green finance, and supporting government policies that encourage proenvironmental business development and discourage harmful practices. These changes can shift whole markets in ways that commercially reward good environmental behavior on the part of firms. When market conditions shift in ways that promote "race to the top" behavior, it generates a "virtuous circle" where firms continue to improve their environmental performance over time, which benefits consumers, innovative firms, and the planet. These changes help transform a dynamic where consumer and corporate behavior leads to the depletion of environmental resources—for example, the "tragedy of the commons"—into a situation where consumer and corporate behavior leads to the restoration of environmental resources, a kind of "replenishing of the commons."

Although they may not be able to influence markets as directly as business and the government, advocates in the NGO sector can work through their networks to engage in activities that help promote proenvironmental shifts in corporate behavior. First, as discussed at length in chapter 5, the most important action that activists can take is to support their allies in business who are trying to make positive change. They can do this by providing allies with technical expertise, facilitating political access, aiding government and business to negotiate legal and regulatory changes, and publicizing successful examples of corporate activities that have benefited the environment.

NGOs can also help by building and reinforcing the transparency infrastructure. While governments have baseline reporting requirements, companies seeking to demonstrate a commitment to environmental and sustainable development goals must do far more than meet the minimum government standard. To enhance public trust, many reporting mechanisms require third-party verification, which means that an entity that is independent of both the buyer and the seller must verify that the reported numbers are true. Frequently, NGOs serve as the third party for this purpose. In some cases, such as with the Carbon Disclosure Project, the NGO may have established an entire platform where the disclosure takes place, gathering the metrics for numerous companies in one place, rather than having them located only in company annual financial or sustainability reports.[112]

Finally, NGOs can help support allies in business by calling out laggards through "name and shame" activities that draw public attention to companies that are harming the environment.[113] Fear of the commercial and legal repercussions of being caught in environmentally damaging behavior can force companies to think twice before engaging in it.[114] NGOs that are savvy can select particular companies or industries to put forward as examples, forcing them to confront their problematic behavior and address it. Once an industry leader is exposed and subsequently reforms, other companies in the same and related industries have both an example of the public relations horror and the commercial damage that can come from ignoring the issue and some concrete steps they can follow to get ahead of the problem, avoiding the bad press while improving their business at the same time (this tactic will be discussed at greater length in chapter 9).[115]

As businesses come to recognize the economic benefits of environmentally responsible behavior and as norms of ESG reporting spread across multiple industries, corporations are becoming more transparent. Greater transparency makes it easier for consumers, activists, and policymakers to reward firms that are improving the environment and punish those that harm it. Markets then shift, generating positive cycles where business innovation leads to better and better environmental outcomes. A "make it work for business" advocacy strategy is especially powerful because it is an area that can be supported by the collaboration of business, NGOs, and the government, and it generates win-win-win-win outcomes that are good for business, good for governments, good for society, and good for the planet.

8   Make It Matter: Using Art to Engage the Heart and the Imagination

More than twenty years ago in *From Art to Politics*, Murray Edelman observed, "In a crucial sense, then, art is the fountainhead from which political discourse, beliefs about politics, and consequent actions, ultimately spring."[1] And yet art continues to be overlooked as an advocacy strategy, and artists are generally not considered to be political actors who intentionally effect political changes through their artwork.

There are intellectual and cultural reasons for this oversight. Political scientists tend to view art as a form of cultural expression rather than as political advocacy. When art is explicitly political, as in the case of protest art, it is commonly considered to be a component of the protest strategy or a tool for social mobilization and publicity, rather than as a strategy in itself.[2] Since artists are generally not the decision makers for policy, and since they commonly operate outside political institutions (advocacy art is considered distinct from government propaganda, which does operate inside political institutions), it has been difficult for political scientists to incorporate them into theoretical models that focus on actors and institutions.

Not surprisingly, cultural scholars, art historians, and artists themselves have more complex and nuanced understandings of the profound influence that artists and their creations can have on politics.[3] However, there is often a tension between art that inspires political action and art as political action. Anthony Downey articulates the dilemma well in his *Art and Politics Now* (2014): "There is a distinction to be had between art practices that engage with politics, and the overall aims of political activism: art as a practice does not necessarily have any clear-cut goals, nor are its end results quantifiable in terms of desirable or fixed outcomes."[4] While art as a practice may be different from political activism, he acknowledges that in the sphere of environmental advocacy in particular, "we witness a deliberate blurring of any distinction between eco-activism and art practice."[5]

For the most part, scholarship about art's role in politics either describes a range of artistic works that are engaged with a specific political issue, such as civil rights, AIDS, or the environment,[6] or focuses on a single or small number of works and discusses how those particular pieces were influential in a specific political battle.[7] The last political scientist to engage theoretically about how art relates to political action was Edelman, and his focus was primarily on art's symbolic value, the way that it can frame public understandings of political issues through the images and narratives it creates.[8]

In this chapter I take a slightly different approach. I am primarily concerned with artists as political actors and art as an effective advocacy strategy, which is sometimes related to art-as-symbolism, but not always. The examples in this chapter aim to illustrate and explain aspects of art's efficacy as an advocacy strategy and focus on a selection of art and artists who have generated proenvironmental change. Individually and collectively, the examples selected here demonstrate how artists can effect political change through their networks by attracting the attention of, eliciting an emotional response from, and inspiring policymakers and the public. These individuals are then activated to make change, and that energy travels through multiple networks, energizing the decision makers to act. Through their influence on multiple stakeholders simultaneously, artists are able to shift the way entire networks of people are responding to the political challenges they face.

The examples set forth here are not intended to be a comprehensive review of political art in the four countries that serve as this book's focus. Furthermore, I do not distinguish between art that emerges organically from the inspiration of the artist and then has political effects (whether intended or not) and art that is commissioned for the specific purpose of generating a desired political or social effect. Both forms represent the use of art as an advocacy strategy.

Instead, this chapter is concerned with the interplay between artists, their audiences, and political responses. While I am sensitive to the concerns of artists that their cultural production not be considered to be valuable only when it serves a political, social, or moral purpose, my intellectual focus here is on examining the ways that art and artists can be political as well as cultural. As in the other chapters, I first ask why art as a strategy is effective and then move to discuss the ways that art's "engage the heart" and "inspire the imagination" strategies can be scaled up—from a single incidence of success to one that generates positive change for a wider set of people and places.

Art as Effective Advocacy: How It Works

When I asked my interview subjects why art was an effective advocacy strategy, they emphasized the power of art to connect with people's emotions and imaginations. As many of them described it, scientists and journalists can fill your head with numbers, but they can't make you care. As Ruby Yang, an Oscar- and Academy Award–winning filmmaker, phrased it to me in a 2011 interview in Beijing, "Reporting and journalism is about stating the facts, but [in filmmaking] we use broader strokes and draw people into the story and characters. We present the facts in a way that people can relate to, so that they say, 'I want to do something,' when they get to the end of the film. And they also ask questions. We give voice to people that don't have voice—orphans, gay men, villagers."

Furthermore, the process of making art in a specific place can draw people directly into their environments, as well as connect individuals and communities that may not realize their ties to one another.[9] This section will focus on four reasons why art is such an effective advocacy strategy: it makes the invisible visible; it makes political problems culturally relevant; it facilitates the imagination; and it enables community connections.

Make the Invisible Visible

Artists can help audiences connect to people and phenomena that they cannot see, or that they do not see in their regular lives. Giving me the example of radioactivity, Eiko Otake, a Japanese dancer who has been active for decades in antinuclear campaigns, said of artists' political role, "We make the invisible visible." Her collaboration with photographer William Johnston, *A Body in Fukushima*,"[10] captures images of Otake's body as it moves through the destroyed, radioactive, and recovering spaces in Fukushima. The ethereal nature of her work allows viewers to see her as embodying the landscape, the radiation, the community and landscape's rebirth, or the viewer's own body, depending on the image and the viewer's own perceptions. Elaborating further on why "making the invisible visible" is so politically powerful, she observed during a 2013 interview with me in

Middletown, Connecticut, "Art can help bridge the gap between the issue and the individual. ... It gives someone 'emotional access' to the issue."

To give a historical example of important artists in Japan's environmental movement, Otake referenced the work of Eugene Smith, whose compassionate and horrific images of the effects of mercury poisoning on the people of Minamata played a critical role in generating national and international outrage about the costs that individuals and communities were paying for economic growth. *Tomoko Uemura in Her Bath* is perhaps the most iconic image of the antipollution movement in Japan and perhaps worldwide. A black-and-white photo published in *Life* magazine in 1972, it captured the agony and compassion of a mother bathing her deformed and emaciated child. The wrenching image touched the hearts of millions, stirring Japanese nationwide to demand solutions from their government.[11] *Tomoko* was intentionally created by Smith to become a symbol of the crisis in Minamata.[12] Not just an emotional expression of pain, it was the deliberate creation of an artist seeking to generate political and social change.

In *Paradise in the Sea of Sorrow: Our Minamata Disease* (1990), Michiko Ishimure accomplishes with words what Smith did with photographs—she makes the stories of victims accessible to the public, arguing that the tragedy of industrial pollution is not "their" problem, but rather "our" disease—a collective problem faced by a whole community. By mixing victims' stories with medical accounts, government reports, and the author's own observations and reactions, Ishimure elicits feelings of anger and admiration for the proud victims who are suffering. Published almost twenty years after the disease devastated Minamata, the work engaged a new generation of Japanese. Its publication contributed to a new wave of activism on the part of victims seeking compensation, as well as city and environmental activists working to commemorate the tragedy while engaging in proactive environmental education and activism.[13] To remember its history and honor the many victims of the disease, Minamata City established an eco-park on reclaimed land built on top of the toxic sludge that remained in the bay. The park contains a stone memorial to the life-forms that died as a result of the pollution. In 1999 the city received ISO (International Standards Organization) 14001 certification to establish Minamata as an eco-city, and it became an official eco-town recognized by the Japanese government in 2001.[14]

More recently, video has become a choice medium for artists to make the invisible visible. Ruby Yang's *Warriors of Qiugang*[15] tells the story of the successful fight of a small village to shut down the factory that was polluting its community. It is full of the drama of all good films—an unlikely hero, a dark and seemingly unbeatable villain, and supporting characters who are relatable to any viewer in any country as neighbors, friends, and classmates. It was nominated for an Academy Award in 2011. Although it is difficult to demonstrate direct political effects, within two months of it being posted online by *Yale Environment 360*, the Chinese government pledged $30 million to clean up the river featured in the film.

A final example of how art can make the invisible visible and generate positive change comes from South Korea. Tree Planet, a social enterprise committed to reforesting the planet (and discussed in greater detail in the next chapter), became concerned when it heard that many of the beautiful gingko trees lining Garosu-gil Avenue (Tree-lined Avenue) in Seoul were disappearing. It made the missing trees visible to the public through a photo exhibition at the entry of Garosu-gil Avenue that portrayed photographs of the trees in pain. Shoppers added their names to a petition to support tree planting, and their names were turned into leaves that gradually filled in a bare tree that was projected onto a screen installed on the avenue.[16]

Tree Planet needed two thousand signatures for its campaign, but over only four days it gathered nearly ten thousand. In March 2013 it planted 413 trees. While it wasn't able to replace the missing trees on the original Garosu-gil Avenue, it was able to create a new *garosu-gil* avenue along a different street in the Dogok-dong neighborhood that previously had no trees. One of the new trees was then photographed and put onto a canvas by photographer Lee Myeong-ho. The artwork sold for $45,000, and the proceeds will be used to plant more trees on more treeless streets in Seoul.[17] The sTreet Campaign won Tree Planet and its collaborators the 2013 Red Dot Design Award for communication design and the 2014 iF Design Award grand prize.[18] Art made the invisible, missing trees visible to the public and generated new trees to purify the air, cool the sidewalks, and provide beauty for the city.

Make the Environment Culturally and Socially Important

Although Ishimure's writings and Yang's documentaries are direct critiques of capitalist systems that promote excessive consumption by people who are far removed from the workers who make their products and the environmental harm caused by their production, capitalist market and social forces can also be harnessed to promote proenvironmental behavior. When a CEO sets a reusable coffee mug on the conference table at the start of a meeting or a big-name celebrity pulls out her reusable water bottle in the middle of a press conference, it sends a powerful signal to those watching that these eco-friendly practices are not just acceptable but also socially desirable. As a result, proenvironmental behaviors and their associated accessories can become trendy status symbols.[19]

The cultural phenomenon of artists influencing proenvironmental consumption patterns is perhaps the most obvious in the area of sustainable fashion. While ideas of sustainable fashion have been around for more than a century,[20] in the twenty-first century it has become a big business supported by industry retail giants like H&M, Gap, and Uniqlo, as well as major designers such as Eileen Fisher.[21] These big-name, mass-producing designers and distributors are becoming more sustainable by increasing the extent and transparency of their environmental, social, and governance reporting, generally seeking to get as close as possible to a circular economy production model that generates low to zero waste and offers safe and dignified working conditions to employees throughout their supply chains. Additionally, niche designers like Hung Weiyu, Ayako Yoshida, and Im Seonoc and new, innovative eco-fashion distributors such as Redress (based in Hong Kong) focus on "upcycling," which takes discarded fabrics and used clothing and turns them into higher-value, luxury items.

As in the foregoing trendy fashion examples, environmental art is frequently crafted to elicit proenvironmental behavior. Occasionally, however, it is designed not to make eco-friendly behavior "cool" but rather to make eco-harmful behavior "uncool." A funny and crass version of this could be found in Hong Kong in spring 2019. Graffiti art around several areas with well-known restaurants had the message "Shark fin makes your penis small. Very small." spray-painted on various surfaces.[22]

In the same way that artists can make the things you put on and in your body "cool," they can also transform cultural practices tied to how we think and move through public space. One of my favorite examples is Green Pedestrian Crossing, designed by Jody Xiong of DDB China, which promoted nonmotorized transportation options. Xiong placed giant white canvases with bare trees in the middle of Shanghai crosswalks. Blotters with eco-friendly green paint were placed on the curbs on either side of the road. Pedestrians would then fill in the "leaves" of the tree as they

crossed the street, helping them understand how their walking was helping the environment. DDB estimates that its campaign reached nearly four
million people, and citizen awareness of environmental protection rose by
86 percent in the city.[23]

Pop music can reach audiences who might not otherwise care about
environmental issues, spurring outrage, hope, longing, and calls to action.
Taiwanese American Wang Leehom's 2007 *Change Me* album focused on the
issue of climate change, not only discussing its problems but also issuing a
call to action among its listeners. The album sold more than a million cop-
ies within its first month and went on to become one of the top ten selling
Mandarin albums of the year.[24] Longtime environmental activist and seven-
time Golden Melody Award winner Lin Shen Xian released *Quit Plastic Poison* in 2016, and critics give him credit for contributing to the political
pressure on political leaders that resulted in Taiwan's 2018 announcement
of a comprehensive ban on single-use straws, cups, bags, and utensils to be
implemented by 2030.[25]

Artists contribute to the meaning, the experience, and sometimes the
security of other advocacy activities, especially protests. Musicians and
other celebrities can draw nonactivists to protests. In post-Fukushima Japan,
musicians played a vital role in crafting the experience of the no-nuke protests, which often took on the feel of a music concert where environmental
issues could be discussed rather than a protest with an occasional band.
Many participants came to the events more because their favorite band
was playing than because they had a commitment to environmental issues.
They would learn about the issues while attending the protest/concert and
leave resolved to become more politically involved.[26]

In Taiwan, martial arts can enhance the cultural relevance and the security of environmental events. In one particularly notable protest in 1987,
activists seeking to stop the development of a fifth naphtha cracker (a large,
industrial petroleum-processing facility) in Houchin, Taiwan, blockaded the
access gate, and confrontations with police and security personnel had frequently been marred by violent clashes. In December the organizers worked
with a local temple to erect a spirit altar at the gate to the complex. During
the planning of the event, it appeared that the ceremony would be highly
contested and perhaps stopped by police. However, as part of the ceremony,
the temple's traditional martial arts group performed, using large swords

and spears. The police, who had assembled for the day, stood down, and the ceremony went off without any violence. In the end, a negotiation with the plant was reached to remove the altar and replace the activists' banner at the gate.[27] Thus, artists can play critical roles in ascribing cultural relevance to advocacy efforts; sometimes it is the presence of not just their art but also their bodies (and their swords) that matters.

Capture the Imagination

Perhaps the most iconic artist to capture our collective, global environmental imagination is Hayao Miyazaki. Beginning with *Nausicaä of the Valley of the Wind* (manga, 1982; film, 1984) and continuing throughout his long and productive career, whose work includes *My Neighbor Totoro*, *Princess Mononoke*, and dozens of other films, Miyazaki has been able to stir millions of viewers around the world. His worlds allow us to imagine environments that are better than our own, where the lush, silent forests overwhelm us with a feeling of peace and awe, and where we can talk with the spirits of the trees.[28]

Miyazaki's films also help us imagine more alarming futures that might come about if we don't take care of our planet. *Nausicaä*'s bleak world is dominated by machines, run by a militaristic state, and haunted by a toxic forest. It is similar enough to our own world to instill fear, allowing us to cheer eco-warrior Nausicaä as she battles and triumphs, and leaving us yearning for a similar hero for our own world. Viewers who have difficulty imagining a future where the sea levels rise when they read scientific studies of glacier runoff will have no trouble picturing themselves living in a world of rising seas once they've watched *Ponyo on the Cliff by the Sea* (2008), where the ocean first floods beaches and roads, and then swallows houses and villages while children and fish seek safety and companionship amid the rising water.

Imagination enables people to inhabit worlds that do not exist in our current reality. Ah Cheng's novella *King of Trees* (1985) is set in the lush jungles of southern Yunnan, where urban youths have been sent by authorities to clear out the "useless" trees of the wild forest to plant "useful" trees that can be harvested. Based on the author's own experience during the Cultural Revolution, the novella reveals the deep divide between urban and rural understandings of nature. Like Miyazaki and many others in this genre, Cheng deeply questions ideas of "progress" when contrasted with conceptions of "primitive," "backward," and "simple." Whereas governments and business seem to be pushing forward some kind of "progress," these novelists and filmmakers require audiences to question their own experience—Is their urban life better than the nature-filled life of their grandparents? Do the city scientists really understand how to identify and grow "useful" trees better than the rural farmers?

For a final example in this section, I turn to Japan immediately after the 3/11 disaster. In June 2011, just three months after the disaster, I was speaking with artist Ozawa Tsuyoshi in Tokyo about a project that he had just finished in Fukushima. He was so saddened to hear that a local Fukushima high school couldn't have its graduation because the school had been destroyed. As a result, all the children in the area were stuck in evacuation centers, bored and with nothing to do. He wanted to do something for them, and came up with the idea of a kite-making workshop. The children would write, draw, or paint something they liked on the traditional paper—pictures, poems, stories—and then they would turn the paper containing their words and images into colorful kites and fly them.

Some of the pictures and stories were so sad—destroyed homes, lost loved ones—but the kites created joy and symbolized hope as they flew high above in the breeze. It was cathartic and fun. The devastated community members who were living in temporary shelters laughed as they ran after the soaring kites. For that afternoon, at least, the grim reality of leaking radiation and flattened houses gave way to sunshine and the laughter of children. The community could play, remembering the world before the disaster and imagining a future when healing would have already happened. The medium of kites requires people to cooperate with nature to fly them high and make them dance. The power of the kite's art came much less from the beauty of the product itself and much more from the process of making and flying them—the sense of joy and freedom that they elicited from participants' imaginations.

Connect Communities

As in the example of the kite making in Fukushima, the process of making art can frequently be even more powerful than the artistic product itself. One of the goals of place-based and community-engaged environmental art is to use the art-making process to connect individuals and communities that are linked through their environment but might not realize it. Different types of individuals are drawn into the project and interact with people very different from themselves, but they discover common ties and understanding by working on a common project together.

A good example of how this can work is Ichi Ikeda's Moving Water project, which occurred from 2006 to 2008 along the Kedogawa River in Kagoshima, Japan. The community around the Kedogawa River is rural, and the river flows down from Sakurayama (Cherry blossom mountain) to the East China Sea. At the top of the river is a forestry school and at the bottom, a fishery school. Although the city has only about twenty thousand residents, the logging community and the fishing community rarely interact. Ikeda's water project made visible, through art, that the river is a common thread that connects everyone together.

The project began in 2006 with Moving Water Days. Participants first created a number of eighty-liter water boxes. The water boxes symbolize water as a human right; the World Health Organization has determined that every human needs eighty liters of water to support daily life—drinking, washing, cooking, and other activities. It is difficult to visualize how much water eighty liters is, so Ikeda made water boxes—eighty centimeters cubed and weighing eighty kilograms. Participants then transferred water from the present to the future by carrying the boxes from the mouth of the river up to its source in the mountains using bamboo backpacks. During the final Moving Water Days, they displayed the boxes in different places and used the backpacks to create a water wheel at the mouth of the river to symbolize the water cycle that brings water to the planet.[29]

The project expanded during its second year. Ikeda worked with a local nongovernmental organization, Eco Link Association, to help him connect with the forestry and fishery schools and recruit a diverse set of participants. As he recounted to me in a 2011 in a Toyko interview, "We had all types of people from the community—farmers, salt makers, teachers, construction workers, people from the forestry school, river school, and ocean school. ... Water follows a cycle: forest → river → ocean → forest. We want to make that kind of society." Participants gathered dead bamboo from the forest and constructed rafts that were connected to the shore at the Kagoshima fisheries high school and extended out about one hundred meters into Kagoshima Bay, which then connects to the East China Sea. The rafts were then released from shore and, while standing on the rafts, the paddlers spelled out SAVE WATER using semaphore (flag letters). They then changed the shape of the assembled rafts—from a line stretching out from the shore to a floating circle. The symbols for the water cycle were placed around the circumference of the circle, emphasizing their relationship to one another. As Ikeda described it, "The fishery school wasn't connected to the community, so I connected the two with my art. Art is a collective enterprise. ... It really raised everyone's awareness."[30]

In the final year, energy surrounding the project was even higher. Earlier in 2008 Ikeda and his water boxes had been featured at a UN conference on the environment in New York City,[31] so the residents were even more excited to take part. They used bamboo and other natural materials to create five floating islands to symbolize the earth's five main landmasses (of the seven continents Europe and Asia are merged, and Antarctica is excluded). Rice shafts were used to mark the desert, and cedar leaves were used to represent the green leaves. As Ikeda described the work to me in 2011, "Actually the [real] desert is bigger [than it is in the piece], but I hope for a smaller desert, more like what we made." Both the community-building process to create the work and the final work itself not only offer a powerful symbol of the importance of our environment, they created a tangible connection between the people, the earth, and the water. A photo of *Five Floating Isles* graces the cover of this book.

A different form of connection can be seen in the work of Kim Young-il, one of the official artists for the 2018 Olympic Games in Pyeongchang, South Korea.[32] Kim has been photographing Pyeongchang's landscapes for thirty years and wanted to do something to help the people in Seoul feel connected to the place that was hosting the Olympics. In *Mountains in PyeongChang & Sound of Korea*, Kim displayed fifteen of his pieces around Seoul's pedestrian walkways to invite residents and visitors to walk through Pyeongchang's landscape before and during the games. The visual works were accompanied by recordings from the area—mountain winds, splashing waves, and the ringing of a temple bell. Taken together, they enabled pedestrians traveling through the megacity to be transported to a more natural place.

Scaling Art Advocacy: Expanding the Effect of Art Advocacy

Of all the advocacy strategies discussed in this book, art is the least scalable. It is the most difficult to extend its efficacy beyond a single place and moment. Before I move on, I would like to clarify how I will be using the word *scale*. As in my discussions of the other strategies examined in this book, the word *scale* is used in this chapter as a verb to discuss how an advocacy effort is expanded or extended beyond its first instance, when it moves beyond its community of origin and beyond the first set of advocates involved. With art, the word *scale* is usually used as an adjective to describe the physical size of artwork (e.g., a large-scale sculpture of an elephant vs. a small-scale sculpture of an ant). However, to remain consistent with the other chapters, I continue to use the word *scale* as a verb to describe the expansion of the scope and extent of the advocacy effort rather than as an adjective describing the size of the artwork.

One of the most powerful aspects of most art—particularly public installations and ephemeral live performances—is precisely that it is not replicable. It must be experienced in a particular place at a particular time, and its meaning would change if the piece were moved, repeated, or copied and placed elsewhere. However, although these art forms cannot usually be expanded past their original installation or performance, their use as an advocacy strategy sometimes can be. This section suggests three ways that art advocacy can scale: go viral via social media, hit the road, and cultivate local art.

In all three cases, the efficacy of art as an advocacy strategy lies in its capacity to create and activate diverse networks of people within a specific community and across multiple communities. By engaging emotion and activating the imagination of diverse stakeholders, networks become energized, propelling policymakers to act and priming them to be open and interested in proenvironmental solutions when they sit down to make policy. Unlike most of the other strategies discussed in this book, art advocacy is less about targeting particular nodes in a network and seeking to influence the most powerful people in the network. Instead, it is more about affecting the entire network by shifting the culture in which the network exists such that hundreds or even millions of individuals connected to a network view the environmental issues we face as urgent problems that must be addressed.

Go Viral

As discussed earlier, new communication technology has enabled some forms of art—photographic images and video especially—to be viewed much more broadly than used to be possible. No longer must someone go to a theater to watch a film or visit a gallery to see an image; we can now access them on our phones. Although the experience of watching a film on a phone may be less intense than watching it in a theater, the ability to share the experience with friends located elsewhere means that millions of people can watch the same film or see the same image almost instantaneously. The individualized experience of looking at something on a phone can become a collective experience shared with one's friends and thousands, sometimes millions, of others.

Perhaps the most extreme example of this occurring in the environmental art space is the multimedia documentary about air pollution *Under the Dome*,[33] created by former CCTV journalist Chai Jing. It was released on February 28, 2015, on an official *People's Daily* website that also contained an interview with the filmmaker and an expression of gratitude from the minister of environmental protection. The film was viewed more than one hundred million times in the first twenty-four hours, and within four days, after having been viewed more than two hundred million times, it was censored and removed from all social media in China.[34]

Although full of horrifying statistics, the film's power comes not from the data it shares but from the way it connects those numbers to personal stories that the viewer can easily relate to. The film opens with the director's own story of discovering a tumor in her unborn child, a tumor she was convinced was caused by air pollution. Later, viewers can feel their hearts crack open when Chai interviews a six-year-old child. "Have you ever seen a real star before?" "No, I haven't." "What about blue sky?" "I've seen one that's a little blue." The story becomes even more personal when she talks about the collective self-delusion that she and so many others have been perpetuating about China's air pollution. "We kept calling it 'fog.'"

The images, stories, and art in this film are what make it powerful. Reports in newspapers and journals have extensively documented the extent of China's pollution problems with plenty of facts, statistics, and analysis. In contrast, the film makes that "fog" clearly visible as threatening pollution, offers an emotional connection to the victims, forces the realization that the viewer is vulnerable, and inspires him or her to be part of the solution. As the film closes, Chai declares, "I'm not going to shirk the responsibility, I'm going to stand up and do something." She then pleads, "If you don't know how to fix it, please stop breaking it." After watching the film online, thousands of people responded, committing to personal actions in their public comments, such as walking rather than driving to work, in order to become a part of the solution.

Why was the film banned? Probably not because it documented China's pollution problems. As discussed extensively in this book, the Chinese government has been working very hard to raise awareness of pollution issues and convince individuals, companies, and local governments to improve environmental outcomes. Likely, the problem was Chai's declaration at the end of the film, "We have the right to know, the right to participate, the right to sue for damages." Gary King, Jennifer Pan, and Margaret Roberts's insightful analysis suggests that most censorship in China is not due to content; rather, "[China's] censorship program is aimed at curtailing collective action by silencing comments that represent, reinforce, or spur social mobilization, regardless of content."[35] It seems likely that, while initially supportive, the government became worried when it became apparent that the film was not being seen by a small group of environmentally minded activists but had gone viral—albeit briefly—to reach millions of angry citizens and perhaps spur them to act.

Chai's documentary exerted political pressure through the networks that it influenced. As a well-known journalist, Chai used her individual influence over people in her own network, but her use of art was not aimed at the powerful "node" actors but rather aimed at activating multiple networks simultaneously by engaging with the broader public. Viewers used social networks to spread their anger, and those same networks ultimately reached policymakers who first acted to shut down the video and then acted to strengthen policy—China's thirteenth Five-Year Plan, which was put into place almost exactly one year after *Under the Dome* was aired, has aggressive air quality targets in addition to other environmental targets that it has strengthened.[36] As with other instances of art activism, it is difficult to draw a direct causal connection between the art and the policy outcome, but in the case of *Under the Dome*, it is easy to see how the art raised public awareness of the issue and increased political pressure on decision makers to act in proenvironmental ways.

Hit the Road

It is common for popular exhibits to travel from one gallery to another, but usually there is no overt relationship between exhibition locations, and audiences are unaware of where the show was previously or where it is going next. The recent trend of "pop-up" exhibits generally follows the same model, only the exhibition time is much shorter and the locations can be unconventional.[37]

However, it does not have to be the case that exhibits remain disconnected from one another even in the case of place-based art. Since it is usually the case that pop-up exhibits are carried by a single truck, multiple venues can be connected to one another if that is part of the artists' vision. The places where the art stops along the way can become connected to one another by a common experience, and that sense of connection can be made more obvious when the visitors interact with and respond to the art in some way such that they leave their own mark on the exhibit. This allows for the possibility of a dynamic relationship among people in different locations who, through the moving exhibit, co-create an unfolding, collective experience or artwork, as well as a network of individuals who have become connected through the art.

An example of how this has worked is the Daylily Art Circus (the daylily represents rebirth in Japanese culture), which connected the victims of the 1995 Kansai earthquake with those of the 2011 earthquake in Tohoku.[38] Organized by Kaihatsu Yoshiaki, the interactive exhibit began in Kobe in August 2011, displaying large air-filled sculptures and combining those with interactive activities for children. The whole exhibit was contained in an Daylily Art Circus truck, which would stop at different destinations every day, starting in Kobe in August and finishing at a village outside Sendai in September. The initial stops were not just opportunities to exhibit the art and engage participants; they also served as fund-raising opportunities. As the exhibit got closer to the epicenter, the flow of money reversed and aid was distributed to communities devastated by the triple disaster.

The Kansai region was physically quite disconnected from the disaster in the Tohoku region—its inhabitants did not feel the earthquake, are on a different power grid so were largely unaffected by the disruptions in power supplies, and are far enough away that even if the Fukushima plant had blown up like Chernobyl, they would have faced little to no nuclear fallout. And yet, thousands of refugees from the Tohoku/Fukushima disaster took refuge in Kobe (including me—me, my husband, and our two small children were living in Tokyo when the earthquake hit, and we moved to the city to stay with friends for several uncertain weeks afterward). Additionally, residents of Kansai also felt especially connected to the victims in Tohoku because the devastating earthquake that hit their region in 1995, which killed about five thousand people and displaced three hundred thousand, was still fresh in the minds of many. Thus, the Kansai region was filled with people who "want to do something," and the Daylily Art Circus helped them engage meaningfully in a common experience with counterparts in Tohoku while contributing financially to communities that needed aid.

During an interview with me in 2011, Kaihatsu described how helpless he felt at the time of the 1995 earthquake in Kobe. "At the time I couldn't do anything. I couldn't go there and was kind of frozen. ... This time after the disaster I was better prepared. In some ways I've been preparing for this [post-Fukushima activism] for fifteen years." As he explained his motivation and inspiration for the traveling exhibit, he said, "I am an artist, so I can give art. If I were a farmer and grew tomatoes or apples, I would send tomatoes or apples to the victims. If I could do construction, I would help fix houses. I can't make tomatoes or fix houses, but I can make joy."

Like the kite workshop described earlier, Kaihatsu's exhibit was designed to elicit joy. The sculptures were made of the same nylon material as the blow-up lawn art that shows up across American lawns over holidays or in front of used-car shops to beckon customers inside. They were large, colorful, funny, and sometimes ridiculous. Some of them flapped around, generating spontaneous laughter from anyone watching. Giant clowns, animals, flowers, and other figures drew in viewers and participants. Children would reach out to hug elephant trunks, and adults would marvel at the pure whimsy of the figures. When the Daylily Art Circus truck stopped, it held art-making workshops for kids—children in the early part of the trip drew pictures and wrote messages of hope to the children who would be visited later.

The Daylily Art Circus ran for several years. In 2011, its first year, it made the journey from Kobe, which was physically unaffected by the disaster, to Sendai, which was devastated. In its early stops in areas unaffected by the disaster, it collected donations and art filled with hope and encouragement to be passed from Kansai residents to their counterparts in Tohoku. Farther along, when it stopped in Tohoku and Fukushima, it frequently held its workshops outside, because people were still living in the school gyms, where the exhibits in Kobe had been held. In those localities, it shared the art of hope created by those in the early part of the trip and conducted its joy-producing workshops in the devastated communities, reassuring residents that others were rooting for them to recover and giving them an afternoon of laughter and smiles with wiggly sunflowers and smiling blow-up elephants before they had to return to school gyms to sleep.

In 2012 it returned to several towns around Fukushima that were still suffering. In 2013 the circus truck once again traveled from sites in Kansai to those in Tohoku, stopping at schools, community centers, and public parks along the way. In 2014 it once again visited recovering communities in Fukushima. In 2019, although it was no longer making the long trek up the coast, a few of the sculptures were displayed in a commemorative exhibit in the Fukushima Prefectural Museum.

In the manner of most art, the Daylily Art Circus did not scale in the regular meaning of the word—the sculptures were not mass-produced and spread around the world; they didn't get bigger physically in some way. However, the advocacy element of the art was scaled. The art was not displayed in a single place or even a dozen, disconnected places. The display and interactive participation that occurred in one place was then connected to exhibits in different communities. It revealed and amplified the connection between those in Kansai and those in Tohoku, their mutual vulnerability to natural disaster, their common reliance on nuclear power, the outpouring of volunteers to support victims, and the capacity of art to facilitate healing. Collectively, the power of the whole tour sutured place and time together. Art as an advocacy strategy could move beyond a single place to affect multiple communities across a longer period of time, linking victims and supporters, citizens and policymakers, together in multiple networks of humanity and art making.

Cultivate Local Art

As discussed earlier, place-based art draws its power from the specific time, place, and community in which it is created.[39] Therefore, it is generally impossible to scale art of this type. However, it is possible to multiply the places where the art is developed and displayed, and this has been occurring around the world as global cities compete to attract the creative class,[40] and rural communities strive to revitalize their communities and avoid extinction.[41] Although arts festivals can sometimes be purely economic ploys by local governments,[42] and their proliferation can spread audiences too thin as well as forcing artists to choose among competing festivals,[43] their proliferation is offering interested artists unprecedented opportunities to engage in environmental advocacy. This section will illustrate how this works with three examples.

Perhaps the most heart-warming story is that of Huang Yung-Fu, a ninety-six-year old grandfather who saved his Taiwanese village from demolition by painting it.[44] Rainbow Village, as it has come to be known, was a village of 1,200 homes that were largely occupied by former soldiers like Huang, who came to Taiwan when the nationalist army retreated from the mainland. As the soldiers and their families aged, died, and moved away, developers began buying up the land. By 2008 only eleven houses remained, and the government began planning to demolish the rest.[45]

Feeling bored, lonely, and a bit helpless in his fight to save his village from demolition, Huang took up painting at the ripe young age of eighty-six, painting a bird inside his house and then moving to the outside. He liked the results. The images were cheery and hopeful. He decided to keep painting, covering his neighbors' walls with bright flowers, whimsical cats, mythical beasts, and other fanciful creations that erupted from his imagination until the entire village was bursting with colors.[46] In 2015 the Defense Ministry pledged to preserve the village as a cultural site,[47] and Rainbow Village now receives more than a million visitors every year.

On a slightly larger scale in South Korea, the Sea Art Festival was held for the first time in 1987 in the run-up to the 1988 Olympics in Seoul, just months after the June Democracy movement, and one month before a new constitution was established that guaranteed a range of new democratic rights to South Korea's citizens. Now a valued tradition, the festival moves to different beaches around Busan, showcasing site-specific installations and bringing local residents and international visitors to less visited areas in South Korea.

The theme of the 2017 festival was "sea+art+fun." While many of the pieces were inspired by nature but not particularly political (e.g., "a slice of summer"),[48] others had strong political messages. For example, PERBOS's *Floride* showcased whimsical, treelike creations that were reminiscent of Dr. Seuss's truffula trees and were designed to make viewers smile and want to play. However, the installation's description read, "A critical outlook on society and environment lies beneath the seemingly light and humorous work. The artist collected and combined waste resources produced from raw petroleum material at a demolition site and produced works in the form of palm trees, throwing a critical outlook on the stance between human civilization and environment."[49]

Rural Japanese villages have been using art festivals as a method to revitalize their communities for nearly twenty years, and the country now boasts the largest number of art festivals in the world.[50] Art festivals gained popularity as a method of revitalization with the success of the Echigo-Tsumari

Art Triennial, which displays art across approximately two hundred villages in the Echigo-Tsumari region of Japan.[51] Like many rural areas around the world, the villages in this mountainous part of Japan had been suffering from depopulation as young people moved to the cities for work and did not return. Begun in 2010, the Echigo-Tsumari Art Triennial sought to create a tourist attraction that would bring young people and visitors to the area, stimulate the economy, and offer opportunities for urbanites to connect with nature.

Inspired by the traditional concept of *satoyama* (mountain village), which evokes images of the landscape and traditional lifestyle of rural Japan, the festival aims to promote the idea that "humans are part of nature." The method of displaying the works, spread across two hundred villages, is intentionally "inefficient" and "deliberately at odds with the rationalization and efficiency of modern society. Wandering among the artworks which emphasise the beauty and richness of *satoyama* and reveal the accumulated temporal layers of human inhabitation opens the senses to the wonder of existence and revives the soul."[52] In addition to reviving the local economy, the festival offers an opportunity for artists to engage in environmental activism, such as when Hong Kong–based artist Ricky Yeung "planted" bright yellow construction beams in a rice field to protest the spread of urban development that displaces farmers.[53]

Echigo-Tsumari's success has inspired others—the Setouchi Triennale draws visitors to twelve islands and two port towns in the Seto Inland Sea.[54] Annual art and music festivals also take place in Miyagi, Sado Island, and Kitakyushu, to name a few.[55] While the other festivals may not place *satoyama* and nature at the center of their themes, artists in Japan have a long history of using these venues as sites for social and environmental activism.[56]

Conclusion

Art is a powerful advocacy strategy. As a cultural practice, it has the tools to engage people's emotions and imaginations in ways that can make an intellectual problem become a personal one. It can make issues matter to people in ways that are difficult or impossible to achieve merely by providing scientific information.

There is a deep paradox at play when artists engage in advocacy. One of the sources of their power is that they are generally not considered to be politically powerful. They don't sit on policymaking committees, and they're not (usually) elected for public office. As a result, they're able to push the envelope on permitted action because they blur the boundary between expression and activism. Artists are permitted to make art that critiques society and the government, art that outrages or disgusts viewers, art that reveals injustice and makes you cry. This is permitted at local art festivals, in small galleries, and sometimes even in public spaces.

However, this unusual level of freedom accorded artists can sometimes lead them to misjudge what they can get away with. This is what happened to one of China's most well-known artists, Ai Weiwei. Ai has been producing art related to a variety of social and political issues throughout his career. In terms of his environmental art, he frequently deals with pollution issues. For example, following a huge oil spill in northern China that threatened the Yellow River, Ai created a ceramic installation, *Oil Spills* (2006), that looked like giant puddles of oil clustered on the gallery floor.[57] *Tree* (2010) was constructed from dead branches gathered from the mountains of southern China and intended to mimic a traditional custom found in Jiangxi Province where curiously shaped trunks and roots are purchased in markets to be displayed at home. The Tate commentary suggests, "*Tree* can be read as a reference to [the] Taoist ideal of harmony—unifying the work of man with nature as well as linking the earth and the sky."[58]

In 2008 Ai was a darling of the Chinese government for his spectacular design of the Olympic stadium, which showcased the rise of China to the world. One year later, a devastating earthquake in Sichuan killed nearly eighty thousand people, and Ai joined the public outrage when it came to light that many of the deaths of children could have been avoided had corrupt officials not pocketed money that had been intended for school construction. He began investigating and creating works that drew on and fed the outrage—*Remembering* was constructed using nine thousand school backpacks to represent the children who died.[59] His activism led to police beatings that landed him in the hospital, the demolition of his studio, and house arrest. In spring 2011, while activists in the Middle East were engaged in overthrowing their authoritarian governments, Ai was arrested on charges of tax evasion. He was held for three months and then released to home arrest. In 2015 Ai left China and moved to Berlin. In 2018 authorities demolished his Beijing studio without warning.[60] He now makes and displays his art around the world, but has not returned to China.[61]

While some have called Ai a "force of nature,"[62] political scientists and other scholars should stop treating artists and their artistic contributions to advocacy as exogenous forces akin to earthquakes. As the examples in this chapter and others have demonstrated, organizations around the world have recognized their importance and systematically incorporate art and artistic expression into their advocacy. Artists themselves are important actors who play influential roles in shaping both public opinion and public policy. Scholars should pay closer attention to the role of artists and incorporate them explicitly into their advocacy models.

9 Be a Game Changer

Occasionally advocates are so innovative and effective that they not only effect positive change in their desired direction, but they transform the advocacy landscape for everyone else. In the course of my research, there were a very small number of individuals and organizations that my interlocutors would refer to repeatedly. These innovators would be talked about as examples to emulate. They would be discussed as pivot points in the narrative of a country's advocacy. They were described as providing the foundations on which other advocates would build.

The efforts of these innovative advocates do not fit neatly into a particular strategy. Their power does not come from a particular tactic or technique that they employed, but rather from their transformational effect. These advocates were able to "think outside the box" of typical advocacy and do something completely different, something that had—and is having—a transformative effect not just on their issue area of interest but on the process of advocacy itself. In particular, these advocates have found ways to catalyze positive interactive effects among individuals, companies, nonprofits, and governments in ways that changes a zero-sum political game where a small group of individuals and companies win at the expense of the broader public and the planet into a positive-sum game where individual actions by consumers and companies contributed to a replenishing of the commons, rather than its degradation.

To use the language of the Connected Stakeholder Model, these game changers were able to make new networks, reconfigure old ones, and energize both. They found innovative ways to connect stakeholders who were previously disconnected, helped them discover common interests, and identified new ways to work together to promote proenvironmental outcomes.

In some cases, they were able to bring new stakeholders into networks connected to environmental policymakers. In other cases, they took on the role of the artists in the last chapter, shifting broader culture in ways that generated proenvironmental attitudes among the public and policymakers, facilitating proenvironmental policy changes.

This chapter seeks to use three stories of game-changing advocates—drawn from three different countries, sectors, and generations—to illustrate how effective advocacy may not necessarily be about developing a winning strategy over an opponent but rather be about changing the advocacy game into one in which everyone wins. Ma Jun in China comes from the nongovernmental organization (NGO) sector. His pollution map has created new platforms of accountability that encourage polluting firms to clean up their act. Koike Yuriko in Japan comes from the government sector. She launched the Cool Biz campaign while she was environmental minister, tapped into the transformative effect that cultural practices can have on environmental outcomes, and has continued to be active in local and global policy transformation as governor of Tokyo. Finally, Jeong Mincheol and Kim Hyungsoo, come from the business sector in South Korea. They founded Tree Planet, an innovative social enterprise that is using gaming and social media technology to reforest the planet.

Ma Jun and the Institute of Public and Environmental Affairs: Building a Collaborative, Proenvironmental Policy Ecosystem

Ma Jun began his career as an investigative journalist for the *South China Morning Post*. In 1999 he wrote *China's Water Crisis*, which has been likened to Rachel Carson's *Silent Spring* in the way that it raised public awareness of China's environmental problems. In 2004 he spent a year at Yale University as part of its World Fellows program, working closely with faculty in the School of Forestry and Environmental Studies to develop his vision of a nonprofit organization that would bring more transparency to China's environmental problems and develop a multistakeholder platform that would encourage collaborative improvement. As he phrased it to me during a 2010 interview in Beijing, "I am convinced that civic participation is the answer to our environmental problems, and that it starts with transparency." His concept was deceptively simple: create a user-friendly website that made official government data about pollution readily available to the public. He incorporated the Beijing-based Institute of Public and Environmental Affairs (IPE) as a nonprofit organization in 2006 and launched its public website (ipe.org.cn) the same year.

With a staff of three people, IPE created the China Water Pollution Map in 2006, and the map has since been expanded to include air and solid waste pollution, as well as governmental transparency measures. The map gathered official, government-reported environmental inspection metrics and made them accessible to everyone. Previously, these factory-level data were publicly available, but people had to know where to find them, and they were required to go to separate websites and government offices to get the records for different towns and companies. IPE collected all of these data in one place, allowing users to click on any part of China and find out the level of various kinds of pollutants in that area, as well as their source. With these data, IPE then listed companies that had pollution violations, making it possible for buyers, the public, and local regulators to put additional pressure on polluting companies to clean up their facilities.

The map and Ma Jun's efforts won him immediate global and local recognition, including being named Green China Man of the Year, as well as one of the one hundred most influential people by *Time* magazine in 2006. His profile, recognition, and influence have only grown over time: In 2012 he won the prestigious Goldman Environmental Prize and was named one of *Foreign Policy*'s Top 100 Global Thinkers. In 2015 he became the first Chinese social entrepreneur to win a Skoll Award, and he is currently a Global Fellow of the China Environment Forum at the Wilson Center in Washington, DC.

Although the pollution map itself was helpful, the "game-changing" aspect of IPE was the way that it integrated the map's data with a change management tool, which enabled multistakeholder collaborations. IPE's pollution data contain the government inspection records of individual facilities, so it is possible to search the database by company name and find out whether a specific company has met or violated China's environmental regulations, and use that information to track changes over time. The database now contains more than 1.6 million inspection records, as well as additional information from companies seeking to explain their environmental records and their plans for remediation.[1] Among the first groups to see the potential of this new tool were global corporations who already had commitments to green their supply chains but were having difficulty finding ways to monitor their suppliers in China.

One of the earliest multinational corporations to approach IPE was Walmart. As discussed in chapter 7, the global retail giant had made sustainability one of its corporate goals in 2005, so it was eager to take advantage of IPE's new capability to help the company green its supply chain in China. Since Walmart had thousands of local suppliers in China, frequent onsite inspections of each facility were logistically impossible. IPE's database offered a practical solution: Walmart could use it to ascertain whether its suppliers were complying with local environmental regulations. If Walmart discovered that one of its suppliers was in violation of national standards, it would send a formal letter putting the supplier on notice that it needed to improve or it would no longer be able to supply Walmart, naming IPE as an organization that could help it clean up its operations. The local supplier would then contact IPE, which would work with the supplier to develop a plan that would enable it to come back into compliance. The supplier could then take the necessary steps to remedy its environmental issue and provide supporting documentation of its cleanup actions to IPE, which would then publicly disclose the company's actions on IPE's website. This process ensured that the company addressed the problem and also increased the transparency of the process.

Through its actions, IPE shifted the compliance calculations for Chinese manufacturers supplying global multinational corporations. Whereas before, it was common for violating companies to pay fines in lieu of making changes to their manufacturing processes—a legal method for addressing the violation—local manufacturers could now be pressured by their buyers to improve their actual environmental performance even in the absence of more stringent governmental enforcement. Furthermore, polluting companies were rewarded by their buyers for improving their environmental performance; those that did not improve were punished with a loss of business.

As corporate interest in cooperating with IPE to help manage supply chains grew, it became apparent that IPE, with its tiny staff in Beijing, would not be able to meet the demand for third-party audits by itself. Therefore, in 2007 it initiated the Green Choice Alliance (GCA), a coalition of NGOs that would partner with IPE in providing information, supervising third-party audits, and reviewing audit reports. The number of NGOs in the GCA more than doubled in five years, growing from the original twenty-one in 2007 to forty-nine by 2012. The GCA also served important cross-checking, burden-sharing, and transparency purposes, since approval of all GCA members was required for company records to be removed from "violator" lists.

Third-party audits are an important but not a perfect fix to the problem of environmental violations. Deception related to environmental audits is a serious problem in China. In fact, in recent years specialty consulting firms have formed to help suppliers find ways to pass inspections and audits without having to address their violations.[2] Although it is nearly impossible to guarantee that no cheating occurs, the GCA model involved many parties and created a collaborative process oriented toward improvement rather than punishment, helping to make compliance more cost effective than cheating. In his study about implementing supplier codes of conduct in China, Bin Jiang found that "most buying companies' auditors or independent third-party auditors do not work with factories together to provide assistance or support regarding how to make the required changes." He goes on to recommend that buyers and suppliers "set practical, step-by-step goals and give credit for incremental change; measure continuous improvement rather than compliance; [and] reward actual changes."[3] This is exactly the approach taken by IPE and the GCA: by allowing suppliers to post additional documentation to the IPE website, incremental improvement is recognized. IPE reports are oriented toward disseminating examples of best practices while also documenting and publicizing success stories of how violating companies have been able to find creative ways to address their environmental problems.

IPE's searchable database and the GCA had successfully created a transparency-based platform that enabled responsible corporations to promote greener supply chain management. However, Ma Jun soon noticed that many of the most highly polluting companies and industries were not taking advantage of the new platform—they didn't seem to care about compliance or pollution. Therefore, starting in 2010, IPE, along with partner NGOs, decided to begin a campaign that targeted pollution of heavy metals, which is one of the worst forms of pollution from a human health standpoint. They found that the information technology (IT) industry was largely responsible for this kind of pollution in China. Strategically, targeting this industry in particular was politically useful because it helped support a national governmental effort to curb heavy metal pollutants, which had led to thirty-two mass incidents in 2009 and were causing significant harm to China's public health, food security, and agricultural sector.

Furthermore, many of the highly polluting factories were supplying major multinational corporations with international brands to protect, making those buyers both more vulnerable to consumer pressure and more financially able to respond to the problem.

IPE's first report on the industry outlined the negative consequences of heavy metal pollution, then discussed industry responsiveness to the NGO inquiries. The report highlighted the good behavior of the most responsive companies (e.g., Panasonic and Sanyo)[4] while also making clear which companies had been unresponsive (twenty of the twenty-nine companies did not respond to the initial inquiry). The first IT report was published in April 2010, and three subsequent reports were published quarterly in 2010. IPE's reports, combined with its international media efforts to draw attention to the issue, were highly effective. By the fourth report, nearly every IT firm had responded, and companies were ranked on six criteria having to do with responsiveness and an additional four criteria that included action plans for not just tier 1 but also tier 2 suppliers.

By the time of the fourth report, in January 2011, only one of the twenty-nine technology companies had still not been responsive to IPE's inquiries about its supply chain: Apple. IPE's campaign then shifted its focus to target that industry leader. *The Other Side of Apple* report, published in August 2011, targeted Apple almost exclusively. It included a compelling web video clip that was circulated widely across the internet, highlighting the plight of workers at Apple supplier companies and the suffering of the residents who lived near those factories. The report received extensive global media coverage immediately after it was published. This was followed by a second report that focused extensively on environmental problems caused by Apple suppliers. The campaign worked. After years of avoidance, Apple executives met with Ma Jun in San Francisco in September 2011, and by January 2012, Apple began to disclose its suppliers in China and take concrete steps to engage in better supply chain management in China.

In keeping with its policy of focusing on improvement and rewarding incremental change, the following IPE report on the IT industry, entitled *Apple Opens Up*, highlighted three major Apple suppliers that had been committing serious environmental violations and documented the ways that they were moving to improve their practices. In an impressive turnaround, by 2019 Apple had become the top-performing company according to IPE's Corporate Information Transparency Index, which reports transparency-related metrics for 384 companies across thirteen different industries.[5]

After its success with Apple, IPE expanded its campaigns to target other industries, and word of the usefulness of its platform for helping companies audit their Chinese suppliers quickly spread to other multinational corporations interested in greening their global supply chains. Companies such as General Electric, Levi's, Nike, Unilever, Coca Cola, and other large multinational corporations that purchase from tens of thousands of suppliers in China are now working closely with IPE to check whether their suppliers are violating Chinese environmental standards and offer those suppliers advice on how to clean up their facilities to move them into compliance.[6]

Although use of IPE's platform had begun to spread among global brands, most local firms continued to conduct business as usual—paying fines rather than cleaning up when they were caught violating environmental regulations. Ma Jun realized that another key pressure point for firms—all firms—is financing, and IPE has been at the forefront of the green finance movement discussed in chapter 7. First, IPE partnered with Chinese banks to encourage them not to provide financing to firms found to be violating environmental regulations. In 2007 the Chinese government became one of the first governments to integrate environmental concerns into its finance policy, and since 2007 it has required all companies seeking loans from the People's Bank of China to demonstrate compliance with environmental regulations.[7] By 2011, China's stock exchange regulators were requiring companies to disclose their environmental information with IPE before doing an initial public offering.[8]

Second, IPE began to target other investors by connecting firm-level environmental data to stock listings. Previously, those using IPE's database had to know the name of a specific firm (often Chinese only) in order to research its inspection records. While this worked for global multinational corporations and other corporations that knew the names of their subcontractors and suppliers, most investors do not know the specific names of suppliers; they just know the parent company's name. IPE began to explore ways to make the environmental records of listed companies more transparent. Through a partnership with the *Securities Times*, IPE developed an online tool that allows investors to research the environmental risk of publicly listed companies. Launched in 2015, the Environmental Index tracked 1,365 key monitored enterprises that were connected to 519 publicly listed companies.[9] This effort has continued to expand, adding companies and supervisory records and linking individual firms to their stock ticker numbers. As of this writing, IPE's enterprise database contains more than six million firms, of which more than one million have supervisory records.[10] Its listed company database has almost four thousand records.[11]

IPE's most recent innovation has been to make all of these data dramatically more accessible through a mobile app. The Blue Map App (as it is now called) puts the information available on IPE's interactive pollution map into the hands of anyone with a phone. The Blue Map App includes several features designed to make it a very useful tool for consumers and citizens, and it also makes the data gathered useful for enforcement authorities. In addition to basic weather and pollution forecasting data, the Blue Map App includes data that help users understand the source of the air and water pollution that they may be experiencing. Users may also use the app to do the same enterprise name and stock ticker number searches available on IPE's website, enabling them to check the environmental records of individual firms and listed companies.[12]

Finally, the Blue Map App has been fully integrated with the Chinese government's Black and Smelly Waters program, which was launched in 2016 to allow individuals to report pollution violations directly to the Ministry of Ecology and Environment. This means that with just a few clicks, users anywhere in China who observe pollution, such as witnessing dumping, smelling stinky discharge, or hearing suspicious drilling, can upload photos and text government officials. They are guaranteed to receive some kind of response from officials within seven days.[13]

When I asked Ma Jun what he thought the keys to his success were, he listed three: (1) IPE relies only on open-source data. (2) IPE has positive intentions. It is not trying to destroy companies. Although it doesn't condone the belief that firms should have as their sole goal the maximization of profits and should accept pollution as a necessary cost of achieving that goal, it offers a clear path to a solution. (3) IPE always tries to collaborate with other stakeholders.[14]

Ma Jun and IPE have been environmental advocacy game changers. They have not only generated positive environmental policy and behavior change among companies, governments, and consumers; they have shifted the entire landscape of environmental advocacy in China and, indeed, the world. IPE's transparency-based platform has created the information infrastructure necessary to enable consumers to pressure corporations to change behavior, even in a context in which government enforcement is still weak.[15]

IPE's actions have empowered others: The public interface and data-gathering capacity of the Blue Map App enable government officials to discover and target the pollution that is bothering citizens the most. The enterprise database enables investors—individuals and corporations—to direct their funds toward companies with better environmental records and away from companies that are violating environmental regulations. IPE's consultancy services, its collaborations with other NGOs, and its improvement-oriented approach enable all firms, from tiny local companies to global multinational corporations, to identify ways that they can improve their operations and then implement those improvements.

IPE's database, app, and web tools have facilitated collaborations among multiple stakeholders. They are creating proenvironmental incentives for companies, allowing consumers and citizens to reward proenvironmental companies and political leaders while pressuring those that have been slower to act. By making the connections among consumers, citizens, companies, banks, and government more visible, IPE has formed proenvironmental networks that did not previously exist. By working with each other through these networks, consumers, suppliers, global multinational corporations, banks, and governments are individually able to take meaningful proenvironmental actions, and collectively their actions are transforming the entire system of environmental governance in China.

In short, Ma Jun has catalyzed environmental advocacy in China by thinking beyond traditional dichotomies of confrontation versus co-optation. His creative interactions with the public, nonprofit, and for-profit sectors have make it possible for him not only to craft a single solution but also to build the informational infrastructure for an advocacy ecosystem that nurtures the development of policies and practices that are generating proenvironmental outcomes that are good for individuals, governments, companies, and the planet.

Koike Yuriko and Cool Biz: Cultural Innovation to Benefit the Climate and the Economy

While many of the other advocates featured in this book are engaged in "bottom-up" advocacy, working first at a local, grassroots level in order to inspire broader change, Koike Yuriko's Cool Biz campaign was a top-down effort to catalyze proenvironmental changes in Japanese (and global) culture that were instituted while she was Japan's environmental minister. Born in 1952 near Kobe, Koike was the daughter of an international businessman who gave her extensive exposure to foreign cultures and encouraged her to travel. Taking his advice to learn more about the world and Arab cultures in particular, she studied Arabic and completed her undergraduate degree in sociology at Cairo University in 1976. After graduation, she worked as an interpreter and in journalism, receiving the Female Broadcaster of Japan award in 1990 for her coverage of Iraq's invasion of Kuwait.[16]

She ran for the House of Councillors in 1992 as a member of the Japan New Party and was elected to the House of Representatives the following year. In 2005 Prime Minister Koizumi Junichiro appointed her to be minister of the environment, and from that position she was able to enact some of her many ideas about how to promote better environmental behavior. One of her biggest and most successful was Japan's Cool Biz campaign, which has since spread globally.

Cool Biz encourages people to dress more casually in the summer, without a tie and jacket, so that they can be more comfortable in warmer temperatures. It also encourages businesses to keep their thermostats set higher than usual—no lower than twenty-eight degrees Celsius during the summer months—to save energy.[17] The primary goal of the campaign was to reduce energy consumption in the summer, which would reduce carbon emissions and save money. There were also anticipated benefits for public health—the Japanese have a specific word, "air-con disease" (*koola-byou*), to describe the symptoms that people get from too much air conditioning, such as headaches, dizziness, dehydration, and metabolic dysfunction.[18] Finally, Koike hoped to spur commercial activities as people shopped for Cool Biz clothing to wear to work.

By all accounts, the campaign has been highly successful. In its first year it reduced carbon emissions by 460,000 tons, and by 2012 it was estimated that annual carbon savings had reached 2.2 million tons.[19] Cool Biz spurred more business than expected. Prime Minister Koizumi wore a short-sleeved Okinawan Kariyushi shirt when he launched the campaign on June 1, 2005, and the company that made it immediately received eight hundred orders for the same shirt.[20] The garment industry saw sales increase by more than $850 million in 2005,[21] with further benefits coming in 2011 with the launch of Super Cool Biz, a reboot of the Cool Biz campaign that followed the 2011 earthquake and nuclear disaster.[22]

In May 2015, I asked Koike what gave her the idea for Cool Biz. Her response was remarkably similar to Sheri Liao's description of her inspiration for the 26 Degree Campaign in Beijing, discussed in chapter 6.

> I had the idea a long time ago. So, when I became [environment] minister, I could do something about it.... The original idea came because I was concerned about the women workers who are in the office all day. The sales people get to come and go, but those clerical women have to stay in the same place all day. They had to put blankets on their lap to stay warm in the middle of the summer. It was ridiculous!
>
> But, you know, I wasn't the fashion minister.
>
> The [Cool Biz] concept is to achieve the ultimate [environmental] goals and policies, but you have to have sympathy from the people. Without support from the people, even with a lot of budget, you're not going to get anywhere. And with support, it can go well even without a budget. It really didn't cost much....
>
> The ministries aren't very good at marketing, but marketing is important....
>
> [The Cool Biz campaign] emphasizes cool, happy, emotional things. The issue is global, meaningful, and urgent. You have to start as soon as possible.

Koike had two important insights into the challenge of environmental policymaking: (1) Many of Japan's toughest environmental challenges could be addressed by cultural change and did not need significant financial investment or new technology if people would change their behavior. (2) Environmental policymaking had a marketing problem—proenvironmental change was seen as a sacrifice rather than a beneficial opportunity. When she took office in 2005, Koike thought that the Ministry of Environment could market proenvironmental behavior that would be attractive to citizens, businesses, and the government, which in turn had the potential to generate enormous environmental benefits for everyone.

Culturally, it is impossible for junior-ranked people in a Japanese office to shed their ties and jackets if senior managers are all wearing suits and ties, so she recruited top businesspeople and senior politicians to promote the idea that more casual summer attire was not just acceptable but also desirable. After Prime Minister Koizumi launched the campaign wearing a short-sleeved shirt on June 1, 2005, he followed this up with a cabinet meeting in which all the cabinet ministers were encouraged to dress in a summer fashion, leaving their ties at home.

On June 5, 2005, several of Japan's top CEOs took part in a Cool Biz fashion show during the 2005 World Exposition.[23] Throughout that summer, the prime minster, cabinet ministers, Supreme Court judges, and top business leaders were frequently seen in public and covered by the press wearing short-sleeved shirts and forgoing a tie.[24] Many businessmen were initially unsure about how to respond, some pocketing ties "just in case," but as the summer of 2005 progressed, the public grew more accustomed to seeing men in more casual dress, and the younger generations celebrated.[25] During the first year, workers reported that fewer than a third of businesses were implementing Cool Biz, but by 2007 the figure had reached 47 percent, and by 2009, 57 percent of those surveyed were working in offices that had adopted Cool Biz.[26]

Koike used her international experience to collaborate with foreign dignitaries in promoting Cool Biz at home and spreading it abroad, and foreign delegations welcomed the opportunity to showcase their traditional clothing. Not surprisingly, given Koike's connections to the region, one of the first delegations to join the Cool Biz campaign came from Arab countries. On June 16, just two weeks after its launch, a group of ambassadors, including those from Kuwait, Saudi Arabia, and Qatar, arrived to their meeting with Prime Minister Koizumi dressed in their traditional attire.[27] The following year, the Environmental Ministry hosted a "Cool Asia" fashion show in Tokyo where top Asian diplomats came in designer-made summer clothing. While Shinzo Abe (who was chief cabinet secretary at the time) sported a Louis Vuitton suit, foreign guests showcased traditional attire made from local fabrics—South Korea's Ra Jong-yil wore a stylish jacket made from *moshi*, a traditional Korean cloth, and China's Wang Yi stole the show in a Tang dynasty–design shirt woven from summer-friendly hemp.[28]

Inspired by Japan's success, in 2006 South Korea's Ministry of Environment cohosted the Cool Biz Fashion Campaign with the Green Fund.[29] Three years later it conducted a public naming contest for South Korea's version of Cool Biz, and "Coolmaepsi," a combination of "cool" and the Korean word for style, *maepsi*, won. Also in 2006, the UK Trades Union Congress began promoting "Cool Work" fashions to allow employees to be healthier and more comfortable in the summer's sweltering heat and encourage employers to reduce electricity use.[30] In 2008 UN secretary-general Ban Ki-moon launched Cool UN, raising the air conditioner settings from twenty-two to twenty-five degrees Celsius in most secretariat buildings.[31] By 2018, Cool Biz had even spread to North Korea.[32]

Japan's efforts to shift culture in environmentally friendly ways also moved beyond Cool Biz. In 2006 Koike launched Mottainai Furoshiki to reduce the use of single-use bags and packaging. A *furoshiki* is a large cloth traditionally used in Japan to wrap gifts, carry lunch boxes, and bring shopping home from the market. *Mottainai* is a Japanese word that means "useless waste" or "wasteful."[33] In 2011, Cool Biz was upgraded to Super Cool Biz, sparked by the energy conservation efforts made necessary by the shutdown of nuclear power plants following the Fukushima disaster. Super Cool Biz advocated even more casual attire—short sleeves and even shorts, not just no tie or jacket—as well as using blinds to block the sun, shifting working hours to the morning, taking more vacation time during the summer, and working from home.[34] The following year, the ministry promoted Cool Share—a campaign that promoted gathering together to share cooler spaces such as family rooms, malls, parks, and the beach rather than cooling many individual spaces. The idea was to promote community development while also reducing energy consumption.[35]

Cool Biz has become an annual event in Japan, celebrating fifteen years in 2019, and has spread across the world. In May 2015 in Tokyo, when I spoke with Doi Kentaro, the Ministry of Environment official who was in charge of implementing Cool Biz, about why the campaign was successful, he credited Koike for both the original idea and its ultimate success. "She came up with the basic ideas of no ties in the summer and then got [Prime Minister] Koizumi and [Toyota president and president of Keidanren] Okuda [Hiroshi] to come on board. It had to come from the top. It really wouldn't have been possible if the top of government and business hadn't been promoting it."

As with our other game changers, Koike did not stop innovating with one successful campaign. After holding several other positions in national politics, including that of defense minister (2007), she ran for governor of Tokyo in 2016 and won a landslide victory even without the support of her party (the Liberal Democratic Party supported a different candidate). As governor, she has made environmental policy one of the cornerstones of her administration, and she is using this to position herself as not just a local but also a global leader on the issue. Every year her influence and ambitions for making proenvironmental changes have expanded.

In 2017, soon after taking office, Koike signed a memorandum of understanding with the mayor of London to promote collaboration between the two cities in a range of areas of mutual interest, including "developing and promoting Environment, Social and Governance (ESG) investment and green finance."[36] The following year she hosted the 2018 Tokyo Forum for Clean City and Clear Sky, which resulted in twenty-two mayors signing the Tokyo Declaration on Realization of Clean Cities and Clear Skies, which committed their cities to reducing waste, promoting zero emissions, and sharing best practices. They also indicated their "aim to create a social movement through comprehensive advancement of various activities, including raising enthusiasm among the citizens, cooperating with private companies that have excellent technologies, and building effective system & policies."[37]

In 2019, in conjunction with the 2019 G20 meetings to be held in Tokyo the following month, Koike hosted the U20 Mayors Summit and Urban Resilience Forum. The final communiqué from that meeting was endorsed by the leaders of thirty cities, which together represented 126 million people. It committed those cities to a number of ambitious targets, including peaking their emissions by 2020 and reaching zero emissions by 2050, achieving 100 percent renewable electricity by 2030 and 100 percent renewable energy by 2050, enacting policies mandating that new buildings operate at zero net carbon by 2030, and phasing out the use of single-use plastic.[38]

Koike inherited an environmentally committed city from her predecessor, and she is continuing to build on that legacy. Tokyo implemented one of the world's first municipal-level cap-and-trade programs in 2010. The goal for carbon emissions reduction in the second compliance period (2015–2019) was 15–17 percent, but under Koike's leadership the city almost doubled that, achieving a 27 percent reduction in CO_2 in 2017 as compared with the base year (average of 2002–2007 emissions) among covered facilities.[39]

As with her Cool Biz campaign, Koike's environmental (and other) initiatives are largely done in partnership with others. In developing and executing the programs, the municipal government works closely with individual citizens, the nonprofit sector, and for-profit companies to enhance the public good. For example, the city has a program to collect unwanted cell phones and small electronics in order to harvest the components to make the five thousand metals needed for the Olympic games; seventy thousand devices had already been collected by December 2017. As part of the governor's effort to promote the development of a hydrogen society, the metropolitan government is cooperating with manufacturers and energy providers to expand the number of hydrogen stations in the city and the use of fuel-cell buses. The Tokyo Financial Big Bang, launched in November

2017, aimed to catalyze growth in financial technology, including green finance and environmental, social, and governance investment in the city.[40]

With all of her initiatives, Koike has worked to energize new stakeholders in the effort to shift consumption patterns in more environmentally friendly ways. She had created new networks that connect business, political, and nonprofit leaders in collective efforts to promote healthier, more environmentally sustainable ways of living. Through multiple efforts at the local, national, and international levels, she has facilitated cultural change, shifting the conception of proenvironmental behavior change from a painful sacrifice made by individuals and corporations to a healthy, fun, and profitable opportunity that can benefit everyone.

In sum, after many years working as an advocate at the grassroots and meta-levels of policymaking, Koike has moved to become an important environmental advocate who is a game changer operating at the top of local, national, and global policymaking. She is helping to change collective ideas about environmental policymaking as a zero-sum game of competition with stakeholders fighting other stakeholders for their small slice of the pie to one in which governments, both national and local, can act as catalysts for change, helping to incentivize proenvironmental behavior and offering models of policies that work for cities and countries that continue to struggle with developing win-win-win environmental solutions.

Jeong Mincheol, Kim Hyungsoo, and Tree Planet: Social Entrepreneurs Reforest the Planet (and Make Peace along the Way)

All of the game changers featured in this chapter have developed innovative ways to direct public and corporate participation toward proenvironmental action within their own countries and across the region. Whereas Ma Jun works primarily in the nonprofit sector and Koike Yuriko acts in the public sector, Jeong Mincheol and Kim Hyungsoo are primarily in the private sector, using social and market forces to spur engagement and fund reforestation projects around the world. Similar to Ma and Koike, Jeong and Kim didn't just come up with one good idea and stop—they continue to find creative new ways to expand public awareness and direct financial capital toward improving environmental outcomes.

The idea for Tree Planet came when Jeong and Kim were bunkmates (along with thirteen others) serving out their mandatory two-year military service in South Korea. They had both studied animation and film during college and were both interested in environmental advocacy. As they talked about nature and dreamed of trees, they began to investigate how they could use their technical and creative skills to do something about the deforestation problem. At the time (2008), the digital games Tamagotchi and Happy Farm were very popular, so the two friends wondered whether they could find a way to make trees into characters (like Tamagotchi) and somehow link a virtual tree-growing process (like Happy Farm) with growing real trees.

They began to explore the idea on their days off—traveling to Seoul and other parts of South Korea to talk with business leaders, environmental activists, arborists, and local government officials.[41] By 2009 their ideas were beginning to gel, and their timing was fortuitous. Two years earlier, the South Korean government had enacted the Social Enterprise Promotion Act, which was designed to support the development of prosocial businesses. Under the terms of the act, the national, city, and provincial governments would all develop social enterprise support plans every five years, and the Social Enterprise Promotion Program.[42]

The Seoul Metropolitan Government was quick to respond. Inspired by the establishment of the Office of Social Innovation and Civic Participation by US president Barack Obama in 2009, and seeking to address the employment fallout from the global financial crisis, the government enacted the Seoul Enterprise Promotion Ordinance,[43] which allowed it to identify and support potential social entrepreneurs.[44] The city's Social Enterprise Promotion Program offered financial support, education, and social networking opportunities for potential social entrepreneurs. The program also provided payroll support for low-income and immigrant employees, as well as matching funds for business development expenses such as market research and advertising.[45]

Jeong and Kim were among the first group of entrepreneurs to take advantage of the social start-up excitement in the city and South Korea's growing social entrepreneurial ecosystem. Ten days after finishing their military service, the two friends incorporated Tree Planet. Although everyone they talked with liked the idea for a game about growing trees, there wasn't clarity about how it could turn into a business model. As Jeong put it to me during a conversation in in Seoul in 2019, "We got to be part of the first generation of social entrepreneurs. [Although people liked the game,] no one believed this crazy business about making a forest. No one cared about the environment, and people didn't think it could be a business model. But then investors saw that it could work, and it took off." Tree Planet garnered immediate attention, placing third in the Global Social Venture Competition in 2011 and winning the Korea Mobile Award Grand Prize in 2012.[46] Its gaming app had more than one million downloads by 2014.[47]

The basic idea behind the Tree Planet gaming app was to connect game players who were taking care of virtual trees with the planting and care of real trees planted in places suffering from desertification. The game made money from advertising (e.g., a Prius might drive past your avatar while it is watering your tree, or a chemical company's logo might appear on the fertilizer you use to feed your tree) and from in-app purchases (e.g., to buy heroes with special powers to help your tree grow). The funds were then given to NGOs that planted the trees.

One of the first projects was a collaboration with Hanwha, a Chinese chemical and solar power company. It began with advertising through Tree Planet and then deepened its collaboration to help Tree Planet site a forest in an area of Ningxi, China, that was suffering from desertification and use Hanwha solar panels to run the irrigation system. Together they ended up planting more than four hundred thousand trees in the forest, which has seen an incredible turnaround from arid desert to rich greenery.[48] The game continued to increase in popularity, and by 2014, 650,000 users had planted more than eight hundred thousand trees in thirteen forests in five different countries.[49]

Many Tree Planet users were also K-Pop fans, and Tree Planet began to receive requests from fans to build forests not just based on their gaming characters but also based on real K-Pop stars. Kim and Jeong responded by establishing "Star Forests." Fans could establish forests for their favorite K-Pop (or other) stars to celebrate the group or mark special events (e.g., the TVXQ forest was established to celebrate the tenth anniversary of the group). Some Star Forests are symbolically meaningful, such the Paul McCartney Beatles Forest located within the Dorasan Peace Park in the demilitarized zone between North and South Korea.[50] Tree Planet would collect donations, work with local government officials to identify a place that needed new trees (often a section of a public park), and then contract with local nonprofits or arborists to plant the trees. The Star Forests became a place where fans could gather and the stars themselves could hold events.[51] There are now more than forty Star Forests, which have been supported by fans from more than twenty countries.[52]

As Tree Planet was expanding its Star Forests, tragedy struck South Korea. On April 16, 2014, more than 300 people, including 250 school children, died when the MV *Sewol* sunk on its way to Jeju Island. One of the individuals who was deeply affected by the disaster was Sean Ferrer, son of Audrey Hepburn, who had spent a year in South Korea in the 1970s while working on a film project and still felt closely connected to the country. Soon after the accident, he reached out to Tree Planet with the idea of building a memorial forest. Ferrer knew of Tree Planet's Star Forests and thought they could help him build a memorial forest for the ferry victims.

Kim and Jeong loved the idea and began to study how memorial forests could work, contacting local government officials to find appropriate land on which to plant their forest and discussing other logistics related to fundraising. During their planning period, the press and the public were flinging accusations at public officials and the ferry company, seeking to assign blame, and Tree Planet realized that a memorial forest had the potential to bypass the negative politics surrounding the tragedy. A memorial forest could offer a healing moment for victims' families as well as an opportunity to contribute something lasting and positive for those, like Ferrer, who were not immediate victims but sought to offer condolences.

Planning was difficult. Not only did they have to find a location that was appropriate—somewhat near the accident, accessible enough for people to visit, in need of trees, and available for reforestation—they also had to design the forest and the memorial. They came up with brilliant branding for the project, "foRest in Peace," and filled the forest with symbolism. The location is 4.16 kilometers from Paengmok Port, and the memorial wall with victims' names is 416 centimeters long to honor the date the ferry sank (April 16). Gingko trees were chosen to populate the forest because of their thousand-year lifespan.[53] Although the number of victims was 304, only 300 trees were planted. The mismatch was intentional because the forest is meant to be symbolic and collective. Furthermore, if the number of trees and the number of victims matched exactly, and a tree later got sick, damaged, or died, those natural occurrences might cause further trauma for the survivors, so the forest was planted with 300 trees.[54]

After almost a year of planning, Tree Planet announced its plans for the memorial forest and opened the donation platform. By the time of the groundbreaking ceremony held on April 9, 2015, about year after the ferry went down, it had raised more than 200 million won (about $185,000) from 2,985 people from South Korea and abroad, which was more than double its initial goal of 100 million won.[55]

The success of its first "foRest in Peace" opened a new chapter in Tree Planet's business model. Kim and Jeong recognized that foRests offer a living, lasting memorial for victims and created an opportunity for a wide range of people to contribute to and be in solidarity with victims, even if they are geographically far away. They also provide a chance for some participants to gather in person for a meaningful ceremony. Efforts to remember a tragedy are frequently stymied because the media and key actors become more focused on assigning blame and avoiding responsibility than on finding a meaningful way to honor victims and care for their loved ones. foRests avoid many of the political problems that arise when a government or company attempts to craft a memorial and ceremony for victims on their own.

For example, one of the most intractable political issues in contemporary Japanese-Korean relations is the issue of comfort women, the women and girls, many of whom were Korean, who were forced to provide sex for Japanese soldiers during World War II. While governments and activists argued about how Japan could offer a sincere apology, Tree Planet planted two foRests (on in Seoul's Peace Park, and the other in China) to honor these women and girls. The foRests are garden forests that include flowers because the "grandmothers like flowers," and some of the victims and their families were able to gather and speak about both pain and healing at the ceremony.[56] In 2016 Tree Planet began working on a "foRest for peace" in North Korea. Although it has been stalled because of political tensions, the company is poised to begin again as soon as it becomes possible.[57]

Through their participation in the foRest ceremonies, Jeong and Kim realized that they wanted to create more opportunities for individuals to connect directly with trees—real trees, not just virtual ones. Jeong read Joan Maloof's *Nature's Temples* (2016) and was moved by its observation that old-growth trees have their own characters. He began to think about how Tree Planet could help people value and understand trees as having their own characteristics, to care for trees rather than seeing them as generic inanimate objects. The Tree Planet game required each player to take a personality test at the beginning of the game and assigned them a tree to care for that had specific features.[58] Jeong and Kim began to think about how they could help people feel connected to the individual characters of real trees.

Their musings coincided with several air-quality scares in Seoul,[59] which led to a corresponding explosion in the market for air-purifying systems.[60] Jeong and Kim thought that rather than adding more machines to our lives, trees and natural plants do a much better job of purifying the air, are pleasant to look at, don't require electricity to run, live much longer, and won't end up in a landfill when they finally give out. They came up with the idea for a "companion tree," a tree that you could buy and care for as a kind of pet, and it would offer companionship and clean air for you.

Tree Planet selected trees that had special characteristics to become companion trees. For example, they were rare, could purify the air, and were indigenous to Korea. Tree Planet would give buyers information about the tree's unique characteristics when they shipped it to them. Furthermore, for every tree that someone bought for their house or garden, another tree would be planted in areas threatened by desertification, or in areas suffering from fine dust pollution, or in botanical gardens devoted to preserving endangered species.[61]

The idea of a special tree as a gift has spread in several ways. One has been to replace the giant plastic wreaths commonly given for weddings and funerals with live plants. Rather than a plastic wreath that would be thrown out at the end of a ceremony, the couple or family would receive a small tree—complete with a beautiful label and appropriate message—that could be planted or kept as a houseplant rather than thrown away. Furthermore, as with the other companion trees, the purchase of the tree for the ceremony would result in another tree being planted in a place that needed it.

Another demand for companion trees that has been expanding is in classrooms. As concerns about air pollution in Seoul grew, so did worries about indoor air quality, especially in schools. In 2018, the government mandated that air purifiers be deployed in classrooms with vulnerable populations like young and special-needs children,[62] and in 2019 it declared the air pollution problem a "social disaster" in order to enable the use of disaster management funds to buy the purifiers.[63] The significant investment is being made even though purifiers take energy to run, are often loud, can't purify all the air in the room, suffer from mechanical problems, and will eventually be added to a landfill.[64] Tree Planet thought that its air-purifying plants would be a better option than commercial air filters. After extensive research, it found that nine of its plants could purify a classroom's air as well as or better than a typical commercial air filter.[65]

In early 2019 Tree Planet reached out to some of its corporate sponsors to suggest a clean air donation program to make these plants available to classrooms. Companies would buy the plants to donate to the schools; Tree Planet would train the teachers to take care of the trees; and the teachers would teach the students. With seventeen to twenty students per elementary school classroom, there would be one plant for every two to three students. The students would name the tree, learn to care for it, and even come to school during holidays to water it. The students would learn the importance of trees in their ecosystem and how to care for them. In only the first six months, Tree Planet created "forests" in more than one thousand classrooms.[66]

As its business grew, Tree Planet needed more office space, so it began to lease space in a WeWork coworking office. Once there, Jeong and Kim noticed that their employees and the other WeWork users drank coffee—a lot of coffee. Since coffee comes from trees, Jeong and Kim wondered whether they could use coffee as another way to connect people to trees. Upon further investigation, they learned that coffee is frequently harvested in environmentally harmful ways and that the coffee farmers did not make much money from the coffee they grew because they sold only the raw beans rather than processed beans. Because of Jeong and Kim's experience in reforestation projects in China and elsewhere, they knew how to build seedling facilities located off the grid using solar power. The solar power in a coffee plantation could not only power the seedling and processing facility, it could also light up the plantation at night, making nighttime picking possible when daytime temperatures rose too high. Tree Planet reached out to WeWork, Korean coffee suppliers, and the Korea International Cooperation Agency to put together a plan to promote shade-grown coffee rather than clear cutting and build local processing plants that would generate more profit for local farmers. So far, they have supported projects in Nepal, Rwanda, and Indonesia.[67]

Jeong and Kim's Tree Planet is less than ten years old, so they have not had as much of a game-changing effect on others when compared with Ma Jun or Koike Yuriko. However, even in this short period of time, it is possible to see the broader transformative effects of their innovations. Their gaming idea has been picked up by others—in 2016 Alipay's mobile client,

Ant Financial Services Group, offered a new game called Ant Forest that allows users to plant virtual trees in virtual forests that would then result in real trees being planted in areas of China in need of reforestation.[68] Between its launch in 2016 and 2019, the game has caused more than fifty million trees to be planted across than more than five hundred square kilometers.[69] The Korea International Cooperation Agency is expanding its efforts to help South Korea's social entrepreneurs go global.[70] And home decorators around South Korea have adopted Tree Planet's model of companion trees, marketing house and garden plants not merely as decorations for your house but rather as special gifts for loved ones and as a replacement for plastic congratulations wreaths.[71]

Tree Planet is also continuing to develop new ways to leverage the market to spread public environmental awareness and craft collaborations that unlock private capital for the good of the public and the planet. It is currently working to reforest a landfill in Seoul[72] and is continuing and expanding its sTreet Campaign[73] to plant trees on city streets, as discussed in the previous chapter. As is the case with the other two game changers profiled in this chapter, Jeong and Kim have found ways to move beyond traditional advocacy "boxes." They are using commercial markets to change the ways that individuals think about trees, giving them character and personality and helping people form relationships with them. They have found creative collaborations that enable private individuals and corporations to work with governments and nonprofit organizations to reforest areas that need more trees. And as with the other two game changers, they are contributing to the construction of an advocacy ecosystem that nurtures collaborative solutions that benefit the planet. They have networked consumers, businesses, nonprofits, and local government officials together in ways that are facilitating proenvironmental behavior change across South Korea.

Conclusion: Transforming a Zero-Sum Game into a Positive-Sum Game

At first glance, the game changers profiled in this chapter may appear to have nothing in common since they come from different countries, different sectors, and different generations. I selected them from among all the advocates I interviewed for this book because they are all having an outsized effect on environmental advocacy that is reaching far beyond their own efforts. They are contributing to a transformation of the entire advocacy landscape by changing what had been a zero-sum game of environmental exploitation into a positive-sum game of environmental restoration.

All of these game changers have found ways to engage the public on a massive scale and to catalyze cooperation across sectors. In many ways, they are transforming the "tragedy of the commons" problem, in which individual use of public resources results in the depletion of those resources and their transformation into private goods, into a "replenishing of commons" solution, in which individual consumption of private goods is contributing to the expansion and enrichment of public resources.

Ma Jun did this by building a platform that enabled consumers and investors to direct their capital toward proenvironmental companies and away from those that are harming the environment. This, in turn, improved the working conditions, lives, and environments of people in and around polluting factories that were encouraged and supported in their cleanup efforts. It supported the dramatic expansion of green finance, spurring the development of innovative proenvironmental firms and catalyzing the restructuring of polluting companies. The Blue Map App has empowered citizens to contribute to the betterment of their neighborhoods and rewarded government officials who make improvements. Individual and corporate actions made out of self-interest are now directed in ways that help the public good and the planet.

Koike Yuriko enabled consumers to direct their consumption toward fashion choices that not only were stylish and more comfortable but also reduced the energy consumption of their homes, offices, and cities. This, in turn, inspired other countries to engage in similar campaigns and helped support new businesses focusing on eco-fashion and green building design. Her work at the national and international levels enabled her to reshape Tokyo's urban policymaking to develop a city where proenvironmental choices made by consumers and companies are contributing to a healthier and a more economically prosperous city.

Jeong Mincheol and Kim Hyungsoo have channeled the energy of video gamers, coffee drinkers, and home decorators toward reforestation projects around the world. They are training children and adults to think about trees with the same love and care that they might pets. They are using forests to heal decades-old emotional wounds. As Ma Jun did with the Blue Map App, they have found ways to target individual and corporate commercial activities to directly improve the air quality of local residents (e.g., those living on a sponsored coffee plantation, along one of Seoul's sTreets, or near a Star Forest), as well as the planet, though their reforestation efforts.

In short, these game changers are shifting the dichotomies that have led to the "tragedy of the commons" of our environment. They have reversed the "individual versus collective" conflict—individual actions such as playing a video game, buying clothing, and making profitable financial investments are replenishing our common environmental resources rather than depleting them. They have eliminated the "environmental benefit versus economic profit" conflict by designing products and markets that allow economic gains for individuals and companies to generate positive environmental outcomes. They have transformed "public versus private" conflicts into opportunities for collaborations by crafting modes of engagement where public, private, and nonprofit sectors can all contribute positively to common, proenvironmental goals. By connecting diverse stakeholders in novel ways, they have fundamentally shifted the political, corporate, and social environment in which everyone is operating. In sum, these environmental activists are transforming a zero-sum game where some individuals and companies win at the expense of our planet into a positive-sum game where individuals, companies, governments, and ecosystems can all win.

Conclusion: Replenishing the Commons

This book[1] began by asking how advocates could be effective in convincing individuals, businesses, and governments to change their behavior in proenvironmental ways. In order to answer this question, it has investigated the strategies that environmental advocates are using to generate successful outcomes in East Asia. East Asia is a difficult place to be an environmental advocate. Compared with those in North America and Europe, East Asia's advocacy sectors are smaller, their green parties weaker, their probusiness governments stronger, and their democratic experience shorter or nonexistent.

And yet China leads the world in producing and using renewable energy. Japan has long had very high emission standards and is a leader in developing green technology. South Korea has reoriented its national industrial policy around green growth and has expanded its public and protected green spaces. And Taiwan boasts some of the highest recycling rates in the world and is active in helping other countries do the same.

Clearly, East Asia's environmentalists must be doing something right. This book has engaged in a process of discovery to find out what these advocates were doing that was effective and why. The research presented here demonstrates that advocates across the region were using a remarkably similar set of strategies in their efforts to generate proenvironmental changes in their societies, despite their very different political systems:

1) Education—Help individuals and policymakers understand environmental problems and why they matter, and provide solutions.

2) Make friends on the inside—Cultivate and empower allies with policy influence.

3) Make it work locally—Successfully implement an environmental solution locally and then disseminate that success to other places.

4) Make it work for business—Identify and support the development of products, services, and markets where proenvironmental behavior can also generate economic profit.

5) Engage the heart and inspire the imagination—Use art to attract attention, emotionally connect people with the environment, and inspire them to act.

6) Be a game changer—Change the context in which other advocates are working by transforming culture and markets in ways that generate positive cycles where the individual consumption of private goods is contributing to the expansion and enrichment of public and ecological resources rather than their degradation.

This book has further shown that these strategies are not just common and effective in East Asia, they are common and effective all over the world. These findings are surprising because they go against two common assumptions that pervade environmental politics and advocacy literatures: (1) East Asia is different—it has an unusual cultural and political history that makes advocacy in the region different from that found in the other parts of the world.[2] (2) Regime type significantly affects advocacy success—advocates in democracies will have a larger and more diverse set of strategies to choose from, and they will generally be more effective than their counterparts in nondemocracies.[3]

In contrast, *Effective Advocacy* has revealed that while societies in East Asia have unique cultural histories and political contexts that affect the precise way that advocacy strategies are implemented, the strategies found to be most common and effective in East Asia are also common and effective around the world. Educating small children about the natural world and instilling in them a sense of responsibility to care for it is as important in Jakarta and Rio de Janeiro as it is in Taipei. Appointing a former Ministry of the Environment official to the board of directors of your environmental nongovernmental organization (NGO) is as useful in Buenos Aires and Moscow as it is in Seoul. While the specific method for developing, implementing, and disseminating a solution to a community's stinky solid waste problem is going to be different according to local conditions, the overall strategy of "making it work locally" is likely to be effective for advocates operating in any community with a functioning government. Essentially, the strategies highlighted as successful by East Asian environmental advocates are effective everywhere.

Why? Why were these strategies the ones that proved to be especially effective? The Connected Stakeholder Model (CSM) offers a new conceptual framework for understanding the process of policymaking that helps explain why these strategies are effective, and why they are effective across such diverse political contexts. Derived from the findings in this book, the model asserts that policymakers' networks—their number, size, and especially their diversity—are the most important factor in determining both the form and the efficacy of policy. Individuals make policy, and those individuals are each connected to multiple personal and professional networks that influence their perspective on the policies they develop. The strategies discussed in *Effective Advocacy* enable advocates to (a) build new networks that connect diverse stakeholders in new ways, (b) link different networks together, (c) strengthen and expand networks, and (d) energize networks, inspiring the individuals, policymakers, companies, and organizations that are part of them to take action.

Scholars have long recognized that decision makers do not make policy in a vacuum but rather have their ideas about problems and solutions shaped by the networks to which they belong. The new insight that the CSM offers is to recognize that these networks are not just a channel through which information can travel; they can exert an independent influence on policymaking. The number, size, and diversity of networks connected to policymakers will influence not just the type of policy that they develop but also its quality.

Previous research on networks connected to policymakers—whether they are called policy networks, advocacy networks, advocacy coalitions, or some other name—have tended to describe the policymaking process as a game with teams. In these models, one set of policymakers, who are connected to their team via one set of networks, competes with another team (or multiple teams) that has its own networks. Based on the team's resources (financial, political, social, informational), one team will triumph over the other.[4]

As chapter 3 describes in more detail, the CSM conceptualizes the process of policymaking and the role of networks differently. Rather than being perceived as a group of political actors who have a clearly defined set of hierarchically organized interests for which they are fighting, policymakers are viewed as individuals connected through a variety of personal and professional networks to multiple stakeholders who have diverse interests. The policymaking process is conceptualized as one in which multidimensional individuals linked to diverse networks seek to collaborate on developing policy solutions to complex problems. Advocates who are not "at the table" with policymakers can change outcomes by influencing the networks of those who are "at the table." To the extent that advocates' ideas reach more decision makers, the more likely it is that they will be heard, their perspective understood, and their recommendations followed. To the extent that more stakeholders are connected to the policymaking process through diverse networks, the better the policy will be.

Advocates—in civil society, in business, and in government—can use the strategies discussed in this book to change the networks to which decision makers are connected. They might do this by targeting particular individuals who act as "network nodes." Those influential people are at the nexus of several networks and have the capacity to transmit ideas to multiple networks simultaneously. Advocates might build new networks, connecting stakeholders who might not have been connected to the policymaking process before. They can act in ways that increase the size (more people connected to the network) or influence (connected to more people in positions of power) of networks by supporting allies who are network members, by introducing more people to the network, or by linking one network to another.

The critical shift in conceptual orientation is from a focus on individual decision makers who are thought to have a single "stake" in the outcome to a focus on the networks connected to those decision makers and the multiple, connected stakeholders contained in those networks. When we think about policymaking as a process where individual people connected to multiple stakeholders make thoughtful decisions together, it becomes easier to understand it as a creative, collaborative process that can produce positive-sum outcomes rather than as a game in which one team wins and another loses. When we understand advocacy to be a process of engagement and connection, it becomes easier for people to know that anyone has the capacity to participate and make a difference. Finally, by focusing on networks rather than individuals or institutions, we are able to expand our vision of policy from a narrow range of options framed by formal institutions and specific individuals with a finite set of job titles to a much wider array of possibilities for creative action that, coordinated or not, can generate positive change that addresses serious public issues.

Implications for Policymakers

What are the implications of the CSM for policymakers? The CSM offers a new conceptualization about the process of policymaking, so there are several concrete recommendations that the model suggests should help policymakers make good policy. Since this model was developed inductively, it is likely that many of the most innovative and effective policymakers are already engaging in policymaking that reflects these recommendations.

1. Select policy advisers connected to diverse networks.

The primary takeaway of the CSM for policymakers is that policy advisers included in the policymaking process should be conceptualized as "nodes" connected to different networks. The primary value of these participants will be to lend insights into the interests, perspectives, and experiences of the many people in their networks. Having diverse perspectives included in the process increases the capacity of policymakers to anticipate problems, find win-win solutions, and generate buy-in from the people most affected by the proposed policy.

While this method of choosing policy advisers is somewhat consistent with a multistakeholder approach, it does have some key differences. The stakeholder approach was originally developed as a business concept to help organizations "manage the relationships with [their] specific stakeholder groups in an action-oriented way."[5] An organization that does this well is one "which understands its stakeholder map and the stakes of each group, which has organizational processes to take these groups and their stakes into account routinely as part of the standard operating procedures of the organization and which implements a set of transactions or bargains to balance the interests of these stakeholders to achieve the organization's purpose."[6]

Consistent with this idea, the CSM suggests that policymakers should do their best to identify all the communities that might have some kind of stake in the policy. However, rather than picking one person to represent each of the identified stakes, the CSM recommends that a constellation of people known to be connected to diverse communities related to the policy area be gathered together. These advisers would not be fighting for their stake but rather collaborating using their diverse experience and expertise to try to generate a policy that is most likely to be successful in achieving the policy objective.

To give a hypothetical example, when trying to develop a new clean water management system, the relevant policymaker is likely to recognize that local environmental groups, local residents, and the local polluting corporation all have a stake in the outcome. The policymaker is likely to invite at least one person from each of these stakeholder groups be part of the policymaking process. However, while the stakeholder approach would suggest that it should not matter which environmental group, community resident, or corporate representative is invited, the CSM suggests that it is critically important that the people selected to be part of the process be those with multiple connections, not just one. Representatives from designated stakeholder groups should not just be fighting for their group's narrow interests but should bring a wealth of experience and perspectives to the table.

To the extent possible, the policymaking process should not be a battle for the relative position of one's stake but rather a collaborative enterprise that involves sharing interests and perspectives and generating creative win-win outcomes for everyone involved. Experienced policymakers understand this very well, especially when it isn't practiced. During a 2019 interview with me, Kim Sungwoo, a senior environment and energy consultant based in Seoul, described a committee he had served on that had been formed using the multistakeholder logic rather than CSM logic.

> There are two problems [in the way the committee was formed]. First, the person who represents the business sector had a science background, but he only represented one part of the business sector. The committee didn't have other business representatives, or even the main voice from that sector, but everyone seemed to think that he represented everyone in the sector. That was a huge problem.
>
> Second, in [this policy case] the government was the referee, while the NGO wanted more, the industry wanted less, and the professors were in the middle. The problem is that while the committee had representation from industry, it was only from part of the industrial sector—the environmental team—so, those folks didn't have investment decision-making authority and didn't really understand how the environmental policy options impacted the balance sheet. As a result, [even though there was a representative from the business sector,] there wasn't anyone on the committee who really understood the perspective of the finance people. All of the committee members were looking at the options only from their own narrow perspective. The policy discussions didn't go very well.

Not everyone who cares about a policy can be in the room making policy. The stakeholder approach assumes that all relevant stakeholders can be included in the process. The CSM recognizes that this is usually impossible.

There may be power brokers who are not physically present who are exercising influence over people in the room. There may be minor stakeholders who are excluded because of space or other concerns. There may be major stakeholders who have very different perspectives from those in the room who are supposedly representing the entire group. There may be stakeholders who were not identified but who can affect or will be affected by the policy. Inviting individuals who are part of many networks that connect them to diverse stakeholders will be the best way to ensure that the people in the room making policy are knowledgeable about the concerns and interests of as many stakeholders as possible, the known and unknown.

2. Recognize that everyone is a political actor.

The CSM recognizes that no one is neutral in policymaking—everyone is acting to benefit their own networks and their own position in those networks, whether they are bureaucrats, activists, businesspeople, or academics. Even if someone is invited to take part as a technical expert, it is important to remember that any given individual will be connected to his or her own networks. The person's perspective will be influenced by his or her connections to unique networks, and he or she can be used as a source of information about diverse stakeholder perspectives, not just as a source of technical expertise.[7]

Presumably, everyone who is invited to participate in the policymaking process is asked to do so because they have some kind of knowledge and experience. Their expertise on the matter at hand will require that they be connected to communities related to that policy area. Even an academic who is brought in as a technical expert will also be a political actor when sharing knowledge. The person should care about the policy and want it to be a good one. Just as stakeholder participants should be selected for their access to diverse networks of people related to the policy, technical experts should also be selected not just for their specific knowledge but also for their connections to others with knowledge and perspectives that might be relevant. Everyone should recognize that all participants are political actors and no one is neutral.

Although the CSM recognizes that "conflicts of interest" can be a problem—it is inappropriate and harmful for those at the table making policy to be able to tailor that policy in ways that can garner them personal benefit—having an interest, indeed several interests, in the outcome of the policy will make those involved in the process more vested in the outcome. Being connected to multiple communities with a stake in the outcome should help the committee care more about crafting a good policy and give its members a broader and more diverse perspective than would be possible if members were truly neutral technical advisers.

3. Recognize institutional constraints and use networks to bridge them.

Institutions, both formal and informal, create constraints and opportunities for policymakers. Working within these constraints is one of the key challenges that policymakers face as they seek to develop effective policy. Some organizations and not others are selected to implement policy. Specific bureaucratic processes are designed to encourage efficient workflow, gather relevant feedback, and ensure accountability. In order to be effective, policies must be designed to conform with these constraints and others.

However, the CSM encourages policymakers to recognize the institutional constraints and identify networks that may have developed to overcome those constraints. Indeed, if networks have not yet formed to overcome known institutional constraints, policymakers may want to use policy to create new networks that help overcome barriers that are hindering policy development and implementation. Network creation and activation can be an important outcome of policy and should be thoughtfully considered as part of the policymaking process.[8]

4. Design policy for flexibility and further innovation.

The CSM recognizes that policy is created and refined through an iterated process. Policies designed by focusing on a fixed set of stakeholders operating within a particular set of institutions can lead to rigid policy outcomes that hinder adaptation to changing circumstances.[9] Especially at the municipal level, we can see that effective policies are frequently designed through a collaborative process that brings the public, private, and nonprofit sectors together in ways that enable experimentation, innovation, and adaptation as policy learning takes place and as circumstances evolve. Key to the flexibility of the policies are large and diverse networks connected both formally and informally to the policymaking process.[10]

5. Recognize clients as cocreators of policy.

The CSM shares with other stakeholder models the idea that clients should be cocreators of policy. In some contexts, it may be difficult to have those most directly influenced by the policy sitting at the table to help craft it, but it should always be possible to have people connected to those affected sitting at the table. Indeed, in contrast to other stakeholder models, which recommend that every community of stakeholders have a representative to fight for their stake, the CSM makes it possible for several of the people directly involved in making policy to be voicing the concerns of affected communities.

Selecting policy advisers because they have multiple connections rather than a single interest or identity has several benefits in terms of representing clients in the process. First, if there are multiple people representing the views of a particular affected community, the policymaking process runs less risk of being hijacked by a single, nonshared view. Similarly, because there should be multiple people in the policymaking process who are connected through their networks to affected and vulnerable communities, those speaking on behalf of the vulnerable are less likely to be viewed as token representatives who are expected to represent some kind of unified perspective on behalf of an entire community of diverse individuals.

Having multiple people in the policymaking process offers multiple perspectives on how a community might be affected by a policy and creates a fuller picture for decision makers as they weigh options. Finally, with multiple people connected to affected communities involved in the policymaking process, implementation should improve, since there will be greater buy-in at the beginning and more channels through which policymakers can communicate the policy change and adjust it as needed.[11]

Implications for Advocates

Most of this book has been concerned with strategies that advocates can use to be effective in their advocacy for change. This section is less concerned with particular strategies for effective advocacy and more focused on the implications that the CSM has for advocates. In particular, to the extent that the CSM accurately describes how policy is made, the following offer some implications for advocates in terms of the way they can think about approaching effective advocacy.

1. Use multiple, diverse networks to gain policy access.

As common sense and much policy analysis has shown, policy access is the most important factor in gaining influence over policy.[12] The easiest way to do this is for a policymaker to invite the advocate to be part of the group of people in the room when policy is being formulated. Short of being in the room themselves, the best option for advocates is to try to access other people who are in the room.

The Advocacy Coalition Framework (ACF) and multistakeholder models suggest that there is likely to be a small set of people in the room with whom an organization shares similar beliefs and goals. Those models recommend connecting with these like-minded people and trying to grow their number to expand their influence when they are sitting at the table and negotiating behind the scenes.

The CSM does not disagree with this recommendation, but it goes a step further. In addition to being connected to like-minded people, one of the most important implications of the CSM for advocates is that any and all connections to policymakers matter—even and maybe even especially those with whom the organization may not share beliefs and goals. Multiple connections to many people who are in the room are better than a single connection to a lone "stakeholder representative."

To the extent that advocates have diverse networks that can enable them to connect to multiple people in the room, their capacity to ensure that their perspectives are considered will increase. This means that advocates should not just attend conferences where they keep running into the same set of people; they should also be seeking to make connections with potential detractors, as well as those whose interests are compatible with but perhaps not central to the missions of the advocates' own organizations. Furthermore, cultivating social, and not just professional, connections can also be very valuable to gaining access. This means that while professional conferences are important networking events, neighborhood festivals, PTA meetings, college reunions, and other events also serve as valuable opportunities to expand an advocate's networks in ways that may prove useful when trying to influence policy.

2. Cultivate long-term relationships.

While access to the people with decision-making authority is important, advocates should not focus exclusively on top leadership. Advocacy usually takes a long time to be effective, and relationships take a long time to form. Ideally, advocates should not be seeking to build new relationships with decision makers. By the time someone has risen to a top decision-making level, it would be best if the advocate already had a deep and multiyear relationship with that person. This means that the relationship-forming process should start when policymakers are still fairly junior, when they are middle managers or heading up small projects. Relationships with decision makers do matter, but advocates seeking long-term effectiveness should be cultivating potential decision makers as well as those already in positions of power.

Indeed, as highlighted in chapter 5 advocates are often in very good positions to help the careers of these lower-level bureaucrats and local politicians. NGOs and others can provide professional opportunities, good publicity, and sometimes resources to bureaucrats and politicians who are just starting out. To the extent that the advocate can be helpful to the policymaker before the latter has decision-making authority, it will make it easier for the advocate to approach that person later. Policymakers are much more likely to take a phone call from someone whom they have known for ten or fifteen years, who has proved to be a supportive, trustworthy person, who was helpful when they needed help, than from someone who is calling for the first time.

Relatedly, people who hold high levels of decision-making authority can be influential allies even after they step down or are removed from a position of power. As discussed in chapters 5, 6, and 9, people have long careers and move around, and the longer and more successful their career, the more networks they are likely to join. It is quite common for retirees to step down from high positions in government and business and spend another decade or two in the advocacy sector, or to move from national-level politics to local-level politics. These individuals have numerous connections that can enhance the efficacy of advocacy. Essentially, advocacy can be a lifelong process, and individuals in all sectors and in all stages of a career can be important allies.

3. Prioritize people and organizations that are network nodes.

The CSM offers a network-based framework for understanding the policymaking process in which the individuals involved in policymaking form a matrix of networks. As advocates attempt to gain influence in this matrix, they should prioritize people and organizations that form nodes in the network. Nodes have connections to multiple other nodes and networks.

It is likely that the most powerful actors will also be nodes in a network, but this is not necessarily the case. There may be individuals or organizations that exert significant power (e.g., they control high levels of funding resources, have decision-making authority) but may have only limited connections to other organizations. Similarly, there may be some organizations (often NGOs or government-organized NGOs) that are involved in forming and supporting numerous networks but may not exert direct decision-making authority or control significant financial resources. However, their capacity to help connect advocates to those with power is a vital asset. When making strategic decisions about which people to cultivate, advocates should not just think about the decision-making power of the individual, or his or her future decision-making power (points number 1 and 2 in this section), but also the capacity of the person to connect the advocate to others—the person's location and role in multiple policy-related networks.

The CSM posits that an advocate's most important resource is his or her network, and growing that network, connecting to new networks, and becoming more influential and important within those networks should be key priorities when advocates make decisions about how to allocate their time and resources.

4. Use networks to overcome and work around institutional barriers.

One of the most difficult aspects of policymaking, for both governmental officials and advocates seeking to influence policy, is that bureaucracies are required to follow specific processes and often have a very narrowly defined scope for their authority, making innovation and change difficult. The CSM helps reveal that bureaucratic barriers are frequently not as insurmountable as they might appear. Any given policymaking process has multiple people involved, and any given person has multiple networks to which he or she is connected. Advocates can use their own networks to connect to multiple policymakers, enhancing the chances that their views are heard by decision makers.

Furthermore, advocates working outside government are frequently able to see the institutional barriers and design networks to work around them. Perhaps the most obvious example of this was the China-US Energy Efficiency Alliance discussed in chapter 5. Local government officials in China and the US would ordinarily never have a chance to meet each other. Additionally, since China is so large, it is also frequently difficult for local government officials to get the attention of Beijing. By facilitating face-to-face meetings between local government officials from Jiangsu and California and connecting them to relevant utility and technology companies, the advocates at the Natural Resources Defense Council were able to work around the institutional barriers that inhibited those communications. Once all the relevant people were able to meet and identify areas of common interest, they were able to design a new network that could enable them to pursue these productive collaborations. Advocates did not remove or even challenge the preexisting institutional barriers; they just networked around them.

5. Network with others to amplify impact.

Perhaps the best examples of how networks can be used to amplify impact can be found in the examples used in chapter 6 to discuss the "make it work locally" strategy. Sheri Liao of Global Village in Beijing had only a handful of volunteers in her organization when she and her fellow environmental activists began discussing their ideas for the 26 Degree Campaign. There was no way that any one group would have enough people to measure the indoor temperatures of any more than a few hotels and offices in the city. By working together, the activists were able to mobilize hundreds of people to carry out "inspections" of indoor air temperature, and now the number of people involved is in the thousands, and their local initiative has become national policy and spread to other countries as well. Especially for small, local groups, working with other groups is an important way of amplifying impact.

Even for large, well-resourced organizations, networking with others can be an important way to amplify the impact of advocacy efforts. The Institute for Global Environmental Strategies did this by forming the KitaQ Composting Network (featured in chapter 6). While it was nice that they were able to help the city of Surabaya address its solid waste problem by developing a new method of household composting and finding a neighborhood-based way to disseminate the new method and support its continued use, it was the international network that was formed that enabled more than thirty cities to adapt the system for use in their own communities. While it is difficult for good public policy to "go viral" in the manner of posts shared over social media, networks formed by advocates can enable good ideas and policies to spread much farther and faster than they would if they were championed by only a single advocate.

6. Do things that matter, and then form networks to support them.

Advocates can spend all their time schmoozing and networking, but it will be a complete waste of time if they are not doing anything that matters. Networks, especially informal, ad hoc networks, will dissolve if they are not useful. The CSM reveals how important it is to the policymaking process to have many diverse networks. *Effective Advocacy* offers examples of strategies that advocates can utilize to influence policymaking. In all cases discussed in this book, the good idea, the innovative person, and the positive outcome came first, and the network to disseminate and amplify that positive outcome followed.

Advocates should not become so enamored of creating networks that connect to people with power that they lose sight of why they are seeking to influence policy. Networks of the kind described in this book are generally organic, and they form and are designed for specific purposes, to engage and connect the people who are relevant for a specific policy goal or task. Therefore, advocates should focus on doing something that matters first, and worry about growing the network later.

It is not uncommon for a network to form around a good idea, but then to have the networked people identify new ideas and form new networks around them or shift and expand the original network to accommodate the new ideas and projects. The increasing number of international networks of cities, such as the C40 Cities Climate Leadership Group, are one example of how this can happen. City mayors first gathered to find ways to support each other in their common challenges related to climate change and formed C40. Once they gathered, they discovered a whole range of issues for which collaboration could be useful—including disaster management, housing, immigration, and public safety. Some of those issue areas were incorporated into C40's activities, and others spun off and formed new networks. Often individuals who help organize new networks remain connected to the original one, even if they are no longer as involved. This is one of the key processes through which networks grow and diversify.

Implications for Scholars

The CSM has several implications for the way that scholars examine actors, institutions, and the policymaking process. The CSM challenges researchers to examine a wider diversity of actors who may be involved in influencing policy. Rather than focusing only on the actors whose institutional affiliations would identify them as active stakeholders, scholars should include all actors in a network matrix who are seeking to influence the policy process, including academics, journalists, artists, and others. No actor should be assumed to be serving purely a technical or cultural role; everyone involved should be assumed to be a political actor in their own right.

Furthermore, this study suggests that scholars should reconceptualize the role of policy actors. Rather than trying to identify an actor's most salient interests and trying to score a competition among divergent interests, scholars should be examining the multitude of interests and perspectives that any given actor brings to the table through his or her networks as a way to determine how those interests are combined to create policy. This perspective will draw scholars' attention to marginal interests that may have low-level salience for multiple actors across different networks. It may also help explain the unexpected policy outcomes that emerge when policies are crafted to support minor interests that no actor was willing to fight for but that many actors were willing to support.

The CSM also conceptualizes a new analogy for the policymaking process and a different role for institutions in the process. It rejects the common analogy of competitive sports. Instead, it adopts computer networks or social networks as its base analogy. Institutions function to provide "rules of the game," guiding participants toward behavior that will facilitate the growth rather than the death of the network (e.g., friending is good, spamming is bad). Some of these rules are formal (e.g., no child pornography), and some of them are informal (e.g., avoid ultra-long posts). Some of the actors are actively trying to expand their influence (e.g., raise their Muckety or Klout score), and there are big players and small players constantly trying to change the rules of the game (e.g., rewriting privacy laws). However, most actors in the system are not actively engaged in trying to change, break, or maintain rules. Furthermore, new technology can open up brand-new ways for actors to interact with each other and the policymaking process itself (e.g., the introduction of the smartphone, the creation of Facebook).

Most importantly, there are no fixed "teams." Individuals and groups are connected to each other in complex ways. Actors might be working together on one policy issue but working against each other on another. There are numerous ways that actors can engage with each other outside the channels provided by formal institutions. In this conceptualization, the primary function of institutions is to create opportunities for actors to connect with one another, to encourage the creation of multiple nodes, and to facilitate innovation—not constrain behavior. In fact, even if the institutions do constrain behavior, it is common for actors to find work-arounds to those constraints. Finally, there is no single "referee" who is ultimately responsible for policing. All actors involved in the network are responsible for employing enforcement mechanisms, with informal, social, and market-based methods of enforcement being utilized far more frequently than formal legal sanctions.[13]

To reiterate, the CSM conceptualization does not deny that institutions create constraints or that actors compete and have conflicting interests. What this network-based conceptualization accomplishes is that it moves us away from models that assume that actors are required to be in competition with one another because they belong to particular teams and that assume that the competitors can be clearly distinguished from teammates or referees. Instead, the model allows for more dynamic and complex interactions among a wide variety of actors seeking to influence policy.

Thus far, policy scholars have given considerable attention to policy actors, the institutions where they reside, and the interests they represent. More attention should now be paid to the networks that these actors form with one another. Scholars utilizing the ACF have begun to do this, but more research is needed to understand the cross-subsystem and cross-sector connections and to tease out the nature and function of these policy-relevant networks. How do they form? How are they maintained? Which kinds of networks are most influential? How do networks work to strengthen or undermine one another in a policy dialogue? Do they strengthen or undermine the policy itself? Do decision makers prioritize one type of network over another? How do policy actors activate their networks for information gathering? How do policymakers identify the nodes of a network and invite those people to take part in decision-making? How do advocates work to create new networks that might increase their influence in policymaking?

Placing networks at the center of policy analysis offers a new perspective on the policymaking process. Rather than a competition between actors on opposing teams fighting for their interests, policymaking is conceptualized as a negotiation among actors who all have multiple interests that they are promoting. This new conceptualization allows for competition and conflict as well as a wide range of other forms of interactions in which different actors work together to craft policies that benefit diverse constituencies.

This view that networks are central to policymaking is in line with the ACF. Where the CSM departs most significantly from the ACF is in the recognition that the networks that form are not necessarily formed by people with similar belief systems and common interests. For the CSM, individuals need only be connected to each other in some way; it can even be a purely social connection. Their beliefs and interests do not need to be aligned for the network to be relevant. Indeed, the more diverse and dynamic the networks to which a decision maker is connected, the more useful it will be for the individual within the network, and the more useful it will be to the policymakers seeking to develop effective policy.

Another important implication of this model is that democracy matters less for the policymaking process than would be expected by pluralist models. Relevant actors might include political parties, but they also might not. It is assumed that relevant exchanges among actors are likely to occur in locations and manners that are hidden from public view. It allows for different actors to have different levels of power in the system. It does not require that actors be clearly defined as public, private, or nonprofit sector—it allows for individual actors to hold multiple identities that may cross sectoral or ideological lines.

It is likely that policymaking in democratic societies will consist of a broader array of actors who are engaged in larger networks that are more horizontally organized, since democratic societies tend to have larger and more independent civil societies.[14] However, the model should still work in societies where the networks are fewer, smaller, and more asymmetric. Therefore, although the model will not apply to societies without sufficient state and societal capacity to implement policies, it should have broader applicability than pluralist-based models that generally assume democratic or democratic-like political contexts.

Examining a different dimension of politics, a network-based approach to policymaking has the potential to improve our understanding of gender politics. Previous approaches to policy have erroneously suggested that women need to gain greater expertise or occupy leadership positions in institutions to gain influence.[15] However, as numerous popular culture writings and more scholarly research attest, it is not primarily the level of expertise but rather access to the right networks, which often requires off-hours commitments (e.g., going out for drinks after work, meeting with constituents on the weekends), that makes it particularly difficult for women to gain leadership positions in business and politics.[16] On the other hand, the CSM offers a strategy and a path for women to gain more influence and a solid justification for their inclusion in policymaking even if their job title does not warrant selection based on a stakeholder model. Women are frequently network nodes, and they are often connected to different networks from men. Including more women in male-dominated policy decision-making and more men in women-dominated policy decision-making should increase the number and diversity of networks connected to decision-making, which should improve the quality of the policy.

In sum, CSM offers scholars a new method for studying policymaking around the world. It incorporates more actors into its model and allows for a new conceptualization of the role of institutions in constraining and enabling those actors to craft policies for their societies. A network-based model is more descriptively accurate and more analytically applicable to policymaking processes found around the world than current models based on interests and institutions. It offers a breakthrough in our attempts to understand and analyze policymaking in an increasingly complex and inter-connected world.

Effective Advocacy and Replenishing the Commons

Ultimately, effective environmental advocacy is about fundamentally trans-forming the political, economic, and social dynamics in our societies that have resulted in a "tragedy of the commons,"[17] where individual and col-lective use of environmental resources has led to their degradation and dis-appearance. Elinor Ostrom won the Nobel Prize in Economics for a career that was devoted to developing design principles that can help regulators and communities manage the use of these common pool resources such that their destruction can be avoided and mitigated.[18] As showcased perhaps most dramatically by the game changers profiled in the previous chapter, East Asia's environmental advocates have gone a step further. They are not just slowing the depletion of our planet's resources, they're enhancing them.

East Asia's environmentalists have found ways to incentivize individuals, companies, and governments to engage in proenvironmental behavior, not just mitigate their environmental harm. As the examples of Toyota, Walmart, and Tree Planet (discussed in chapters 7 and 9) show, they have created and supported new markets where greener products, renewable energy, and less waste promotes not just a cleaner planet and happier and healthier people but also wealthier investors. The KitaQ network, discussed in chapter 6, and

Taiwan's Rainbow Village, highlighted in chapter 8, demonstrate how East Asia's environmental solutions often don't just solve a single, narrow environmental problem like solid waste management (KitaQ network) or overdevelopment (Rainbow Village); they can contribute positively to the social, economic, ecological, and aesthetic conditions of their communities.

East Asia's political systems—Japan's mature democracy, South Korea's and Taiwan's newer democracies, and China's authoritarian state—are all rewarding leaders with good environmental records and extensive experience by electing and appointing them to powerful political positions with the authority to integrate environmental concerns into broader policymaking. Tokyo's Koike Yuriko and Taipei's Hau Lung-pin were both elected mayor of their capital cities after demonstrating their commitment and effectiveness when serving as their country's environmental minister, and now their cities regularly rank among the greenest cities in the world. Lee Myung-bak was elected president of South Korea largely as the result of the success of the Cheonggyecheon River restoration project he conducted while mayor of Seoul, and once he became president he championed South Korea's new "green growth" policy. Chinese president Xi Jinping developed Fujian as an "ecological province" while he was governor, and he now promotes the expansion of an "ecological civilization" to all of China.

East Asia is a region that is hostile to political advocacy, and yet its environmentalists have found ways to replenish our common environmental resources. They are teaching their children and leaders about the importance of our planet and how to care for it. They are networking around institutional barriers to influence policymakers. They are implementing solutions locally and then spreading them to other communities. They are directing the power of the market toward healing and enhancing the planet rather than destroying it. They are helping us to imagine a world that we want to live in and then inspiring us to create that world. They are changing a planet-killing competition in which we all lose into a creative, collaborative process in which individuals, companies, governments, and ecosystems can all win. They have many lessons to teach us. If we are lucky and work hard, perhaps we can learn them before it is too late.

Appendix A: Interviews

The interviews for this book were conducted over the course of ten years, 2009–2019. Most were held with advocates where they lived—in China, Japan, South Korea, or Taiwan—while some were held in other locations, such as at international conferences. Therefore, while the location of the interview is not a perfect measure of how much time I spent in each place or how many advocates I spoke with who were from that country, it offers a reasonable approximation (figure A.1).

As explained in chapter 1, in order to obtain a diverse set of perspectives on effective advocacy, I sought to speak with as many environmental advocates and as many different kinds of advocates as I could. In each country, I actively sought out, at a minimum, activists from each of the three advocacy sectors—nonprofit (nongovernmental organizations [NGOs], including grassroots organizations, and international NGOs), for-profit (business), and government (national and local). As chapter 1 and especially chapter 8 highlight, many people belonged to more than one sector—an NGO advocate might also serve on a national government advisory panel, and a businessperson might also be on the board of an NGO. Figure A.2 is a chart of how many people I spoke with who were part of each type of organization. Because many people had more than one organizational affiliation, the numbers do not add up to 105 (the total number of interviews in the dataset).

Finally, figure A.3 illustrates the frequency with which different strategies were mentioned in the course of my interviews. As chapter 1 explained, the focus of this study was on effective advocacy, so the vast majority of my conversations were spent discussing strategies that advocates thought were effective and listening to their ideas about why they thought they were effective. Not surprisingly, occasionally conversations about effective advocacy

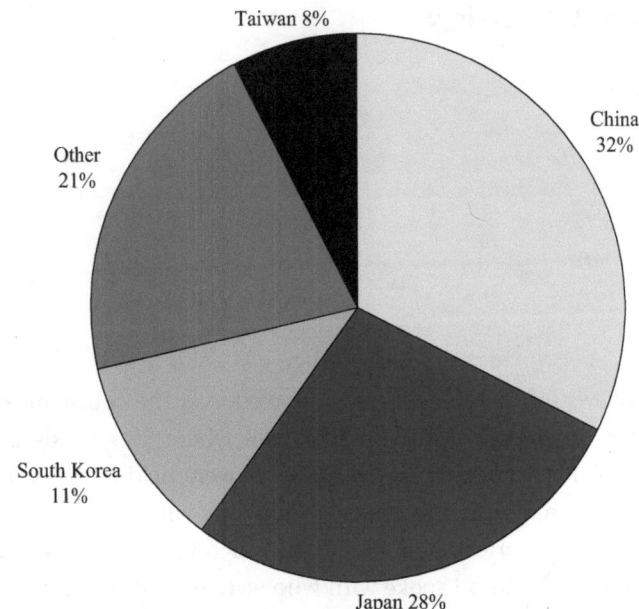

Figure A.1
Location of interviews.

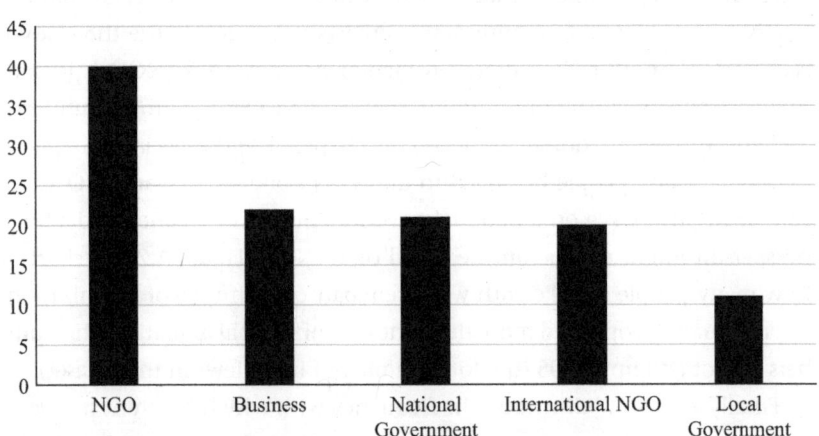

Figure A.2
Organizational affiliation of interview subjects (count).

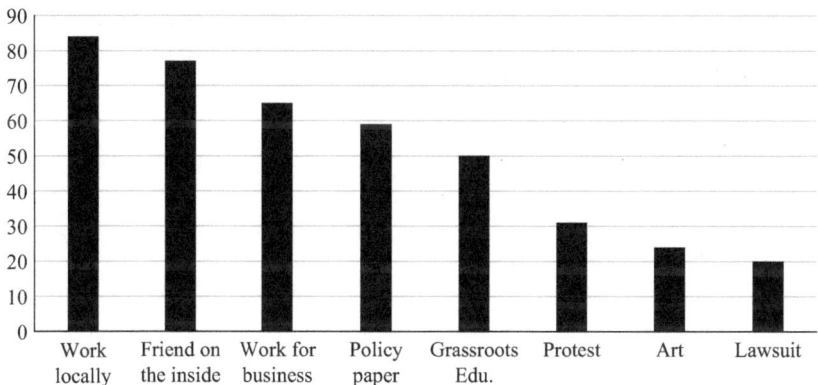

Figure A.3
Strategy type (count of interviews that mention).

would also involve discussions of failed advocacy, as well as advocacy with neutral or mixed results. Figure A.3 indicates the number of interviews that contained a reference to any advocacy strategy, whether the interlocutor viewed it as generating a successful, unsuccessful, neutral, or mixed result.

Because of the nonrandom and noncontrolled pattern of my interviews, these data are not useful for any broader conclusions about frequency of strategy use or effectiveness—the next set of analyses were intended for that purpose. I provide them here because they do indicate which strategies my interlocutors thought were important to discuss, which, in turn, influenced the strategies that I sought to investigate. Because almost all of my interlocutors referenced more than one strategy, the numbers add up to more than the 105 interviews in the dataset.

Appendix B: Datasets

Data from interviews were supplemented by creating two original datasets and conducting a series of analyses on those data. To create the first dataset, my research assistants and I assembled a list of approximately one hundred environmental organizations in each of the four countries (China, Japan, South Korea, and Taiwan) and then added organizations from the US for comparative purposes. The goal was to capture in the database (1) the most influential environmental organizations in the country, and (2) a semirepresentative sampling of the remaining environmental organizations in the country. In all cases, I worked with capable native-speaking research assistants to help with the collection and coding of organizations for the five countries.

For three of the five countries in the database, I was able to begin with a handful of influential organizations and then populate the bulk of the dataset with a random sample of organizations. The US, Japan, and South Korea all had official lists of environmental organizations that I could use to build my database. For the US groups, I began with the oldest and most influential groups as identified by Christopher Bosso in *Environment Inc.* (2005), then supplemented these with a random sampling of organizations registered with the IRS that list environment as a core mission, for a total of 105 US environmental organizations.[1]

For Japan, the first five organizations were included based on my knowledge, and an additional one hundred groups were added using NPO Hiroba (nonprofit organization forum), a list of all the registered nonprofit organizations in Japan.[2] There were 3,597 organizations in the database that included "environmental protection" as one of their focal areas. In order to create a dataset of approximately one hundred groups, I sampled every thirty-sixth organization listed in the output, which was organized according to the prefecture in which the organizations were registered. This methodology helped ensure geographically proportionate sampling (Tokyo has a disproportionately large number of organizations, and I wanted to ensure that all prefectures had proportionate representation). Two of the five original groups were already in the database, resulting in 103 groups total.

For South Korea, the database began with seven organizations that I knew to be highly influential. The South Korean Ministry of Environment publishes an online list of nonprofit organizations, nongovernmental organizations, and social cooperatives related to the environment.[3] The list contained 373 organizations. We randomly selected 100 groups to include in the dataset. For about 30 of the organizations, we could find no additional information, so additional groups were randomly selected from the full list until we had a total of 100 environmental groups about which we could code information.

For organizations in the US, Japan, and South Korea, organizational websites (especially annual reports when available), government reports, and media coverage were used to gather information about the organizations, their membership, and their activities, which were coded and added to the database. For all three of these countries, this search methodology biased the dataset against all-volunteer groups that may be actively engaged in environmental activities but were not officially registered as nonprofit organizations. This bias is less of an issue in the US, where the requirements to file for and maintain 501(c)(3) status are relatively simple, and the tax benefits are significant, creating strong incentives for all organizations, even small ones with no paid staff, to register. However, for Japan and South Korea, the barriers to becoming registered as a nonprofit organization are significant, resulting in fewer registered organizations and biasing the dataset against the all-volunteer, nonregistered groups that constitute the majority of civil society in these two countries. Although the dataset has this limitation, it still is able to offer a portrait of registered environmental groups and their activities, even if it cannot claim to be as representative of all environmental groups.

I could not find comparable official lists of environmental groups for either China or Taiwan. For those two countries, I did my best to follow the spirit of the data collection for the previous countries. I began with a short list of the environmental groups that I knew to be influential. Native research assistants combed the internet for the names of and information about as many environmental groups as they could find. Once the lists were compiled, I circulated them among scholars and environmental leaders in the two countries who were familiar with the environmental groups active in their countries to see whether I was missing any important groups and whether the lists I had developed appeared to these local experts to be fairly representative of environmental groups in their countries. In the end, I was able to include 108 groups from China and 32 groups from Taiwan. As was the case for the sampling method in Japan and South Korea, this search methodology required that the groups be sufficiently well resourced to afford a website in order for us to find them, again biasing the results against local all-volunteer groups. However, the local experts who were consulted assured me that the lists we generated included all of the most important groups and were fairly representative samples of the others.

In order to discover whether there were systematic differences in the boards of directors, we coded information about the background of members of boards of directors for the organizations in the dataset. We were able to obtain information on boards of directors for about half of the organizations in the dataset, usually from annual reports or links on organizational websites. This information was not evenly distributed. All of the Japanese organizations had it publicly available, while it was harder to find for Chinese, South Korean, Taiwanese, and US organizations. For all groups, we coded a wide range of information, including their founding dates, the types of issue areas in which they were active, the advocacy strategies they employed, their budget and staff, and characteristics of the members of their boards of directors.

The second dataset was created using the Factiva media database. The search for cases was conducted in the following manner: Factiva's major news and business publications were searched, and the search was limited to (a) articles with word counts of greater than 1,500 to ensure that there was sufficient information to identify a case, (b) articles that mentioned the environment (or environmental, environmentalism, or other variants) five or more times, and (c) articles published between January 1, 2005, and December 31, 2009. These five years were chosen because they are recent enough to be able to capture advocacy strategies used in contemporary environmental politics, the focus of this study, and they are old enough that there would be a good chance that the outcome (success or failure) of the advocacy could be determined. The search generated 3,567 relevant articles with 177 duplicates, for a final pool of 3,390 articles. Articles were then randomly selected until the dataset contained 200 cases of environmental advocacy.

Media sources inevitably introduce bias, since more exciting advocacy is more likely to be covered than more mundane advocacy efforts, even though the latter may be both more prevalent and more effective. Thus, there is likely a bias in the data in favor of some strategies, such as public protests, which are often deemed to be more newsworthy than local clean-the-river campaigns or similarly low-profile activities.[4]

The dependent variable in the analysis was success. An advocacy effort was coded as successful if the goal of the action as articulated by the advocates was achieved. Thus, a public protest to close a factory was coded as a success if the factory was closed; a local clean-the-river event was coded as a success if the river was cleaned. The outcome was coded as a failure if the advocacy did not result in the desired outcome. Using the prior examples, if the factory was not closed, or if the clean-the-river event was canceled because of rain, it would be coded as a failure. Success was coded as mixed if the effort was partially successful, such as if the factory was closed for a while but then reopened after some cleanup had occurred, or if the clean-the-river event was originally scheduled to clean three riverbanks but only cleaned one. If the outcome could not yet be determined—for example, if discussions about factory closure were ongoing, or if the clean-the-river event was rescheduled but had not yet occurred—then the outcome was coded as "undetermined."

For most advocacy events in the dataset, the goals of the advocates were clear and quite specific, as in the examples of closing a factory or cleaning a local river. However, for a small subset of events (7 percent), the stated goals of the advocates were broad rather than specific—for example, "improve understanding about climate issues." In that small number of cases, the event was coded as successful if the advocacy took place and people participated.

The relative impact of the success was not coded because I could not find an objective way of measuring relative importance. Is the successful prevention of the construction of a single new petrochemical facility more important than the successful creation of a regional watershed management plan? The closest I could come to measuring the scale of the impact was to measure the scope of the advocacy.[5] Scope was measured according to whether the advocacy was directed at the local, regional, national, or global level. It was coded as local if the goal was specific to a particular community—for example, closing a local power plant or conducting a local river cleanup effort. The advocacy was coded as regional if it included multiple communities, such as a watershed protection effort. It was coded as national if the goal was nationwide, such as a new national regulatory standard. It was coded as global if it crossed national boundaries, such as a multinational effort to preserve international fisheries.

Similarly, I was unable to create a consistent method to code failures in the specific advocacy effort that contributed to long-term success on a broader cause. For example, it is somewhat common for a lawsuit to fail in court but succeed in raising sufficient political pressure that it ultimately contributes to new policy.[6] However, it becomes very difficult to link a particular "failed" lawsuit (or protest) with a later policy or legislative success on a similar issue. For the purposes of this dataset, an advocacy effort was coded as a success if the specified goal as articulated by the advocates was achieved completely. It was coded as "mixed" if it was achieved partially. It was coded as "failure" if the goal was not achieved at all. Thus, if a lawsuit failed in court, or a protest against a factory failed to close the factory, those efforts were coded as failures even if, ultimately, those failures were the foundation on which other successes were built. This limitation of the study is discussed in greater detail chapter 2 and 3.

Each case was coded for the strategies that were present in the advocacy effort. Public protests were coded as present if there was a gathering of people in a public place for the purpose of protesting some cause. Online-only demonstrations, protests, and campaigns were not counted unless they also had a physical component. Lawsuits were coded as present if there was evidence that a lawsuit was filed or threatened. Media campaigns were coded as present if there was evidence that the activity was part of a broader media strategy designed to elicit widespread media attention. Letter-writing campaigns were coded as present if there was an effort to have members of the public write to public officials. Lobbying was coded as present if activists were targeting politicians (legislators, mayors, and others) to promote changes in law that favored their causes. International networks were coded as present if international organizations were engaged in the advocacy effort.

Turning to the advocacy strategies highlighted by the policy and bureaucratic politics literature, the effort to cultivate personal networks with policymakers, called the "friend on the inside" strategy, was coded as present if there was evidence that such a connection existed in the advocacy effort. Evidence of this kind would include the presence of a policy-relevant public official at an event associated with the advocacy effort, the presence of a former high-ranking government official on the sponsoring advocacy organization's board of directors, or a similar factor. The strategy of cultivating a "friend on the inside" was distinguished analytically from lobbying based on the type of policymaker. The former strategy targets nonelected policymakers and the latter targets politicians. The distinction between the two is essentially that between "legislative lobbying" and "administrative lobbying" as conceptualized by Jeffrey Berry and David Arons (2003).

The "make it work for business" strategy was coded as present if the environmental advocacy was directed toward profit making in a market. Examples could include energy conservation campaigns that highlighted money saved for businesses or efforts to promote new, eco-friendly technologies. Public education included all efforts designed to educate the public, often children, about environmental issues. Education directed toward elites was measured by the creation of policy papers. Art included any use of environmental art, such as gallery exhibits, public art installations, dance performances, or film releases. Local environmental projects, the "make it work locally" strategy in the results, included concrete local efforts to address the environmental issue in a particular place. Examples of this kind of strategy include a local clean-the-river project or a campaign to ban the use of plastic bags in one particular town. Local networking as a strategy was coded as present if there was evidence that three or more local organizations were cooperating in the advocacy effort.

In addition to the strategies, a number of control variables were also included in the analysis. Since I was interested in testing the common assumption that the level of democracy matters for advocacy strategy selection and effectiveness, I included a measure of democracy. Although there are a number of well-accepted measures of regime type, I chose to use the Freedom House data because they go back to 1973 and enabled me to test the effects of political and civic rights separately. Contextual information that was coded included the world region where the effort was concentrated (using UN regional designations) and the year the advocacy was initiated.

Descriptive and contextual information about the advocacy effort was also included. The issue type (pollution, environmental justice, energy, conservation, and waste) was coded. Whether the advocacy was primarily a NIMBY (not in my backyard) activity was coded, as well as whether there was violence. Violence was coded as present if there was any violence associated with the advocacy, irrespective of whether the activists were initiators or victims. Finally, the actors involved in the advocacy effort were coded (grassroots nongovernmental organizations, business, government, international organizations), as well as which actors initiated the advocacy effort.

Descriptive statistics were used to determine how common different strategies were and what their relative success rates were. To gain greater analytic leverage on which strategies were more successful, ordinary least squares regressions and recursive partitioning were also employed to tease out the relationships between advocacy type, outcome, issue, region, and regime type.

Statistical analyses were performed using the R statistical language.[7] Random forest was used to identify explanatory strategies that differentiated between successful and unsuccessful cases of advocacy. The random forest classification algorithm is an extension of classification and regression trees. Classification and regression trees have been used in multiple disciplines to group observations based on a number of predictor variables. Classification is achieved through recursive partitioning of the dataset into successively more homogeneous groups. If the results are perfect, all nodes will result in completely homogeneous groups. The splits in the data are made using all of the predictor variables, and the best tree structure is determined by the Gini Index. The cforest function in the R party package was used to build the random forest model using conditional permutation importance. The advantage of using the ctree function in the party package as compared with the original random forest implementation by Leo Breiman (2001) is that it produces unbiased individual trees. [8] Informative predictor variables were determined following Carolin Strobl et al. (2008), who determined that variable importance value should be above the absolute value of the lowest negative-scoring variable.[9]

Notes

Chapter 1

1. Peter Evans, *Embedded Autonomy: States and Industrial Transformation* (Princeton, NJ: Princeton University Press, 1995); Meredith Woo-Cumings, ed., *The Developmental State* (Ithaca, NY: Cornell University Press, 1999); Joseph Wong, "The Adaptive Developmental State in East Asia," *Journal of East Asian Studies* 4, no. 3 (September–December 2004): 345–362.

2. Muthiah Alagappa, ed., *Civil Society and Political Change in Asia: Expanding and Contracting Democratic Space* (Stanford, CA: Stanford University Press, 2004); Jeffrey Broadbent and Vicky Brockman, eds., *East Asian Social Movements: Power, Protest, and Change in a Dynamic Region* (New York: Springer, 2011); Paul G. Harris and Graeme Lang, eds., *Routledge Handbook of Environment and Society in Asia* (New York: Routledge, 2015).

3. Peter Ho, "Embedded Activism and Political Change in a Semiauthoritarian Context," *China Information* 21 (2007): 187–209; Jessica C. Teets, *Civil Society under Authoritarianism: The China Model* (New York: Cambridge University Press, 2014); Susan L. Shirk, "China in Xi's 'New Era': The Return to Personalistic Rule," *Journal of Democracy* 29, no. 2 (2018): 22–36.

4. The 2016 Improper Solicitation and Graft Act bans most lobbying activity in South Korea, and the 2008 Lobbying Act significantly restricts political lobbying in Taiwan.

5. Robert Joseph Pekkanen, *Japan's Dual Civil Society: Members without Advocates* (Stanford, CA: Stanford University Press, 2006).

6. Doug McAdam, John D. McCarthy, and Mayer N. Zald, eds., *Comparative Perspectives on Social Movements: Political Opportunities, Mobilizing Structures, and Cultural Framings* (New York: Cambridge University Press, 1996); Sidney Tarrow, *Power in Movement: Social Movements and Contentious Politics*, 2nd ed. (Ithaca, NY: Cornell University Press, 1998); Charles Tilly and Sidney Tarrow, *Contentious Politics* (Boulder, CO: Paradigm, 2006).

7. Marco Giugni, *Social Protest and Policy Change: Ecology, Antinuclear, and Peace Movements in Comparative Perspective* (New York: Rowman and Littlefield, 2004); Dana R. Fisher, "The Broader Importance of #Fridaysforfuture," *Nature Climate Change* 9, no. 6 (2019): 430–431.

8. Russell J. Dalton, *The Green Rainbow: Environmental Interest Groups in Western Europe* (New Haven, CT: Yale University Press, 1994); Raymond Dominick, *The Environmental Movement in Germany: Prophets and Pioneers, 1871–1971* (Bloomington: Indiana University Press, 1992); Ellis Krauss and Bradford Simcock, "Citizens' Movements: The Growth and Impact of Environmental Protest in Japan," in *Political Opposition and Local Politics in Japan*, ed. Kurt Steiner, Ellis Krauss, and Scott Flanagan (Princeton, NJ: Princeton University Press, 1980); Jacqueline Vaughn Switzer, *Environmental Activism: A Reference Handbook* (Santa Barbara, CA: ABC-CLIO, 2003).

9. Francis O. Adeola, "Cross-national Environmental Injustice and Human Rights Issues: A Review of Evidence in the Developing World," *American Behavioral Scientist* 43, no. 4 (2000): 686–706; Arun Agrawal and Elinor Ostrom, "Collective Action, Property Rights, and Decentralization in Resource Use in India and Nepal," *Politics and Society* 29, no. 4 (2001): 485–514.

10. Helen M. Poulos and Mary Alice Haddad, "Violent Repression of Environmental Protests," *SpringerPlus* 5, no. 230 (2016): 1–12; Steve Vanderheiden, "Eco-terrorism or Justified Resistance? Radical Environmentalism and the 'War on Terror,'" *Politics and Society* 33, no. 3 (2005): 425–447.

11. Hilal Elver, "International Environmental Law, Water and the Future," *Third World Quarterly* 27, no. 5 (2006): 885–901; Michael Ewing-Chow and Darryl Soh, "Pain, Gain, or Shame: The Evolution of Environmental Law and the Role of Multinational Corporations," *Indiana Journal of Global Legal Studies* 16, no. 1 (2009): 195–222; Jedediah Purdy, "The Politics of Nature: Climate Change, Environmental Law, and Democracy," *Yale Law Journal* 119, no. 6 (2010): 1122–1209; Patricia Steinhoff, ed., *Going to Court to Change Japan* (Ann Arbor, MI: Center for Japanese Studies, 2014).

12. Simon Avenell, "Legal Experts and Environmental Activism in Japan: Fighting for 'Environmental Rights,'" in *Greening East Asia: The Rise of the Eco-developmental State*, ed. Ashley Esarey et al. (Seattle: University of Washington Press, forthcoming); Raul Lejano, Mrill Ingram, and Helen Ingram, *The Power of Narrative in Environmental Networks* (Cambridge, MA: MIT Press, 2013); Steinhoff; Carol Hager and Mary Alice Haddad, eds., *NIMBY Is Beautiful: Cases of Local Activism and Environmental Innovation around the World* (New York: Berghahn Books, 2015).

13. Norman Miller, *Environmental Politics: Interest Groups, the Media, and the Making of Policy* (New York: Taylor and Francis, 2002); Moti Nissani, "Media Coverage of the Greenhouse Effect," *Population and Environment* 21, no. 1 (1999): 27–43; Helen M. Poulos, "The Media and NIMBY: How Do Grassroots Environmental Protests Incite

Innovation?," in Hager and Haddad; Yuezhi Zhou and Sun Wusan, "Public Opinion Supervision: Possibilities and Limits of the Media in Constraining Local Officials," in *Grassroots Political Reform in Contemporary China*, ed. Elizabeth Perry and Merle Goldman (Cambridge, MA: Harvard University Press, 2007).

14. Louise Chawla, "Research Priorities in Environmental Education," *Children's Environments* 9, no. 1 (1992): 68–71; Edward Johnson and Michael Mappin, eds., *Environmental Education and Advocacy: Changing Perspectives of Ecology and Education* (New York: Cambridge University Press, 2009); United Nations Department of Economic and Social Affairs, *People Matter: Civic Engagement in Public Governance—World Public Sector Report* (New York: United Nations, 2009).

15. Noriko Manabe, "The No Nukes 2012 Concert and the Role of Musicians in the Anti-nuclear Movement," *Asia-Pacific Journal* 10, no. 29 (2012): article 2, https://apjjf.org/-Noriko-Manabe/3799/article.pdf.

16. Robert Glenn Ketchum, *The Tongass: Alaska's Vanishing Rain Forest* (New York: Aperture, 1987).

17. Murray Edelman, *From Art to Politics* (Chicago: University of Chicago Press, 1995); Timothy George, *Minamata: Pollution and the Struggle for Democracy in Postwar Japan* (Cambridge, MA: Harvard University Press, 2002).

18. Mary Alice Haddad, "From Backyard Environmental Advocacy to National Democratisation: The Cases of South Korea and Taiwan," in Hager and Haddad; James Reardon-Anderson, *Pollution, Politics, and Foreign Investment in Taiwan: The Lukang Rebellion* (New York: M. E. Sharpe, 1997).

19. Jessica C. Teets and William Hurst, eds., *Local Governance Innovation in China: Experimentation, Diffusion, and Defiance*, Routledge Contemporary China Series (New York: Routledge, 2015); Tadayoshi Terao and Kenji Otsuka, eds., *Development of Environmental Policy in Japan and Asian Countries* (New York: Palgrave Macmillan and IDE-JETRO, 2007); Kent Portney and Jeffrey Berry, "Civil Society and Sustainable Cities," *Comparative Political Studies* 47, no. 3 (2014): 395–419.

20. Hsin-Huang Michael Hsiao, "Environmental Movements in Taiwan," in *Asia's Environmental Movements: Comparative Perspectives*, ed. Yok-shiu F. Lee and Alvin Y. So (New York: M. E. Sharpe, 1999); Do-Wan Ku, "The Structural Change of the Korean Environmental Movement," *Korea Journal of Population and Development* 25, no. 1 (1996): 155–180; Lei Xie and Peter Ho, "Urban Environmentalism and Activists' Networks in China: The Cases of Xiangfan and Shanghai," *Conservation and Society* 6 (2008): 141–153; Frank Fischer, *Citizens, Experts, and the Environment: The Politics of Local Knowledge* (Durham, NC: Duke University Press, 2000); DeWitt John, *Civic Environmentalism: Alternatives to Regulation in States and Communities* (Washington DC: Congressional Quarterly Press, 1994); Simon Avenell, *Transnational Japan in the Global Environmental Movement* (Honolulu: University of Hawai'i Press, 2017).

21. Aseem Prakash and Matthew Potoski, *The Voluntary Environmentalists: Green Clubs, ISO 14001, and Voluntary Environmental Regulations* (New York: Cambridge University Press, 2006); Marc Eisner, *Governing the Environment* (New York: Lynne Rienner, 2006); Neil Gunningham, Robert Kagan, and Dorothy Thornton, *Shades of Green: Business, Regulation, and Environment* (Stanford, CA: Stanford Law and Politics, 2003); Norman Vig and Michael Kraft, eds., *Environmental Policy: New Directions for the Twenty-First Century*, 8th ed. (Washington, DC: SAGE, 2013).

22. Bryan D. Jones, *Reconceiving Decision-Making in Democratic Politics: Attention, Choice, and Public Policy* (Chicago: University of Chicago Press, 1994).

23. Jeffrey Berry et al., "Power and Interest Groups in City Politics" (working paper, Rappaport Institute for Greater Boston, Kennedy School of Government, Harvard University, Cambridge, MA, 2006); Christopher Bosso, *Environment, Inc.*, Studies in Government and Public Policy (Lawrence: University Press of Kansas, 2005); Peter M. Haas, "Banning Chlorofluorocarbons: Epistemic Community Efforts to Protect Stratospheric Ozone," *International Organization* 46, no. 1 (Winter 1992): 187–224; Patricia J. Libby, *The Lobby Strategy Handbook: 10 Steps to Advancing Any Cause Effectively* (New York: Sage, 2011).

24. Sofie Bouteligier, *Cities, Networks, and Global Environmental Governance: Spaces of Innovation, Places of Leadership* (New York: Routledge, 2013); Haas, "Banning Chlorofluorocarbons"; Paul G. Harris and Chihiro Udagawa, "Defusing the Bombshell? Agenda 21 and Economic Development in China," *Review of International Political Economy* 11, no. 3 (2004): 618–640; Paul Sabatier and Hank C. Jenkins-Smith, eds., *Policy Change and Learning: An Advocacy Coalition Approach* (Boulder, CO: Westview, 1993); Avenell, *Transnational Japan*.

25. Bouteligier; Kemi Fuentes-George, *Between Preservation and Exploitation: Transnational Advocacy Networks and Conservation in Developing Countries* (Cambridge, MA: MIT Press, 2016); Maria Guadalupe Moog Rodrigues, *Global Environmentalism and Local Politics: Transnational Advocacy Networks in Brazil, Ecuador, and India* (Albany: State University of New York Press, 2003); Karen T. Litfin, "Advocacy Coalitions along the Domestic-Foreign Frontier: Globalization and Canadian Climate Change Policy," *Policy Studies Journal* 28, no. 1 (2000): 236–252.

26. Bouteligier; Per-Olof Busch, Helge Jörgens, and Kerstin Tews, "The Global Diffusion of Regulatory Instruments: The Making of a New International Environmental Regime," *Annals of the American Academy of Political and Social Science* 598 (March 2005): 146–167; Margaret Keck and Kathryn Sikkink, *Activists beyond Borders: Advocacy Networks in International Politics* (Ithaca, NY: Cornell University Press, 1998); Lejano, Ingram, and Ingram.

27. Neil Carter and Arthur Mol, *Environmental Governance in China* (New York: Routledge, 2008); Elizabeth Economy, *The River Runs Black: The Environmental Challenge to China's Future* (Ithaca, NY: Cornell University Press, 2004); Peter Ho and Richard

Edmonds, eds., *China's Embedded Activism: Opportunities and Constraints of a Social Movement* (New York: Routledge, 2007); Anna Lora-Wainwright, *Resigned Activism: Living with Pollution in Rural China* (Cambridge, MA: MIT Press, 2017); Andrew Mertha, *China's Water Warriors: Citizen Action and Policy Change* (Ithaca, NY: Cornell University Press, 2008); Lei Xie, *Environmental Activism in China* (New York: Routledge, 2009).

28. W. Puck Brecher, *An Investigation of Japan's Relationship to Nature and Environment* (Lewiston, NY: Edwin Mellen, 2000); Jeffrey Broadbent, *Environmental Politics in Japan: Networks of Power and Protest* (New York: Cambridge University Press, 1998); George; Hidefumi Imura and Miranda Schreurs, eds., *Environmental Policy in Japan* (Northampton, MA: Edward Elgar, 2005); Margaret McKean, *Environmental Protest and Citizen Politics in Japan* (Berkeley: University of California Press, 1981); Brett Walker, *Toxic Archipelago: A History of Industrial Disease in Japan*, Weyerhaeuser Environmental Books (Seattle: University of Washington Press, 2011); Kenneth Wilkening, *Acid Rain Science and Politics in Japan: A History of Knowledge and Action toward Sustainability* (Cambridge, MA: MIT Press, 2004); Avenell, *Transnational Japan*.

29. Young-Bae Chang, Jae-Kak Han, and Woo-Hyun Kim, "Green Growth and Green New Deal Policies in Korea: Are They Creating Decent Green Jobs?" (paper presented at the International Trade Union Conference / Global Unions Research Network workshop "A Green Economy That Works for Social Progress," Brussels, Belgium, 2011); Sung-Young Kim and Elizabeth Thurbon, "Developmental Environmentalism: Explaining South Korea's Ambitious Pursuit of Green Growth," *Politics and Society* 43, no. 2 (2015): 213–240; Dowan Ku, "The Korean Environmental Movement: Green Politics through Social Movement," in Broadbent and Brockman; Su-Hoon Lee, "Environmental Movements in Korea," in Yok-shiu F. Lee and So.

30. Simona A. Grano, *Environmental Governance in Taiwan: A New Generation of Activists and Stakeholders* (New York: Routledge, 2015); Ming-sho Ho, "Lukang Anti-Dupont Movement (Taiwan)," in *The Wiley-Blackwell Encyclopedia of Social and Political Movements*, ed. David Snow et al. (Malden, MA: Wiley-Blackwell, 2013); Hsiao; Reardon-Anderson; Shui-Yan Tang and Ching-Ping Tang, "Democratization and Environmental Politics in Taiwan," *Asian Survey* 37, no. 3 (March 1997): 281–294.

31. Harris and Lang; Arne Kalland and Gerard Persoon, eds., *Environmental Movements in Asia*, Man and Nature, vol. 4 (New York: Curzon, 1998); Yok-shiu F. Lee and So, *Asia's Environmental Movements*; Terao and Otsuka.

32. Evans, *Embedded Autonomy*; Ziya Onis, "The Logic of the Developmental State," *Comparative Politics* 24, no. 1 (1991): 109–126; Joseph Wong, "The Adaptive Developmental State in East Asia," *Journal of East Asian Studies* 4, no. 3 (September–December 2004): 345–362; Woo-Cumings.

33. Alagappa; Broadbent and Brockman; Stephen Osborne, ed., *The Voluntary and Non-profit Sector in Japan*, Nissan Institute / RoutledgeCurzon Japanese Studies Series (New York: RoutledgeCurzon, 2003); Robert Joseph Pekkanen, "Japan's Dual Civil

Society: Members without Advocates" (PhD diss., Harvard University, 2002); Kim Reimann, *The Rise of Japanese NGOs* (New York: Routledge, 2009); Karla Simon, *Civil Society in China: The Legal Framework from Ancient Times to the "New Reform Era"* (New York: Oxford University Press, 2013).

34. Peter Evans, ed., *State-Society Synergy: Government and Social Capital in Development* (Berkeley: University of California Press, 1997); Richard C. Feiock et al., "Local Level Collaborations on Environmental Issues in China through the Lens of Institutional Collective Action" (paper presented at the Annual Meeting of the Association of Public Policy Analysis and Management, Miami, 2015); Peter Ho and Edmonds; Teets; Terao and Otsuka.

35. Esarey et al.; Chang, Han, and Kim; Mark Elder, "Regional Governance for Environmental Sustainability in Asia in the Context of Sustainable Development: A Survey of Regional Cooperation Frameworks," in *Routledge Handbook of Sustainable Development in Asia*, ed. Sara Hsu (New York: Routledge, 2018); Ekaterina Zelenovskaya, *Green Growth Policy in Korea: A Case Study* (International Center for Climate Governance, 2012); Miranda Schreurs, *Environmental Politics in Japan, Germany, and the United States* (New York: Cambridge University Press, 2002).

36. Dafydd Fell, "The Evolution of the Anti-nuclear Movement in Taiwan since 2008," in *Taiwan's Social Movements under Ma Ying-Jeou: From the Wild Strawberries to the Sunflowers*, ed. Dafydd Fell (London: Routledge, 2017); Grano; Lora-Wainwright; Eiko Maruko Siniawer, *Waste: Consuming Postwar Japan* (Ithaca, NY: Cornell University Press, 2018); Haikun Wang et al., "Trade-Driven Relocation of Air Pollution and Health Impacts in China," *Nature Communications* 8, no. 1 (2017): article 738; Walker; Hsu; Gerald McBeth and Tse-Kang Leng, *Governance of Biodiversity Conservation in China and Taiwan*, Environmental Governance in Asia (Northampton, MA: Edward Elgar, 2006); Esarey et al.

37. Economy; George; Dowan Ku, "Environmental Movement and Policies During High Economic Growth in Korea," in *Environment and Our Sustainability in the 21st Century: Understanding and Cooperation between Developed and Developing Countries*, ed. Yuko Arayama (Nagoya, Japan: Nagoya University, 2002), 65–87; Lora-Wainwright; McKean; Reardon-Anderson.

38. Japanese Constitution of 1947. An official English translation can be found online here: "The Constitution of Japan," promulgated November 3, 1946, came into effect May 3, 1947, Prime Minister of Japan and His Cabinet, accessed August 24, 2018, https://japan.kantei.go.jp/constitution_and_government_of_japan/constitution_e .html.

39. Mary Alice Haddad, *Building Democracy in Japan* (New York: Cambridge University Press, 2012).

40. In 1979 prodemocracy protests in Kaohsiung, Taiwan, were violently repressed, leading to the massacre of the Lin family and the arrest and imprisonment of many of the leaders. Tun-jen Cheng, "Democratizing the Quasi-Leninist Regime in

Taiwan," *World Politics* 61, no. 4 (1989): 471–499; Robert Edmondson, "The February 28 Incident and National Identity," in *Memories of the Future: National Identity Issues and the Search for a New Taiwan*, ed. Stéphane Corcuff (Armonk, NY: M. E. Sharpe, 2002). In 1980 a prodemocracy student protest in Gwangju, South Korea, was violently repressed, resulting in hundreds of deaths. Georgy Katsiaficas, *South Korean Democracy: Legacy of the Gwangju Uprising* (New York: Routledge, 2013); Kim Yong Cheol, "The Shadow of the Gwangju Uprising in the Democratization of Korean Politics," *New Political Science* 25, no. 2 (2003): 225–240. Both incidents served to spark what would become successful, nationwide democracy movements.

41. Samuel P. Huntington, "Democracy's Third Wave," *Journal of Democracy* 2, no. 2 (Spring 1991): 12–34.

42. Ming-sho Ho and Chen-Shuo Hong, "Challenging New Conservative Regimes in South Korea and Taiwan," *Asian Survey* 52, no. 4 (July/August 2012): 643–665.

43. Melanie Manion, "Chinese Democratization in Perspective: Electorates and Selectorates at the Township Level," *China Quarterly* 163 (2000): 764–782; Perry and Goldman, *Grassroots Political Reform*.

44. Guobin Yang, *Power of the Internet in China* (New York: Columbia University Press, 2009); Yuezhi Zhou, "Watchdogs on Party Leashes? Contexts and Implications of Investigative Journalism in Post-Deng China," *Journalism Studies* 1, no. 2 (2000): 577–595; Zhou and Wusan; Gary King, Jennifer Pan, and Margaret E. Roberts, "How Censorship in China Allows Government Criticism but Silences Collective Expression," *American Political Science Review* 107, no. 2 (2013): 326–343.

45. Lora-Wainwright; Yeling Tan, "Transparency without Democracy: The Unexpected Effects of China's Environmental Disclosure Policy," *Governance* 27, no. 1 (2014): 37–62; Teets; China Development Brief, *A Report on the Policy Environment for Chinese NGOs* (Hong Kong: China Development Brief, 2018); Xiaoguang Kang and Qun Wang, eds., "Nonprofit Policymaking in China," special issue, *Nonprofit Policy Forum* 9, no. 1 (2018).

46. John A. Daly, *Advocacy: Championing Ideas and Influencing Others* (New Haven, CT: Yale University Press, 2012), 6.

47. Paul Sabatier, "An Advocacy Coalition Framework of Policy Change and the Role of Policy-Oriented Learning Therein," *Policy Sciences* 21, no. 2/3 (1988): 133.

48. Lora-Wainwright; Tan; Teets; China Development Brief; Kang and Wang.

49. Lora-Wainwright; Tan; Teets; China Development Brief; Kang and Wang; Rob Efird, "Closing the Green Gap: Policy and Practice in Chinese Environmental Education," in *Schooling for Sustainable Development across the Pacific*, ed. John Chi-Kin Lee and Rob Efird (Dordrecht: Springer, 2014); John Chi-Kin Lee and Rob Efird, "Introduction: Schooling and Education for Sustainable Development (ESD) across the Pacific," in John Chi-Kin Lee and Efird.

Chapter 2

1. Factiva's Major News and Business Publications subset of news sources was selected because it offered the best geographic diversity of sources and stories. If I used the full set of available articles or other publication subsets, results were even more skewed toward North American news sources and stories.

2. Robert Pekkanen, Steven Rathgeb Smith, and Yutaka Tsujinaka, eds., *Nonprofits and Advocacy: Engaging Community and Government in an Era of Retrenchment* (Baltimore: Johns Hopkins University Press, 2014).

3. Daniel Gillion, *The Political Power of Protest: Minority Activism and Shifts in Public Policy*, Cambridge Studies in Contentious Politics (New York: Cambridge University Press, 2013), 28.

4. Austin Sarat and Stuart Scheingold, eds., *Cause Lawyers and Social Movements* (Stanford, CA: Stanford Law and Politics, 2006); Steinhoff.

5. Peter A. Groothuis and Gail Miller, "Locating Hazardous Waste Facilities: The Influence of NIMBY Beliefs," *American Journal of Economics and Sociology* 53, no. 3 (1994): 335–346; Lora-Wainwright; Eileen Maura McGurty, "From NIMBY to Civil Rights: The Origins of the Environmental Justice Movement," *Environmental History* 2, no. 3 (July 1997): 301–323; Barry Rabe, *Beyond NIMBY: Hazardous Waste Siting in Canada and the United States* (Washington, DC: Brookings Institution, 1994); Daniel Sherman, *Not Here, Not There, Not Anywhere: Politics, Social Movements, and the Disposal of Low-Level Radioactive Waste* (Washington, DC: RFF, 2011).

6. Adeola; McGurty; Poulos and Haddad.

7. Leo Breiman, "Random Forests," *Machine Learning* 45, no. 1 (2001): 5–32.

8. Russell J. Dalton, Steve Recchia, and Robert Rohrschneider, "The Environmental Movement and the Modes of Political Action," *Comparative Political Studies* 36, no. 7 (2003): 768.

9. Poulos and Haddad.

10. Global Witness interactive map "Killings by Country 2010–2015," in "On Dangerous Ground," Global Witness, June 20, 2016, https://www.globalwitness.org/en/reports/dangerous-ground/.

11. Robert Paul Weller, *Alternate Civilities: Democracy and Culture in China and Taiwan* (Boulder, CO: Westview Press, 1999), 116.

12. Bosso; Dalton; Giugni; John; Ferdinand Muller-Rommel and Thomas Poquntke, *Green Parties in National Government* (New York: Routledge, 2002); Aseem Prakash and Mary Kay Gugerty, eds., *Advocacy Organizations and Collective Action* (New York: Cambridge University Press, 2010).

13. Ho, Peter; Yok-shiu F. Lee and So; Terao and Otsuka.

14. Poulos and Haddad.

15. Giugni; Timothy Hildebrandt and Jennifer Turner, "Green Activism? Reassessing the Role of Environmental NGOs in China," in *State and Society Responses to Social Welfare Needs in China: Serving the People*, ed. Jonathan Schwartz and Shawn Shieh (New York: Routledge, 2009); Mertha, *China's Water Warriors*; Muller-Rommel and Poquntke; Switzer.

Chapter 3

1. Robert A. Dahl, *Polyarchy: Participation and Opposition* (New Haven, CT: Yale University Press, 1971).

2. Douglass C. North, *Institutions, Institutional Change and Economic Performance* (New York: Cambridge University Press, 1990), 3.

3. Frank R. Baumgartner and Bryan D. Jones, *Agendas and Instability in American Politics* (Chicago: University of Chicago Press, 1993); Paul Burstein, "Policy Domains: Organization, Culture, and Policy Outcomes," *Annual Review of Sociology* 17 (1991): 327–350; Miller.

4. Eisner.

5. Gunningham, Kagan, and Thornton.

6. Bosso; Mertha, *China's Water Warriors*; Sarah Pralle, "Shopping Around: Environmental Organizations and the Search for Policy Venues," in Prakash and Gugerty; Fischer, *Citizens*; Jessica C. Teets, "Reforming Service Delivery in China: The Emergence of a Social Innovation Model," *Journal of Chinese Political Science* 17 (2012): 15–32; Jeffrey Berry and David F. Arons, *A Voice for Nonprofits* (Washington, DC: Brookings Institution, 2003); Herrington Bryce, *Players in the Public Policy Process: Nonprofits as Social Capital and Agents* (New York: Palgrave Macmillan, 2005); John.

7. Jeffrey Berry and Kent Portney, "The Group Basis of City Politics," in Pekkanen, Smith, and Tsujinaka; Genia Kostka and Arthur P. J. Mol, "Implementation and Participation in China's Local Environmental Politics: Challenges and Innovations," *Journal of Environmental Policy and Planning* 15, no. 1 (2013): 3–16; Ralph Litzinger, "In Search of the Grassroots: Hydroelectric Politics in Northwest Yunnan," in Perry and Goldman; Groothuis and Miller; John; Portney and Berry; Rabe; Ran Ran, "Perverse Incentive Structure and Policy Implementation Gap in China's Local Environmental Politics," *Journal of Environmental Policy and Planning* 15, no. 1 (2013): 17–39; Sherman; Paul Waley, "Ruining and Restoring Rivers: The State and Civil Society in Japan," *Pacific Affairs* 78, no. 2 (2005): 195–215.

8. Carter and Mol; Dalton; Dominick; Wilkening; Giugni; James Mahoney and Kathleen Thelen, *Explaining Institutional Change: Ambiguity, Agency, and Power* (New York: Cambridge University Press, 2009).

9. Arun Agrawal and Maria Carmen Lemos, "A Greener Revolution in the Making? Environmental Governance in the 21st Century," *Environment: Science and Policy for Sustainable Development* 49, no. 5 (2007): 36–45; Graeme Auld and Lars H. Gulbrandsen, "Transparency in Nonstate Certification: Consequences for Accountability and Legitimacy," *Global Environmental Politics* 10, no. 3 (August 2010): 97–119; Busch, Jörgens, and Tews; David Antony Detomasi, "The Multinational Corporation and Global Governance: Modelling Global Public Policy," *Journal of Business Ethics* 71, no. 3 (2007): 321–334; Emerson Kirk, Tina Nabatchi, and Stephen Balogh, "An Integrative Framework for Collaborative Governance," *Journal of Public Administration Research and Theory* 22 (2011): 1–29; Maria Carmen Lemos and Arun Agrawal, "Environmental Governance," *Annual Review of Environmental Resources* 31 (2006): 297–325; Ronald Mitchell, "Transparency for Governance: The Mechanisms and Effectiveness of Disclosure-Based and Education-Based Transparency Policies," *Ecological Economics* 70, no. 11 (2011): 1882–1890; Elinor Ostrom, *Governing the Commons: The Evolution of Institutions for Collective Action* (Cambridge: Cambridge University Press, 1990); Prakash and Potoski, *Voluntary Environmentalists*; Yves Tiberghien, "The Battle for the Global Governance of Genetically Modified Organisms: The Roles of the European Union, Japan, Korea, and China in a Comparative Context," *Les Etudes du CERI*, no. 124 (June 2006) 1–49; Oran Young, "Governance for Sustainable Development in a World of Rising Interdependencies," in *Governance for the Environment: New Perspectives*, ed. Magali Delmas and Oran R. Young (New York: Cambridge University Press, 2009).

10. Detomasi; Archon Fung, Mary Graham, and David Weil, *Full Disclosure: The Perils and Promise of Transparency* (New York: Cambridge University Press, 2007); Robert Kagan, Dorothy Thornton, and Neil Gunningham, "Explaining Corporate Environmental Performance: How Does Regulation Matter?," *Law and Society Review* 37, no. 1 (March 2003): 51–90.

11. Of course, if there are multiple bureaucrats participating, they will be expected to be promoting the interests of their units (e.g., a Finance Ministry bureaucrat will be expected to fight with a bureaucrat from the Environmental Ministry, each promoting his or her ministry's preferred outcome) rather than generally facilitating.

12. Hugh Heclo, "Issue Networks and the Executive Establishment," in *The New American Political System*, ed. Anthony King (Washington, DC: American Enterprise Institute, 1978).

13. Frans Van Waarden, "Dimensions and Types of Policy Networks," *European Journal of Political Research* 21, no. 1–2 (1992): 29–52.

14. David Marsh and Roderick Arthur William Rhodes, *Policy Networks in British Government* (Oxford: Clarendon Press, 1992).

15. Hugh Compston, *Policy Networks and Policy Change: Putting Policy Network Theory to the Test* (Springer, 2009).

16. Sabatier; Sabatier and Jenkins-Smith.

17. Hank C. Jenkins-Smith and Paul A. Sabatier, "Evaluating the Advocacy Coalition Framework," *Journal of Public Policy* 14, no. 2 (1994): 175–203; Litfin; Jennifer Hadden, "Explaining Variation in Transnational Climate Change Activism: The Role of Inter-movement Spillover," *Global Environmental Politics* 14, no. 2 (2014): 7–25; Keck and Sikkink; Rodrigues.

18. Mertha, *China's Water Warriors*; Schreurs, *Environmental Politics*.

19. Evans, *Embedded Autonomy*; Haddad, *Building Democracy in Japan*; Peter Ho and Edmonds; Terao and Otsuka.

20. Joel Migdal, *State in Society: Studying How States and Societies Transform and Constitute One Another*, Cambridge Studies in Comparative Politics (New York: Cambridge University Press, 2001); Joel S. Migdal, "The State in Society: An Approach to Struggles for Domination," in *State Power and Social Forces: Domination and Transformation in the Third World*, ed. Joel S. Migdal, Atul Kohli, and Vivienne Shue (New York: Cambridge University Press, 1994); Saad Eddin Ibrahim, "Civil Society and Prospects of Democratization in the Arab World," in *Civil Society in the Middle East*, vol. 1, ed. Augustus Richard Norton (New York: E. J. Brill, 1995); David A. Lake and Matthew A. Baum, "The Invisible Hand of Democracy: Political Control and the Provision of Public Services," *Comparative Political Studies* 34, no. 6 (2001): 587–621; Jenny B. White, *Islamist Mobilization in Turkey: A Study in Vernacular Politics* (Seattle: University of Washington Press, 2002); Quintan Wiktorowicz, "Civil Society as Social Control: State Power in Jordan," *Comparative Politics* 33, no. 1 (2000): 43–62.

21. Broadbent and Brockman; Haddad, "From Backyard Environmental Advocacy"; Hildebrandt and Turner; Hsiao; Hyuk-Rae Kim, "The State and Civil Society in Transition: The Role of Non-governmental Organizations in South Korea," *Pacific Review* 13, no. 4 (2010): 595–613; See-Jae Lee, "The Environmental Movement and Its Political Empowerment," *Korea Journal* 40, no. 3 (2000): 131–160; Litzinger; Reardon-Anderson; Andrew Wells-Dang, *Civil Society Networks in China and Vietnam: Informal Pathbreakers in Health and the Environment* (New York: Palgrave Macmillan, 2012); Xie and Ho.

22. Broadbent and Brockman; Haddad, "From Backyard Environmental Advocacy"; Hildebrandt and Turner; Hsiao; Hyuk-Rae Kim, "State and Civil Society"; See-Jae Lee; Litzinger; Reardon-Anderson; Wells-Dang; Xie and Ho.

23. Peter Ho, "Embedded Activism and Political Change in a Semiauthoritarian Context," *China Information* 21 (2007): 187–209; Ho.

24. Broadbent and Brockman; Peter Ho and Richard Edmonds; Terao and Otsuka.

25. Fengshi Wu, "New Partners or Old Brothers? GONGOs in Transnational Environmental Advocacy in China," *China Environment Review*, no. 5 (2002): 45–58.

26. Jang Jip Choi, "The Democratic State Engulfing Civil Society: The Ironies of Korean Democracy," *Korean Studies* 34 (2010): 1–24; Hildebrandt and Turner; Peter Ho and Edmonds; Jonathan Schwartz, "Environmental NGOs in China: Roles and Limits," *Pacific Affairs* 77, no. 1 (2004): 28–49; Teets, *Civil Society under Authoritarianism.*

27. Chalmers Johnson, *MITI and the Japanese Miracle: The Growth of Industrial Policy, 1925–1975* (Stanford, CA: Stanford University Press, 1982).

28. Kenneth Lieberthal and Michel Oksenberg, *Policy Making in China: Leaders, Structures, and Process* (Princeton, NJ: Princeton University Press, 1988).

29. Andrew Mertha, "'Fragmented Authoritarianism 2.0': Political Pluralization in the Chinese Policy Process," *China Quarterly* 200, no. 1 (2009): 995–1012; Yves Tiberghien, "The Global Governance of Biotechnology: Mediating Chinese and Canadian Interests," *China Papers*, no. 13 (2010):111–127; Peter Lorentzen, Pierre Landry, and John Yasuda, "Undermining Authoritarian Innovation: The Power of China's Industrial Giants," *Journal of Politics* 76, no. 1 (January 2014): 182–194; Teets, *Civil Society under Authoritarianism*; Woo-Cumings; Keiko Hirata, "Whither the Developmental State? The Growing Role of NGOs in Japanese Aid Policymaking," *Journal of Comparative Policy Analysis: Research and Practice* 4 (2002): 165–188; Evans, *Embedded Autonomy.*

30. Jeffrey Broadbent and Vicky Brockman, 2010; Mary Alice Haddad, "Environmental Advocacy: Insights from East Asia,"*Asian Journal of Political Science* 25 (2017): 401–419; Paul G. Harris and Graeme Lang, eds., Routledge Handbook of Environment and Society in Asia (New York: Routledge, 2015); Yok-shiu F. Lee and Alvin Y. So, eds., *Asia's Environmental Movements: Comparative Perspectives* (New York: ME Sharpe, 1999).

31. John Creighton Campbell, *How Policies Change: The Japanese Government and the Aging Society* (Princeton, NJ: Princeton University Press, 1992); Chalmers Johnson; Richard J. Samuels, *The Politics of Regional Policy in Japan: Localities Incorporated?* (Princeton, NJ: Princeton University Press, 1983); Lieberthal and Oksenberg; Wu; Teets and Hurst.

32. Linda Jakobson and Dean Knox, "New Foreign Policy Actors in China" (policy paper, Stockholm International Peace Research Institute, Solna, Sweden, 2010); Teets, *Civil Society under Authoritarianism*; Katherine R. Xin and Jone L. Pearce, "Guanxi: Connections as Substitutes for Formal Institutional Support," *Academy of Management Journal* 39, no. 6 (1996): 1641–1658.

33. Stanley Wasserman and Katherine Faust, *Social Network Analysis: Methods and Applications*, Structural Analysis in the Social Sciences 8 (Cambridge: Cambridge University Press, 1994); John Scott, *Social Network Analysis: A Handbook* (New York: Sage, 2000); Albert-László Barabási and Márton Pósfai, *Network Science* (New York: Cambridge University Press, 2016).

34. Linton Freeman, *The Development of Social Network Analysis: A Study in the Sociology of Science* (Vancouver, BC: Empirical, 2004).

35. Barabási and Pósfai.

36. Those outside China can find a version of the film with English and Mandarin subtitles here: https://www.youtube.com/watch?v=T6X2uwlQGQM (accessed March 18, 2020).

37. Qiang Zhang et al., "Drivers of Improved PM2.5 Air Quality in China from 2013 to 2017," *Proceedings of the National Academy of Sciences* 116, no. 49 (2019): 24463–24469.

38. Avenell, "Legal Experts."

39. Mary Alice Haddad, "Working with and around Strong States: Environmental Networks in East Asia," in *Civil Society and the State in Democratic East Asia: Between Entanglement and Contention in Post High Growth*, ed. David Chiavacci, Simona A. Grano, and Julia Obinger (Amsterdam: Amsterdam University Press, 2020).

40. Campbell; William Easterly, *The Tyranny of Experts: Economists, Dictators, and the Forgotten Rights of the Poor* (New York: Basic Books, 2014); Frank Fischer, *Technocracy and the Politics of Expertise* (Newbury Park, CA: Sage, 1989); Fischer, *Citizens*; Peter M. Haas, "Introduction: Epistemic Communities and International Policy Coordination," *International Organization* 46, no. 1 (1992): 1–35; Connie P. Ozawa, "Science in Environmental Conflicts," *Sociological Perspectives* 39, no. 2 (Summer 1996): 219–230; Carol Hager, *Technological Democracy: Bureaucracy and Citizenry in the German Energy Debate* (Ann Arbor: University of Michigan Press, 1995).

41. Jakobson and Knox; Cheng Li, ed., *Bridging Minds across the Pacific: U.S.-China Educational Exchanges, 1978–2003* (Lexington, MA: Lexington Books, 2005); Debra C. Minkoff, Silke Aisenbrey, and Jon Agone, "Organizational Diversity in the U.S. Advocacy Sector," *Social Problems* 55, no. 4 (2008): 525–548; Schreurs, *Environmental Politics*; Dominick; Zhidong Hao and Henry A. Giroux, eds., *Intellectuals at a Crossroads: The Changing Politics of China's Knowledge Workers* (Albany: State University of New York Press, 2003); Yiqi Zhang et al., "The Roles of Scientific Research and Stakeholder Engagement for Evidence-Based Policy Formulation on Shipping Emissions Control in Hong Kong," *Journal of Environmental Management* 223 (October 2018): 49–56.

42. Peter Ho; Richard J. Samuels, *The Business of the Japanese State: Energy Markets in Comparative and Historical Perspective* (Ithaca, NY: Cornell University Press, 1987); Jacob M. Schlesinger, *Shadow Shoguns: The Rise and Fall of Japan's Postwar Political Machine*, 2nd ed. (Stanford, CA: Stanford University Press, 1999).

43. Detomasi; Dominick; Haas, "Introduction"; Keck and Sikkink; Lester M. Salamon, *Partners in Public Service: Government-Nonprofit Relations in the Modern Welfare*

State (Baltimore: Johns Hopkins University Press, 1995); Frank J. Schwartz, *Advice and Consent: The Politics of Consultation in Japan* (New York: Cambridge University Press, 1998); Xie and Ho.

44. Berry and Arons; Jones; Pekkanen, Smith, and Tsujinaka; Paul Pierson, "When Effect Becomes Cause: Policy Feedback and Political Change," *World Politics* 45, no. 4 (1993): 595–628.

Chapter 4

1. Carter and Mol; Economy; Norman Eder, *Poisoned Prosperity: Development, Modernization, and the Environment in South Korea* (New York: M. E. Sharpe, 1997); George; Harris and Lang; Kalland and Persoon; Ming-sho Ho, "Environmental Movement in Democratizing Taiwan (1980–2004): A Political Opportunity Structure Perspective," in Broadbent and Brockman.

2. McKean; Takehiro Watanabe, "Talking Sulfur Dioxide: Air Pollution and the Politics of Science in Late Meiji Japan," in *Japan at Nature's Edge: The Environmental Context of a Global Power*, ed. Ian Jared Miller, Julia Adeney Thomas, and Brett Walker (Honolulu: University of Hawai'i Press, 2013).

3. Walker.

4. Kurt Steiner, "Toward a Framework for the Study of Local Opposition," in Steiner, Krauss, and Flanagan, 1.

5. McKean; Frank Upham, "Litigation and Moral Consciousness in Japan: An Interpretive Analysis of Four Japanese Pollution Suits," *Law and Society Review* 10, no. 4 (Summer 1976): 579–619.

6. Interviews with senior managers from Keidanren, Toyota, and Hitachi, Tokyo, 2011.

7. For more about the Kiko Forum, see Kim Reimann, "Building Global Civil Society from the Outside In? Japanese International Development NGOs, the State, and International Norms," in *The State of Civil Society in Japan*, ed. Frank Schwartz and Susan Pharr (New York: Cambridge University Press, 2003).

8. Internet World Statistics, https://www.internetworldstats.com/asia/jp.htm (accessed July 26, 2020).

9. Interviews in Beijing, Hong Kong, Taipei, Seoul, and Tokyo, 2010–2011. For example, in 2010 Greenpeace had fifty staff in its Beijing office, forty staff in Hong Kong, and only fifteen in Tokyo. Interviews in Beijing and Hong Kong offices of Greenpeace, November 2010; email communication with the Tokyo office, October 2012.

10. Jeff Kingston, ed., *Natural Disaster and Nuclear Crisis in Japan: Response and Recovery after Japan's 3/11*, Nissan Institute / Routledge Japan Studies (New York: Routledge, 2012).

11. Interviews with Ministry of Environment staff, Tokyo, June 2011. See also "IAEA Action Plan on Nuclear Safety, International Atomic Energy Agency, accessed October 9, 2012, http://www.iaea.org/newscenter/focus/actionplan/; United Nations Environment Programme, *Managing Post-disaster Debris: The Japanese Experience* (Geneva: United Nations Environment Programme, 2012), https://postconflict.unep .ch/publications/UNEP_Japan_post-tsunami_debris.pdf; and Japanese Ministry of Foreign Affairs, "Future Cities We Want" (2012), https://www.mofa.go.jp/policy/envi ronment/warm/cop/rio_20/pdfs/pamph_01.pdf (accessed 07/26/2020).

12. For the Japan Renewable Energy Institute that Son created, see https://www .renewable-ei.org/en/images/pdf/REI_brochure_E_final_web.pdf (accessed 07/26/2020); for news coverage, see Mitsuru Obe, "First the iPhone. Now Renewables," *Wall Street Journal*, June 18, 2012, http://online.wsj.com/article/SB10001424052702304371504577 404343259051300.html.

13. No Nukes Japan website [in Japanese], accessed October 9, 2012, http://nonukes .jp/wordpress/; Sayonara (Goodbye) Nukes has an English website, accessed October 9, 2012, http://sayonara-nukes.org/english/. For coverage of one of their biggest rallies, see "Japan's Anti-nuclear Protests: The Heat Rises," *Economist*, July 21, 2012, http://www.economist.com/node/21559364.

14. "Nuclear Power in Japan," World Nuclear Association, accessed November 5, 2018, http://www.world-nuclear.org/information-library/country-profiles/countries -g-n/japan-nuclear-power.aspx.

15. Ministry of Economy, Trade and Industry, *Japan's Energy White Paper 2017: Japan's Energy Landscape and Key Policy Measures* (Ministry of Economy, Trade and Industry, n.d., accessed November 5, 2018), http://www.enecho.meti.go.jp/en/category/white paper/pdf/whitepaper_2017.pdf.

16. Bureau of Environment, Tokyo Metropolitan Government, *Final Energy Consumption and Greenhouse Gas Emissions in Tokyo (FY 2015)* (Tokyo: Bureau of Environment, Tokyo Metropolitan Government, March 2018), http://www.kankyo.metro .tokyo.jp/en/climate/index.files/GHG2015.pdf.

17. Tang and Tang, 284.

18. Reardon-Anderson, 11–12.

19. Reardon-Anderson, x.

20. See Tang and Tang.

21. Interviews in Taipei, November 2010.

22. Interviews in Taipei, November 2010.

23. Interview with Echo Lin, Taiwan Environmental Action Network, Taipei, November 2010; interview with Kuang-Jung Hsu, professor of atmospheric sciences, Taiwan

National University, and former chair, Taiwan Environmental Protection Union, Taipei, November 2010.

24. Korean Federation for Environmental Movements, homepage [in Korean], accessed October 9, 2012, http://www.kfem.or.kr/; interview with a KFEM leader, Seoul, November 2010.

25. Jennifer S. Oh, "Strong State and Strong Civil Society in Contemporary South Korea Challenges to Democratic Governance," *Asian Survey* 52, no. 3 (May–June 2012): 528–549; Sunhyuk Kim, "Civic Engagement and Democracy in South Korea," *Korean Observer* 40, no. 1 (Spring 2009): 1–26.

26. See-Jae Lee.

27. Su-Hoon Lee, 92.

28. See-Jae Lee, 143.

29. See-Jae Lee, 144; Duwan Ku, "Environmental Movement and Policies during High Economic Growth in Korea," in *Environment and Our Sustainability in the 21st Century: Understanding and Cooperation between Developed and Developing Countries*, ed. Yuko Arayama (Nagoya, Japan: Nagoya University, 2002), 76.

30. Su-Hoon Lee, 93.

31. Duwan Ku (2002), 76.

32. Duwan Ku (2002), 211–213; Lee, 94.

33. See-Jae Lee, November 26, 2010.

34. Jennifer S. Oh, 547; Choi, 17–22; Hyuk-Rae Kim, "Dilemmas in the Making of Civil Society in Korean Political Reform," *Journal of Contemporary Asia* 34, no. 1 (2004): 55–69.

35. Chang, Han, and Kim.

36. Ming-sho Ho and Hong; Ming-sho Ho, *Challenging Beijing's Mandate of Heaven: Taiwan's Sunflower Movement and Hong Kong's Umbrella Movement* (Philadelphia: Temple University Press, 2019).

37. Choe Sang-Hun, "Park Geun-hye, South Korea's Ousted President, Gets 24 Years in Prison," *New York Times*, April 6, 2018, https://www.nytimes.com/2018/04/06/world/asia/park-geun-hye-south-korea.html.

38. Judith Shapiro, *Mao's War against Nature: Politics and the Environment in Revolutionary China* (New York: Cambridge University Press, 2001).

39. Interviews in Beijing, November 2010 and April 2011.

40. Bin Xu, *The Politics of Compassion: The Sichuan Earthquake and Civic Engagement in China* (Stanford, CA: Stanford University Press, 2017).

41. Carin Zissis and Jayshree Bajoria, "China's Environmental Crisis," *Washington Post*, August 7, 2008, https://www.washingtonpost.com/wp-dyn/content/article/2008/08/07/AR2008080702003.html.

42. Xie and Ho; Wells-Dang.

43. Peter Ho and Edmonds; Wu.

44. For example, in 2019 the Nature Conservancy had $7.7 billion file:///Users/mahaddad/Downloads/TNC-Financial-Statements-FY19.pdf (accessed July 27, 2020) and the WWF collected more than $300 million in revenue http://assets.worldwildlife.org/financial_reports/37/reports/original/WWF-AR2019-FINALPAGES.pdf?1582917951&_ga=2.82725206.1560703302.1595849720-2124463079.1595849720 (p. 33) (accessed July 28, 2020).

45. Interview, Beijing, November 2010.

46. Johannes Friedrich, Mengpin Ge, and Andrew Pickens, "This Interactive Chart Explains World's Top 10 Emitters, and How They've Changed," World Resources Institute, April 11, 2017, https://www.wri.org/blog/2017/04/interactive-chart-explains-worlds-top-10-emitters-and-how-theyve-changed.

47. China Development Brief.

48. Mary Alice Haddad, "Paradoxes of Democratization: Environmental Politics in East Asia," in Harris and Lang.

49. Choi; See-Jae Lee, "Environmental Movement."

50. Harrell et al.

51. Teruyuki Shimizu et al., "A Region-Specific Analysis of Technology Implementation of Hydrogen Energy in Japan," *International Journal of Hydrogen Energy*, published ahead of print, December 15, 2017, https://doi.org/10.1016/j.ijhydene.2017.11.128; Ministry of the Environment, "White Paper on the Environment 2017" (Ministry of the Environment, Tokyo, 2018); Hideko Yonetani, "Construction and Demolition Waste Management in Japan," slide presentation, accessed November 8, 2018, http://www.uncrd.or.jp/content/documents/2661Parallel%20Roundtable(2)-Presentation(4)-Hideko%20Yonetani.pdf.

52. Ministry of the Environment Korea, *Ecorea: Environmental Review 2015, Korea* (Seoul: Ministry of the Environment, 2017); Zelenovskaya; Grano; Robert Paul Weller, *Discovering Nature: Globalization and Environmental Culture in China and Taiwan* (Cambridge: Cambridge University Press, 2006).

53. "China Meets 2020 Carbon Target Three Years ahead of Schedule," United Nations Climate Change, March 28, 2018, https://unfccc.int/news/china-meets-2020-carbon-target-three-years-ahead-of-schedule; International Energy Agency, Solar PV (2019) https://www.iea.org/reports/solar-pv (accessed July 27, 2020).

54. Jun Morikawa, *Whaling in Japan: Power, Politics, and Diplomacy* (New York: Columbia University Press, 2009); Hiroyuki Watanabe, *Japan's Whaling: The Politics of Culture in Historical Perspective* (Victoria, Australia: Trans Pacific, 2008).

55. Rebecca W. Y. Wong, *The Illegal Wildlife Trade in China* (Cham: Springer, 2019).

56. Chun-Chieh Chi, "Capitalist Expansion and Indigenous Land Rights: Emerging Environmental Justice Issues in Taiwan," *Asia Pacific Journal of Anthropology* 2, no. 2 (2001): 135–153.

57. Kyung-Min Baek et al., "Monitoring of Particulate Hazardous Air Pollutants and Affecting Factors in the Largest Industrial Area in South Korea: The Sihwa-Banwol Complex," *Environmental Engineering Research*, published ahead of print, December 17, 2019, https://doi.org/10.4491/eer.2019.419. *Environmental Engineering Research* 25, no. 6 (2020): 908–923.

Chapter 5

1. Sabatier and Jenkins-Smith; Eisner.

2. Busch, Jörgens, and Tews; Haas, "Banning Chlorofluorocarbons."

3. Yuanchao Xu and Woody Chan, "Ministry Reform: 9 Dragons to 2," China Water Risk, April 18, 2018, http://www.chinawaterrisk.org/resources/analysis-reviews/ministry-reform-9-dragons-to-2/.

4. "Organization of the Ministry of the Environment," Ministry of the Environment (Japan), November 2012, https://www.env.go.jp/en/aboutus/organization/organization.pdf.

5. Korean Ministry of the Environment, http://www.me.go.kr/home/web/index.do?menuId=10433 (accessed July 28, 2020).

6. According to an Environmental Protection Agency source, there are about 1,600 network nodes in the ministry, but since it seems likely that there would be no more than one node per person (indeed, it is likely to be less, since shared printers probably take up their own node), there are likely to be fewer than 1,600 staff in that agency. "Automation of Laboratory Procedures," Environmental Analysis Laboratory, Environmental Protection Administration (Taiwan), last updated February 12, 2019, https://www.epa.gov.tw/niea-en/DACD61CDA02CD766.

7. "EPA's Budget and Spending," United States Environmental Protection Agency, accessed March 15, 2019, https://www.epa.gov/planandbudget/budget.

8. "Enforcement Annual Results Numbers at a Glance for Fiscal Year 2017," United States Environmental Protection Agency, accessed July 5, 2019, https://archive.epa.gov/epa/enforcement/enforcement-annual-results-numbers-glance-fiscal-year-2017.html.

9. "The Ministry: Tasks and Structure," Federal Ministry for the Environment, Nature Conservation and Nuclear Safety (Germany), accessed March 15, 2019, https://www.bmu.de/en/ministry/tasks-and-structure/.

10. "About Us," Department for Environment, Food and Rural Affairs, accessed March 15, 2019, https://www.gov.uk/government/organisations/department-for-environment-food-rural-affairs/about.

11. "Directorate-General for Environment: About Us," European Commission, accessed March 15, 2019, http://ec.europa.eu/dgs/environment/index_en.htm.

12. John W. Kingdon, *Agendas, Alternatives, and Public Policies* (New York: Harper-Collins, 1984).

13. Bosso; Carter and Mol; Eisner; Imura and Schreurs; Lemos and Agrawal; Vig and Kraft.

14. Steven J. Balla and John R. Wright, "Interest Groups, Advisory Committees, and Congressional Control of the Bureaucracy," *American Journal of Political Science* 45, no. 4 (2001): 799–812; Berry and Arons; Hildebrandt and Turner.

15. Grano; Muller-Rommel and Poquntke; Michael O'Neill, *Green Parties and Political Change in Contemporary Europe: New Politics, Old Predicaments* (Burlington, VT: Ashgate, 1997).

16. Fuentes-George; Grano; Haddad, "From Backyard Environmental Advocacy"; Switzer; Wilkening.

17. Jones; Kingdon.

18. Bouteligier; Fischer, *Citizens*; Craig M. Kauffman, *Grassroots Global Governance: Local Watershed Management Experiments and the Evolution of Sustainable Development* (New York: Oxford University Press, 2017); Taedong Lee, *Global Cities and Climate Change: The Translocal Relations of Environmental Governance*, Cities and Global Governance (New York: Routledge 2015); Lemos and Agrawal; Rodrigues.

19. Tokyo Metropolitan Government, "Creating A sustainable City: Tokyo's Environmental Policy September 2019, https://www.kankyo.metro.tokyo.lg.jp/en/about_us/videos_documents/documents_1.files/creating_a_sustainable_city_2019_e.pdf (accessed July 28, 2020); Department of Environmental Protection (Taiwan), *Policy Objectives and Implementations in 2018*, accessed February 13, 2019, https://www-ws.gov.taipei/Download.ashx?u=LzAwMS9VcGxvYWQvMzY0L3JlbGZpbGUvMTg2ODUvNzY2NDUzNC85YmRhODQzYS1lMjJkLTQ4YjctYjU0YS0xNWM1MGYxYmE1NTQucGRm&n=UG9saWN5IE9iamVjdGl2ZXMgYW5kIEltcGxlWVudGF0aW9ucyBpbiAyMDE4LnBkZg%3d%3d&icon=.pdf; Solidiance, "Asia Pacific's Top 10 Green Cities," SlideShare, September 26, 2011, https://www.slideshare.net/dduhamel/top-10-asia-green-cities-asian-green-cities-wwwsolidiancecom-2011.

20. Fischer, *Citizens*; Hager; Kingdon; Schreurs, *Environmental Politics*; David Knoke et al., *Comparing Policy Networks: Labor Politics in the U.S., Germany, and Japan*, Cambridge Studies in Comparative Politics (Cambridge: Cambridge University Press, 1996); Sabatier.

21. Harris and Lang; Sunhyuk Kim, *Politics of Democratization in Korea: The Role of Civil Society* (Pittsburgh: University of Pittsburgh Press, 2000); Duwan Ku, "The Korean Environmental Movement: Green Politics through Social Movement," in *East Asian Social Movements*, eds. Jeffrey Broadbent and Victoria Brockman (New York: Springer, 2011, 205–235); Yok-shiu F. Lee and So; McKean; Tang and Tang.

22. Keck and Sikkink; Lemos and Agrawal; Litfin; Robert Rohrschneider and Russell J. Dalton, "A Global Network? Transnational Cooperation among Environmental Groups," *Journal of Politics* 64, no. 2 (2003): 510–533; Young; Haas, "Banning Chlorofluorocarbons"; Knoke et al.

23. Prakash and Gugerty; Zelenovskaya; Haas, "Banning Chlorofluorocarbons"; Fung, Graham, and Weil.

24. Fuentes-George; Jessica Teets, "The Power of Policy Networks in Authoritarian Regimes: Changing Environmental Policy in China," *Governance* 31, no. 1 (2017): 125–141; Detomasi; Richard C. Feiock, "The Institutional Collective Action Framework," *Policy Studies Journal* 41, no. 3 (2013): 397–425.

25. Interview in Taipei, November 2010.

26. Daniel P. Aldrich, "It's Who You Know: Factors Driving Recovery from Japan's 11 March 2011 Disaster," *Public Administration* 94, no. 2 (2016): 399–413; Feiock et al.; Peter Ho; Teets, *Civil Society under Authoritarianism*; Xin and Pearce.

27. The workshop was hosted by IGES in 2018. See "Workshop on the Joint Crediting Mechanism (JCM) and Low-Carbon Technologies in the Philippines," Institute for Global Environmental Strategies, accessed February 12, 2019, https://www.iges .or.jp/en/climate-energy/20180220.html.

28. Bouteligier; Dalton, Recchia, and Rohrschneider; Hildebrandt and Turner; Kristine Kern and Harriet Bulkeley, "Cities, Europeanization and Multi-level Governance: Governing Climate Change through Transnational Municipal Networks," *JCMS: Journal of Common Market Studies* 47, no. 2 (2009): 309–332; Wells-Dang.

29. Eugene Bardach, *A Practical Guide for Policy Analysis: The Eightfold Path to More Effective Problem Solving* (Washington, DC: Sage, 2012); Bouteligier; Lukas Hakelberg, "Governance by Diffusion: Transnational Municipal Networks and the Spread of Local Climate Strategies in Europe," *Global Environmental Politics* 14, no. 1 (2014): 107–129; International Institute of Environment and Development and World Business Council for Sustainable Development, *Breaking New Ground: Mining, Minerals, and Sustainable Development—Report of the Mining, Minerals and Sustainable Development*

Project (London: International Institute of Environment and Development and World Business Council for Sustainable Development, 2002); McKenzie F. Johnson et al., "Network Environmentalism: Citizen Scientists as Agents for Environmental Advocacy," *Global Environmental Change* 29 (2014): 235–245; Teets and Hurst.

30. Alagappa; Charles Armstrong, ed., *Korean Society: Civil Society, Democracy and the State* (New York: Taylor and Francis, 2002); Broadbent and Brockman; Sunhyuk Kim, *Politics of Democratization in Korea*; McKean; William T. Rowe, "The Problem of 'Civil Society' in Late Imperial China," *Modern China* 19, no. 2 (1993): 139–157; Simon.

31. Makoto Imada, "The Voluntary Response to the Hanshin Awaji Earthquake: A Trigger for the Development of the Voluntary and Non-profit Sector in Japan," in Osborne; Jessica C. Teets, "Post-earthquake Relief and Reconstruction Efforts: The Emergence of Civil Society in China?," *China Quarterly* 198 (June 2009): 330–347.

32. Roger Goodman, Gordon White, and Huck-ju Kwong, eds., *The East Asian Welfare Model: Welfare Orientalism and the State* (New York: Routledge, 1998); Salamon; Lily Tsai, *Accountability without Democracy: How Solidary Groups Provide Public Goods in Rural China* (New York: Cambridge University Press, 2007); Tadashi Yamamoto, ed. *The Nonprofit Sector in Japan* (New York: Manchester University Press, 1998).

33. Haddad, *Building Democracy in Japan*; Jessica C. Teets, "Let Many Civil Societies Bloom: The Rise of Consultative Authoritarianism in China," *China Quarterly* 213 (March 2013): 19–38; Joo, Sungsoo, Seonmi Lee, and Youngjae Jo. "The Explosion of CSOs and Citizen Participation: An Assessment of Civil Society in South Korea 2004." CIVICUS http://www.civicus.org/media/CSI_South_Korea_Country_Report.pdf (accessed July 28, 2020).

34. Seth Faison, "5 Greenpeace Leaders Detained at Beijing Protest," *New York Times*, August 15, 1995, https://www.nytimes.com/1995/08/15/world/5-greenpeace-leaders -detained-at-beijing-protest.html.

35. "Achievements," Greenpeace East Asia, accessed February 16, 2019, http://www .greenpeace.org/eastasia/about/achievements/.

36. Greenpeace East Asia, Renewable Energy 2005 report, http://www.greenpeace .org/eastasia/publications/reports/climate-energy/2005/renewable-energy/ (accessed Feb. 1, 2019).

37. Dominic Dudley, "China Is Set to Become the World's Renewable Energy Superpower, according to New Report," *Forbes*, January 11, 2019, https://www .forbes.com/sites/dominicdudley/2019/01/11/china-renewable-energy-superpower /#8c56c745a28c.

38. Rob Gifford, "Yellow River Pollution Is Price of Economic Growth," *All Things Considered*, NPR, December 11, 2007, https://www.npr.org/templates/story/story .php?storyId=16951806.

39. http://www.greenpeace.org/eastasia/Global/eastasia/publications/reports /climate-energy/2005/yellow-river-at-risk.pdf (accessed Feb. 1, 2019).

40. CCTV Report, "[Direct Attack on the Yellow River Pollution] Shanxi Fenhe: Industrial Pollution Affects Agriculture" June 15, 2005, http://www.cctv.com/news /china/20050615/100877.shtml (accessed Feb. 17, 2019).

41. http://en.classora.com/reports/t24369/ranking-of-the-worlds-richest-countries -by-gdp?edition=1997 (accessed February 17, 2019).

42. Classora Data, "Ranking the World's Richest Countries by GDP (1997)" http://en .classora.com/reports/t24369/general/ranking-of-the-worlds-richest-countries-by -gdp?edition=2010&fields= (accessed February 17, 2019).

43. Economy.

44. Various interviews in 2010, 2011, 2015, and 2019 conducted in Tokyo, Seoul, Taipei, and Beijing. The names and affiliations of the interview subjects are not being revealed to protect their identities.

45. http://www.greenpeace.org/eastasia/Global/eastasia/publications/reports/toxics /2010/swimming-in-poison-yangtze-fish.pdf (accessed Feb. 17, 2019).

46. APEC, "New Chemical & New Regulation 'Provisions on the Environmental Administration if New Chemical Substances" (MEP Order No. 7)'" http://mddb.apec .org/documents/2010/CDSG/WKSP1/10_cd_wksp1_005.pdf (accessed July 28, 2020).

47. Interviews with bureaucrats-turned-NGO-leaders in Tokyo, Beijing, and Seoul, 2010, 2011, 2015, and 2018.

48. Kingdon.

49. Reza Hasmath, Timothy Hildebrandt, and Jennifer YJ Hsu, "Conceptualizing Government-Organized Non-governmental Organizations" (paper presented at the Development Studies Association Annual Meeting, Oxford, UK, 2016); Jonathan Schwartz; David L. Blaney and Mustapha Kamal Pasha, "Civil Society and Democracy in the Third World: Ambiguities and Historical Possibilities," *Studies in Comparative International Development* 28, no. 1 (1993): 3–24; John M. Bryson, Barbara C. Crosby, and Melissa Middleton Stone, "The Design and Implementation of Cross-sector Collaborations: Proportions from the Literature," in "Collaborative Public Management," special issue, *Public Administration Review* 66 (2006): 44–55; David Horton Smith, *Grassroots Associations* (Thousand Oaks, CA: Sage, 2000); Robert Wuthnow, ed., *Between States and Markets: The Voluntary Sector in Comparative Perspective* (Princeton, NJ: Princeton University Press, 1991); Steven Rathgeb Smith and Michael Lipsky, *Nonprofits for Hire: The Welfare State in the Age of Contracting* (Cambridge, MA: Harvard University Press, 1993).

50. Peter Ho and Edmonds.

51. Teets 2013; Teets, *Civil Society under Authoritarianism*.

52. China Development Brief.

53. "The 20-Year History of IGES," Institute for Global Environmental Strategies, accessed March 6, 2019, https://www.iges.or.jp/20th/en/index.html.

54. Peter Ho and Edmonds.

55. IGES-YCU Joint Seminar on Low-Carbon and Smart Cities: Seeking Local Energy Solutions after the Nuclear Crisis Report, https://www.iges.or.jp/en/publication_documents/pub/conferenceproceedings/en/2594/IGES-YCUJointSeminar_Low-CarbonSmartCities.pdf (accessed March 10, 2019).

56. See "Strengthening Japan's Environmental Cooperation Strategy as a Leader to Promote Green Markets in East Asia," https://www.iges.or.jp/en/publication_documents/pub/policyreport/en/2301/report+on+green+market+promotionfrom+cai+s1+study_final+version++5+july+2011.pdf (accessed March 10, 2019).

57 IGES 2017 Annual Report, https://www.iges.or.jp/en/publication_documents/pub/annual/en/6627/annual2017_e.pdf (accessed July 28, 2020), p. 3.

58. "About Us," Citizens' Coalition for Economic Justice, accessed June 10, 2019, http://ccej.or.kr/eng/who-we-are/about-us/.

59. Choi; Sunhyuk Kim, "Civic Engagement"; Ku 2011; See-Jae Lee, "Environmental Movement."

60. Myung-Rae Cho, "Emergence and Evolution of Environmental Discourses in South Korea," *Korea Journal* 44, no. 3 (2004): 138–164; Myung-Rae Cho, "The Politics of Urban Nature Restoration: The Case of Cheonggyecheon Restoration in Seoul, Korea," *International Development Planning Review* 32, no. 2 (2010): 145–165; David Von Hippel, Sun-Jin Yun, and Myung-Rae Cho, "The Current Status of Green Growth in Korea: Energy and Urban Security," *Asia-Pacific Journal* 9, no. 44 (2011): article 4, https://apjjf.org/-Myung-Rae-Cho--Sun-Jin-YUN--David-von-Hippel/3628/article.pdf.

61. Lee In-Keun, "Cheong Gye Cheon Restoration Project: A Revolution in Seoul" (presentation to ICLEI, 2006), https://seoulsolution.kr/sites/default/files/policy/%5BEN%5DCheong%20Gye%20Cheon%20Restoration%20Project.pdf.

62. "Case Study: Cheonggyecheon; Seoul, Korea," Global Designing Cities Initiative, accessed June 12, 2019, https://globaldesigningcities.org/publication/global-street-design-guide/streets/special-conditions/elevated-structure-removal/case-study-cheonggyecheon-seoul-korea/.

63. "Revitalizing a City by Reviving a Stream," Development Asia, accessed June 12, 2019, https://development.asia/case-study/revitalizing-city-reviving-stream.

64. Sujoyini Mandal, "Seoul Goes Local in Development," World Bank Blogs, November 18, 2013, https://blogs.worldbank.org/sustainablecities/seoul-goes-local-development.

65. Cho, "Politics of Urban Nature Restoration."

66. Shim Myung Pil, "Human-Friendly River Restoration and Management in Korea: The Four Rivers Restoration Project" (presentation at the Japan Society of Civil Engineers 2014 International Forum, November 20, 2014), http://jsce100.com /international_conf/pdf/forum08.pdf.

67. Editorial, "Four-River Project," *Korea Times*, July 5, 2018, http://www.koreatimes .co.kr/www/opinion/2018/07/202_251780.html.

68. Feiock et al.; Michael Kraft, Mark Stephan, and Troy Abel, *Coming Clean: Information Disclosure and Environmental Performance* (Cambridge, MA: MIT Press, 2011); Miller; Xueyong Zhan, Carlos Wing-Hung Lo, and Shui-Yan Tang, "Contextual Changes and Environmental Policy Implementation: A Longitudinal Study of Street-Level Bureaucrats in Guangzhou, China," *Journal of Public Administration Research and Theory* 24, no. 4 (2014): 1005–1035; Teets, "Let Many Civil Societies Bloom"; Teets, *Civil Society under Authoritarianism*.

69. "2018 Research Report—Fair Winds Charter: How Civic Exchange Influenced Policymaking to Reduce Ship Emissions in Hong Kong 2006–2015," Civic Exchange, September 19, 2018, https://civic-exchange.org/report/fair-winds-charter-2018/.

70. Simon Ng, *Fair Winds Charter: How Civic Exchange Influenced Policymaking to Reduce Ship Emissions in Hong Kong 2006–2015* (Hong Kong: Civic Exchange, 2018), 23.

71. Kai-Hon Lau et al., *Significant Marine Sources for SO_2 Levels in Hong Kong* (Hong Kong: Civic Exchange, 2005).

72. Ng, 16; Yiqi Zhang et al., 52.

73. Ng, 22.

74. Agrawal and Lemos; Bouteligier; Peter Ho; Morgen Johansen and Kelly LeRoux, "Managerial Networking in Nonprofit Organizations: The Impact of Networking on Organizational and Advocacy Effectiveness," *Public Administration Review* 73, no. 2 (2013): 355–363; John; Pekkanen, Smith, and Tsujinaka; Benjamin Van Rooij et al., "From Support to Pressure: The Dynamics of Social and Governmental Influences on Environmental Law Enforcement in Guangzhou City, China," *Regulation and Governance* 7, no. 3 (2013): 321–347.

75. In 2017, with the accomplishment of its original mission, the China-US Energy Efficiency Alliance transitioned into the China-US Energy Innovation Alliance.

76. Bo Shen, Barbara Finamore, and Mona Yew, "Promoting Energy Efficiency as a Cost-Effective Resource in China: A Review of Jiangsu's Efficiency Power Plant Pilot," *ACEEE Summer Study on Energy Efficiency in Industry* 4 (2009): 114–125.

77. Kingdon.

Chapter 6

1. Hildebrandt and Turner; Kostka and Mol; Mertha, *China's Water Warrior*; Teets, *Civil Society under Authoritarianism*; Xie; Choi; Jennifer S. Oh; Pekkanen 2006.

2. The committee's name is sometimes translated as "Citizens' Committee for Green Seoul" and sometimes as "Citizens' Green Seoul Committee," depending on the source. I will use the latter throughout, with the exception that when citing an interview with one of the committee's leaders, I will use the name as listed on the person's business card. They both refer to the same group.

3. "Seoul Metropolitan Government Ordinance on Establishment and Operation of Citizens' Green Seoul Committee," comparison of Partial Amendment No. 6744, January 4, 2018, and Amendment of Other Laws No. 6851, March 22, 2018, Seoul Legal Administration Services, accessed July 5, 2019, https://legal.seoul.go.kr/legal /english/front/page/law.html?pAct=lawComparison&pPromNo=3409.

4. United Nations Sustainable Development, *Agenda 21: United Nations Conference on Environment & Development, Rio de Janerio [sic], Brazil, 3 to 14 June 1992*, accessed July 5, 2019, https://sustainabledevelopment.un.org/content/documents/Agenda21.pdf.

5. Interview with Seong Hwan Min, secretary general of the Citizens' Committee for Green Seoul, Seoul, June 20, 2019; Sunhyuk Kim, "Civic Engagement"; See-Jae Lee, "Environmental Movement."

6. "Seoul Metropolitan Government Ordinance on Establishment and Operation of Citizens' Green Seoul Committee," comparison of Amendments of Other Laws No. 6851, March 22, 2018, and Partial Amendment No. 6910, October 4, 2018, Seoul Legal Administration Services, accessed July 7, 2019, https://legal.seoul.go.kr/legal /english/front/page/law.html?pAct=lawComparison&pPromNo=3706.

7. Won-Ju Kim, "Changes in Park & Green Space Policies in Seoul," Seoul Solution, June 25, 2015, last updated May 8, 2018, https://seoulsolution.kr/en/content /3497.

8. Interview with Seong Hwan Min, secretary general of the Citizens' Committee for Green Seoul, Seoul, June 20, 2019; Won-Ju Kim, "Changes."

9. Jong Youl Lee, "Theory and Application of Urban Governance: The Case of Seoul," *Journal of Urban Technology* 10, no. 2 (2003): 81.

10. Lee.

11. Lee, 74, 75, 82.

12. Myung-Rae Cho, "A Progressive City in the Making? The Seoul Experience," in *The Rise of Progressive Cities East and West*, ed. Mike Douglass, Romain Garbaye, and K. C. Ho (Singapore: Springer, 2019), 54.

13. "Seoul Population," World Population Review, accessed July 8, 2019, http://worldpopulationreview.com/world-cities/seoul-population/.

14. Seoul Metropolitan Government, *Reframing Urban Energy Policy: Challenges and Opportunities in the City Seoul* (Seoul: Seoul Metropolitan Government, August 2017), https://www.ieac.info/IMG/pdf/2017smg-olnpp-book-lr-c.pdf.

15. "South Korea Steps Up Shift to Cleaner Energy, Sets Long-Term Renewable Power Targets," Reuters, April 18, 2019, https://www.reuters.com/article/us-southkorea-energy/south-korea-steps-up-shift-to-cleaner-energy-sets-long-term-renewable-power-targets-idUSKCN1RV06P.

16. Arcadis, *Citizen Centric Cities: The Sustainable Cities Index 2018* (Arcadis, 2018), https://www.arcadis.com/media/1/D/5/%7B1D5AE7E2-A348-4B6E-B1D7-6D94FA7D7567%7DSustainable_Cities_Index_2018_Arcadis.pdf.

17. Daniel P. Aldrich, *Site Fights: Divisive Facilities and Civil Society in Japan and the West* (Ithaca, NY: Cornell University Press, 2008); Groothuis and Miller; Poulos; Rabe.

18. Hager and Haddad.

19. McGurty; Dalton, Recchia, and Rohrschneider; George; Reardon-Anderson.

20. Ming-sho Ho, "Resisting Naphtha Crackers: A Historical Survey of Environmental Politics in Taiwan," *China Perspectives* 2014, no. 3 (September 2014): 5.

21. Ho, 7.

22. Reardon-Anderson.

23. "Taiwan Real GDP Growth," CEIC, accessed April 17, 2019, https://www.ceicdata.com/en/indicator/taiwan/real-gdp-growth.

24. Formosa Plastics, *Formosa Plastics Group: Introduction* (Taipei City: Formosa Plastics, 2018), https://www.fpg.com.tw/uploads/images/media-center/ebook-top/FPG%20Introduction2018_en.pdf.

25. Ming-sho Ho, "Resisting Naphtha Crackers," 8.

26. Tang and Tang, 291; Interview with Echo Lin and Millie Lee in Taipei, November 5, 2010.

27. Ming-sho Ho, "Resisting Naphtha Crackers," 8.

28. Ho, 7–8; Robert Paul Weller and Hsin-Huang Michael Hsiao, "Culture, Gender and Community in Taiwan's Environmental Movement," in *Environmental Movements in Asia*, ed. Arne Kalland and Gerard Persoon (Richmond, UK: Curzon, 1998), 94.

29. Hager and Haddad; Rabe.

30. For a detailed discussion of how companies and governments in Japan and France targeted specific communities for projects that often generate NIMBY

protests and develop location-specific packages to woo residents, see Aldrich, *Site Fights*.

31. Ming-sho Ho, "Resisting Naphtha Crackers."

32. https://www.fpg.com.tw/uploads/images/media-center/ebook-top/FPG%20 Introduction2018_en.pdf (pp. 14–22; accessed April 17, 2019).

33. Ming-sho Ho, "Resisting Naphtha Crackers," 9–10; Haddad, "From Backyard Environmental Advocacy."

34. Ming-sho Ho, "Resisting Naphtha Crackers," 10.

35. Grano, 93.

36. Interview with Tu Wen-ling, Taipei, March, 2019.

37. Ming-sho Ho, "Resisting Naphtha Crackers," 13–14.

38. Grano, 102–105; Ming-sho Ho, "Resisting Naphtha Crackers," 11.

39. Brandon Gaille, "20 Taiwan Chemical Industry Statistics and Trends," BrandonGaille.com, February 8, 2019, https://brandongaille.com/20-taiwan-chemical-industry-statistics-and-trends/.

40. Ming-sho Ho, "Resisting Naphtha Crackers"; Fell; Grano, chap. 3.

41. Interview with Yong He, Green Earth Volunteers, Beijing, 2015.

42. "'26-Degree Campaign' Saves Energy in Beijing," *China Daily*, June 27, 2004, http://www.chinadaily.com.cn/english/doc/2004-06/27/content_343184.htm; "26°C Campaign Launched in China," World Wildlife Fund, June 26, 2004, http://wwf.panda.org/?13951/26C-campaign-launched-in-China/.

43. Liming Qiao and Peng Wang, "The 26 Degree Campaign: Saving Energy," in *The China Environment Yearbook (2005)*, ed. Congjie Liang and Dongping Yang (London: Brill, 2005).

44. Interviews, Beijing, 2015.

45. "Beijing 'Brown-Out' to Save Power," BBC News, July 22, 2004, http://news.bbc.co.uk/2/hi/asia-pacific/3916789.stm.

46. Qiao and Wang, 334.

47. Qiao and Wang, 335–336.

48. Qiao and Wang, 336.

49. Interview with Friends of Nature staff, Beijing, May 2015.

50. Institute for Global Environmental Strategies, Charter for the Establishment of the Institute for Global Environmental Strategies, accessed April 1, 2019, p. 1, https://www.iges.or.jp/files/about/PDF/charter-e.pdf.

51. Institute for Global Environmental Strategies, 1.

52. Institute for Global Environmental Strategies, *Towards a Sustainable Asia and the Pacific: Report of Eco Asia Long-Term Perspective Project Phase II* (Hayama, Kanagawa: Institute for Global Environmental Strategies, 2001).

53. Atsushi Terazono et al., "Waste Management and Recycling in Asia," *International Review of Environmental Strategies* 5, no. 2 (2005): 477–498.

54. Simon Gilby et al., *Planning and Implementation of Integrated Solid Waste Management Strategies at Local Level: The Case of Surabaya City* (Osaka: UN Environment; Hayama, Kanagawa: Institute for Global Environmental Strategies, 2017), 5.

55. D. G. J. Premakumara, "Kitakyushu City's International Cooperation for Organic Waste Management in Surabaya City, Indonesia and Its Replication in Asian Cities" (discussion paper, Kitakyushu Urban Centre, Institute for Global Environmental Strategies, Kitakyushu City, Japan, 2012), 2.

56. Institute for Global Environmental Strategies, Waste Reduction Model of Surabaya City (Institute for Global Environmental Strategies, 2009), https://kitakyushu.iges.or.jp/publication/Takakura/Surabaya_Experience_Full.pdf. See also Toshizo Maeda, "Reducing Waste through the Promotion of Composting and Active Involvement of Various Stakeholders: Replicating Surabaya's Solid Waste Management Model" (policy brief, Institute for Global Environmental Strategies, Kitakyushu, Japan, 2009), 2; "Takakura Composting Method," Japan International Cooperation Agency, accessed April 1, 2019, https://www.jica.go.jp/english/our_work/thematic_issues/management/study_takakura.html.

57. Maeda, 4.

58. Maeda, 2–3.

59. Premakumara, 5.

60. Premakumara, 8.

61. "IGES Signs Academic Partnership Agreement with PT Sarana Multi Infrastruktur (PT SMI) Indonesia," Institute for Global Environmental Strategies, February 20, 2019, https://www.iges.or.jp/en/announcement/20190220.html.

62. Premakumara, 10.

63. "KitaQ System Composting," accessed April 3, 2019, https://kitaq-compost.net/.

64. Fritz Akhmad Nuzir, *Development Model of Takakura Composting Method (TCM) as an Appropriate Environmental Technology (AET) for Urban Waste Management* (Hiroshima: IGES Kitakyushu Urban Centre, 2018), 8.

65. Kitakyushu Asian Center for Low Carbon Society. Presentation on Low-Carbon Development of Hai Phong City, http://asiangreencamp.net/eng/pdf/68.pdf (accessed July 30, 2020); Nguyen Van Thanh, "Haiphong City—Vietnam, Becoming Green Port

City" (presentation, November 2017), https://lcs-rnet.org/pdf/loCARNet_6th_presen tations/P3_4_Do_Quang_Hung-Bangkok_the_6th_LoCARNet_3_11_2017.pdf.

66. IGES, "Collaboration between Asia and the City of Kitakyushu" pp. 44-53. https:// www.iges.or.jp/en/publication_documents/pub/researchreport/en/6655/Actions _for_a_Sustainable_Society_e.pdf (accessed July 30, 2020).

67. Yatsuka Kataoka et al., *Actions for a Sustainable Society: Collaboration between Asia and the City of Kitakyushu* (Kitakyushu City: Institute for Global Environmental Strategies, 2018), 46–53; Trung Hieu Nguyen, "Low-Carbon City Development in Hai Phong and Projects in Collaboration with Kitakyushu City" (presentation, Tokyo, January 23, 2017), http://www.iges.or.jp/files/research/sustainable-city/PDF/20170123 /08_Hai_Phong.pdf; City of Kitakyushu and City of Hai Phong, *Green Growth Promotion Plan in the City of Hai Phong* (May 2015), http://www.asiangreencamp.net/pdf /green_en.pdf.

Chapter 7

1. McGurty; Poulos and Haddad; David Schlosberg, *Defining Environmental Justice: Theories, Movements, and Nature* (Oxford: Oxford University Press, 2009).

2. Neal D. Woods, "Interstate Competition and Environmental Regulation: A Test of the Race-to-the-Bottom Thesis," *Social Science Quarterly* 87, no. 1 (2006): 174–189; Wang et al.; Detomasi.

3. Peter Dauvergne, *Shadows in the Forest: Japan and the Politics of Timber in Southeast Asia* (Cambridge, MA: MIT Press, 1997); Jennifer Farley Gordon and Colleen Hill, *Sustainable Fashion: Past, Present and Future* (New York: Bloomsbury Academic, 2015); International Institute of Environment and Development and World Business Council for Sustainable Development; Lora-Wainwright.

4. David M. Konisky, "Regulatory Competition and Environmental Enforcement: Is There a Race to the Bottom?," *American Journal of Political Science* 51, no. 4 (2007): 853–872; Aseem Prakash and Matthew Potoski, "Racing to the Bottom? Trade, Environmental Governance, and ISO 14001," *American Journal of Political Science* 50, no. 2 (2006): 350–364.

5. Eri Saikawa, "Policy Diffusion of Emission Standards Is There a Race to the Top?," *World Politics* 65, no. 1 (2013): 1–33; Baomin Dong, Jiong Gong, and Xin Zhao, "FDI and Environmental Regulation: Pollution Haven or a Race to the Top?," *Journal of Regulatory Economics* 41, no. 2 (2012): 216–237; Kent E. Portney, *Taking Sustainable Cities Seriously: Economic Development, the Environment, and Quality of Life in American Cities* (Cambridge, MA: MIT Press, 2013).

6. Christina L. Anderson and Rebecca L. Bieniaszewska, "The Role of Corporate Social Responsibility in an Oil Company's Expansion into New Territories," *Corporate Social Responsibility and Environmental Management* 12, no. 1 (2005): 1–9; Christine

Bader, *The Evolution of a Corporate Idealist: When Girl Meets Oil* (New York: Biblio-motion, 2014); Ng; Ray Cheung, "Making Green from Green—How Improving the Environmental Performance of Supply Chains Can Be a Win-Win for China and the World" (brief, China Environment Forum, 2011); Tan.

7. Suzanne Berger and Ronald Dore, eds., *National Diversity and Global Capitalism* (Ithaca, NY: Cornell University Press, 1996); Stephan Haggard, *Pathways from the Periphery: The Politics of Growth in the Newly Industrializing Countries* (Ithaca, NY: Cornell University Press, 1990); Walter Hatch and Kozo Yamamura, *Asia in Japan's Embrace: Building a Regional Production Alliance* (New York: Cambridge University Press, 1996).

8. Marc Epstein and Marie-Josée Roy, "Managing Corporate Environmental Perfor-mance: A Multinational Perspective," *European Management Journal* 16, no. 3 (1998): 284–296; Busch, Jörgens, and Tews; Orly Lobel, "Sustainable Capitalism or Ethical Transnationalism: Offshore Production and Economic Development," *Journal of Asian Economics* 17, no. 1 (2006): 56–62.

9. Peter Dauvergne and Jane Lister, "Big Brand Sustainability: Governance Prospects and Environmental Limits," *Global Environmental Change* 22, no. 1 (2012): 36–45; Mary Alice Haddad, "Increasing Environmental Performance in a Context of Low Governmental Enforcement: Evidence from China," *Journal of Environment and Devel-opment* 24, no. 1 (March 2015): 3–25; Institute of Public and Environmental Affairs and Natural Resources Defense Council, *Greening the Global Supply Chain*, CITI Index and Annual Report (Beijing: Institute of Public and Environmental Affairs and Natural Resources Defense Council, 2014); International Institute of Environment and Development and World Business Council for Sustainable Development; Bin Jiang, "Implementing Supplier Codes of Conduct in Global Supply Chains: Process Explanations from Theoretic and Empirical Perspectives," *Journal of Business Ethics* 85 (2009): 77–92; Keck and Sikkink; Kraft, Stephan, and Abel; Lobel; Amanda Murdie and Johannes Urpelainen, "Why Pick on Us? Environmental INGOs and State Sham-ing as a Strategic Substitute," *Political Studies* 63, no. 2 (2015): 353–372.

10. Shojiro Tokunaga, "Japan's FDI-Promoting Systems and Intra-Asian Networks: New Investment and Trade Systems Created by the Borderless Economy," in *Japan's Foreign Investment and Asian Economic Interdependence: Production, Trade, and Finan-cial Systems*, ed. Shojiro Tokunaga (Tokyo: University of Tokyo Press, 1992); Steven K. Vogel, *Freer Markets, More Rules: Regulatory Reform in Advanced Industrial Countries* (Ithaca, NY: Cornell University Press, 1996).

11. "Our Products," Shell, accessed July 1, 2019, https://www.shell.com.hk/en_hk /business-customers/commercial-fuels/our-products.html; Gammon Construction, "Press Release: The first contractor in Hong Kong to use B5 biodiesel on plant & equipment." https://gammonconstruction.com/en/html/press/press-bbf29e5aa7b44 163a9f4755a7ec59d71.html (accessed July 31, 2020).

12. "As Others Struggle, New Hong Kong Biodiesel Plant Powers Ahead," *Wall Street Journal*, October 4, 2013, https://blogs.wsj.com/chinarealtime/2013/10/04/as-others -struggle-new-hong-kong-biodiesel-plant-powers-ahead/.

13. "Shell Makes Early Move on Biodiesel Launch in Hong Kong," *South China Morning Post*, accessed July 1, 2019, https://www.scmp.com/business/companies/article /1652517/shell-makes-early-move-biodiesel-launch-hong-kong.

14. Mark Sharp, "Biodiesel: The Fuel of the Future?," *South China Morning Post*, March 29, 2013, https://www.scmp.com/lifestyle/technology/article/1201931/biodiesel-fuel -future.

15. "Shell, Maxim's Group Partner in Hong Kong Biodiesel Pilot Program," *Biodiesel Magazine*, March 12, 2019, http://www.biodieselmagazine.com/articles/2516611 /shell-maximundefineds-group-partner-in-hong-kong-biodiesel-pilot-program.

16. Ronald Dore, *British Factory, Japanese Factory: The Origins of National Diversity in Industrial Relations* (Berkeley: University of California Press, 1973); Berger and Dore; Xin and Pearce.

17. "Business in the Blood: Family Firms," *Economist*, November 1, 2014; Yin-hua Yeh, Tsun-siou Lee, and Tracie Woidtke, "Family Control and Corporate Governance: Evidence from Taiwan," *International Review of Finance* 2, no. 1–2 (2001): 21–48.

18. Dore; Daniel I. Okimoto, *Between MITI and the Market: Japanese Industrial Policy for High Technology* (Stanford, CA: Stanford University Press, 1989); Toshimitsu Shinkawa and T. J. Pempel, "Occupational Welfare and the Japanese Experience," in *The Privatization of Social Policy? Occupational Welfare and the Welfare State in America, Scandinavia and Japan*, ed. Michael Shalev (New York: Routledge, 1996); Margarita Estévez-Abe, *Welfare and Capitalism in Postwar Japan* (New York: Cambridge University Press, 2008).

19. Sangjin Yoo and Sang M. Lee, "Management Style and Practice of Korean Chaebols," *California Management Review* 29, no. 4 (1987): 95–110; Greg J. Bamber and Chris J. Leggett, "Changing Employment Relations in the Asia-Pacific Region," *International Journal of Manpower* 22, no. 4 (2001): 300–317; Roger Goodman and Ito Peng, "The East Asian Welfare States: Peripatetic Learning, Adaptive Change, and Nation-Building," in *Welfare States in Transition: National Adaptations in Global Economies*, ed. Gosta Esping-Andersen (London: Sage, 1996).

20. John Hassard et al., "China's State-Owned Enterprises: Economic Reform and Organizational Restructuring," *Journal of Organizational Change Management* 23, no. 5 (2010): 500–516; Varouj A. Aivazian, Ying Ge, and Jiaping Qiu, "Can Corporatization Improve the Performance of State-Owned Enterprises Even without Privatization?," *Journal of Corporate Finance* 11, no. 5 (2005): 791–808.

21. Takeo Hoshi and Anil Kashyap, *Corporate Financing and Governance in Japan: The Road to the Future* (Cambridge, MA: MIT Press, 2004); Takeo Hoshi, Anil Kashyap, and

David Scharfstein, "Corporate Structure, Liquidity, and Investment: Evidence from Japanese Industrial Groups," *Quarterly Journal of Economics* 106, no. 1 (1991): 33–60.

22. Hoshi and Kashyap; Won Yong Oh, Young Kyun Chang, and Aleksey Martynov, "The Effect of Ownership Structure on Corporate Social Responsibility: Empirical Evidence from Korea," *Journal of Business Ethics* 104, no. 2 (2011): 283–297; Stephan Haggard, "The Political Economy of Regionalism in Asia and the Americas," in *The Political Economy of Regionalism*, ed. Edward D. Mansfield and Helen V. Milner (New York: Columbia University Press, 1996); Gregory W. Noble and John Ravenhill, *The Asian Financial Crisis and the Architecture of Global Finance* (New York: Cambridge University Press, 2000).

23. Eri Nakamura, "Does Environmental Investment Really Contribute to Firm Performance? An Empirical Analysis Using Japanese Firms," *Eurasian Business Review* 1, no. 2 (2011): 91–111; Chin-Huang Lin, Ho-Li Yang, and Dian-Yan Liou, "The Impact of Corporate Social Responsibility on Financial Performance: Evidence from Business in Taiwan," *Technology in Society* 31, no. 1 (2009): 56–63.

24. The 2004 figures estimate that 80–90 percent of the products in US stores are made in China. Sam Hornblower, "Wal-Mart & China: A Joint Venture," Frontline, November 23, 2004, https://www.pbs.org/wgbh/pages/frontline/shows/walmart/secrets/wmchina.html.

25. "Walmart China Factsheet," Walmart, accessed June 15, 2019, http://www.wal-martchina.com/english/walmart/index.htm.

26. James Brian Quinn and Frederick G. Hilmer, "Strategic Outsourcing," *MIT Sloan Management Review* 35, no. 4 (1994): 43.

27. Lee Scott's 2005 remarks: https://corporate.walmart.com/_news_/executive-viewpoints/twenty-first-century-leadership (accessed July 31, 2020).

28. Debbie Kalish et al., "Integrating Sustainability into New Product Development," *Research-Technology Management* 61, no. 2 (2018): 37–46.

29. Walmart, *2018 Global Responsibility Report* (Bentonville, AR: Walmart, 2019), 65, 59.

30. Walmart, 28.

31. "China Sustainability Summit: Fact Sheet," Wal-Mart, accessed May 2, 2019, https://cdn.corporate.walmart.com/47/07/368e013b430fb2b978caab7fccbe/china-2008-sustainability-summit-fact-sheet_129945454252825155.pdf.

32. "Environmental Certification Center of Ministry of Environmental Protection China and Walmart Sign MOU to Increase Cooperation in Environmental Sustainability," Walmart, 2008, https://corporate.walmart.com/_news_/news-archive/2008/12/08/environmental-certification-center-of-ministry-of-environmental-protection-china-walmart-sign-mou-to-increase-cooperation-in-environmental-sustainability.

33. "Environmental Certification Center."

34. Cheung; Jun Ma et al., "Greening Supply Chains in China: Practical Lessons from China-Based Suppliers in Achieving Environmental Performance" (working paper, World Resources Institute, Washington, DC, 2010).

35. Institute of Public and Environmental Affairs and Natural Resources Defense Council, 25.

36. "Walmart Continues to Strengthen Global Supply Chain Sustainability; Announces New Commitment to Advance Factory Energy Efficiency in China," Walmart, August 27, 2014, https://corporate.walmart.com/_news_/news-archive/2014/08/27/walmart -continues-to-strengthen-global-supply-chain-sustainability-announces-new -commitment-to-advance-factory-energy-efficiency-in-china.

37. "Walmart Announces New Commitments to Drive Sustainability Deeper into the Company's Global Supply Chain," Walmart, October 25, 2012, https://corporate.walmart .com/_news_/news-archive/2012/10/25/walmart-announces-new-commitments-to-drive -sustainability-deeper-into-the-companys-global-supply-chain.

38. "Walmart Commits to Reduce Emissions by 50 Million Metric Tons in China," Walmart, March 29, 2018, https://news.walmart.com/2018/03/29/walmart-commits -to-reduce-emissions-by-50-million-metric-tons-in-china.

39. Walmart Sustainability Hub, accessed June 30, 2019, https://www.walmartsustain abilityhub.com/.

40. Echo Huang, "Ten Years after China's Infant Milk Tragedy, Parents Still Won't Trust Their Babies to Local Formula," Quarts, July 15, 2018, https://qz.com/1323 471/ten-years-after-chinas-melamine-laced-infant-milk-tragedy-deep-distrust-re mains/.

41. "China's Persistent Food and Drug Safety Problem," Medical Xpress, July 24, 2018, https://medicalxpress.com/news/2018-07-china-persistent-food-drug-safety.html.

42. Walmart, 54.

43. Walmart, 218–221.

44. Jennifer Chait, "Largest Organic Retailers in North America," Balance, November 20, 2019, https://www.thebalancesmb.com/organic-retailers-in-north-america -2011-2538129.

45. "Walmart Upbeat after China Success," Inquirer, December 4, 2018, https:// business.inquirer.net/261643/walmart-upbeat-after-china-success.

46. Melanie Lee, "Wal-Mart's Pork Scandal Highlights Struggles in China," Reuters, October 14, 2011, https://www.reuters.com/article/us-walmart-china/wal-marts-pork -scandal-highlights-struggles-in-china-idUSTRE79D1S020111014.

47. Adam Jourdan, "Wal-Mart Recalls Donkey Product in China after Fox Meat Scandal," Reuters, January 1, 2014, https://www.reuters.com/article/us-walmart-china/wal-mart -recalls-donkey-product-in-china-after-fox-meat-scandal-idUSBREA0103O20140102.

48. Tom Schoenberg, "Walmart Deadlocked with U.S. Over Bribery Probe," Bloomberg, August 2, 2018, https://www.bloomberg.com/news/articles/2018-08-02/walmart -is-said-to-be-deadlocked-with-u-s-over-bribery-probe.

49. Andrew Spicer and David Graham Hyatt, "Walmart Tried to Make Sustainability Affordable. Here's What Happened," Conversation, August 13, 2018, https:// theconversation.com/walmart-tried-to-make-sustainability-affordable-heres-what -happened-76771.

50. Siniawer.

51. Yasuhiro Monden, *Toyota Production System: An Integrated Approach to Just-in-Time* (Portland, OR: Productivity, 2011).

52. Berger and Dore; William E. Bryant, *Japanese Private Economic Diplomacy: An Analysis of Business-Government Linkages* (New York: Praeger, 1975); Dore; Robert Wade, *Governing the Market: Economic Theory and the Role of Government in East Asian Industrialization* (Princeton, NJ: Princeton University Press, 1990); Richard J. Samuels, *"Rich Nation, Strong Army": National Security and the Technological Transformation of Japan* (Ithaca, NY: Cornell University Press, 1994).

53. "Development History of Japanese Automobile Industry," Cars from Japan, last updated November 22, 2017, https://carfromjapan.com/article/industry-knowledge /development-history-japanese-automotive-industry/.

54. "Top Vehicle Manufacturers in the US Market, 1961–2016," Knoema, accessed May 6, 2019, https://knoema.com/infographics/floslle/top-vehicle-manufacturers-in -the-us-market-1961-2016.

55. "Toyota Introduces Global Environment Charter," Toyota, January 16, 1992, https://global.toyota/en/detail/7839331.

56. "ISO 14000 Family: Environmental Management," ISO, accessed May 6, 2019, https://www.iso.org/iso-14001-environmental-management.html.

57. "Chronology of Environmental Initiatives by Field," Toyota, accessed May 6, 2019, https://www.toyota-global.com/company/history_of_toyota/75years/data/comp any_information/social_contribution/environmental/chronology_of_environmen tal_initiatives.html.

58. Toyota Corporation, *Environmental Report 2018* (Toyota Corporation, 2018), 7.

59. Toyota Corporation, 15–18.

60. Hiroshi Ito and Nobuo Kawazoe, "A Review of Toyota City's Eco-policy: Changes in Citizens' Awareness between 2012 and 2015," *Urban Research and Practice* 11, no. 1 (2018): 19–36.

61. "Global Automotive Market Share in 2019, by Brand," Statista, accessed May 6, 2019, https://www.statista.com/statistics/316786/global-market-share-of-the-leading -automakers/.

62. Toyota Corporation, 5.

63. Lídia Simão and Ana Lisboa, "Green Marketing and Green Brand—the Toyota Case," *Procedia Manufacturing* 12 (2017): 183–194.

64. OECD (Organisation for Economic Co-operation and Development), *Environmental Labelling and Information Schemes* (OECD, 2016), 3.

65. OECD, 7.

66. Prakash and Potoski, *Voluntary Environmentalists*; Auld and Gulbrandsen; Tomasz Kijek, "Modelling of Eco-innovation Diffusion: The EU Eco-label," *Comparative Economic Research* 18, no. 1 (2015): 65–79; Seo-Hyeon Min, Seul-Ye Lim, and Seung-Hoon Yoo, "Consumers' Willingness to Pay a Premium for Eco-labeled LED TVs in Korea: A Contingent Valuation Study," *Sustainability* 9, no. 5 (2017): 814–825.

67. Frieder Rubik, Dirk Scheer, and Fabio Iraldo, "Eco-labelling and Product Development: Potentials and Experiences," *International Journal of Product Development* 6, no. 3–4 (2008): 409; Frieder Rubik and Paolo Frankl, *The Future of Eco-labelling: Making Environmental Product Information Systems Effective* (London: Routledge, 2017), 245–247.

68. Vangelis Vitalis, "Private Voluntary Eco-labels: Trade Distorting, Discriminatory and Environmentally Disappointing." Background paper for the Round Table on Sustainable Development meeting "Eco-labelling and Sustainable Development," Organisation for Economic Co-operation and Development Headquarters, Paris, December 6, 2002, https://www.oecd.org/sd-roundtable/papersandpublications/3936 2947.pdf (accessed July 31, 2020).

69. Asian Institute of Technology, *Regional Collaboration on Ecolabelling—Asia Pacific* (United Nations Environment Programme and European Union, 2016), 24.

70. OECD; Patrik Sörqvist et al., "The Green Halo: Mechanisms and Limits of the Eco-label Effect," *Food Quality and Preference* 43 (2015): 1–9; Vitalis; Rubik, Scheer, and Iraldo.

71. OECD; Rubik, Scheer, and Iraldo.

72. Nannette Lindenberg, "Definition of Green Finance," (2014); "Green Financing," UN Environment Programme, accessed May 9, 2019, https://www.unenvironment .org/regions/asia-and-pacific/regional-initiatives/supporting-resource-efficiency /green-financing.

73. United Nations Environment, *Sustainable Finance Progress Report* (Geneva: United Nations, 2019), 19.

74. Green Climate Fund, *GCF in Brief: About the Fund* (May 1, 2018), https://www .greenclimate.fund/sites/default/files/document/gcf-brief-about-fund_0.pdf.

75. "Project Portfolio," Green Climate Fund, accessed May 9, 2019, https://www
.greenclimate.fund/what-we-do/portfolio-dashboard.

76. "National Environmental Funds," Convention on Biological Diversity, May 3,
2017, https://www.cbd.int/financial/0006.shtml.

77. "Japan Green Fund for Renewable Energy Established," Japan for Sustainability,
January 4, 2003, https://www.japanfs.org/en/news/archives/news_id025143.html.

78. U.S.-China Green Fund homepage, accessed May 9, 2019, http://www.uschin
agreenfund.com/home/index/index.html.

79. George Kell, "The Remarkable Rise of ESG," *Forbes*, July 11, 2018, https://www
.forbes.com/sites/georgkell/2018/07/11/the-remarkable-rise-of-esg/#2b9a7986
1695.

80. Julie Ayling and Neil Gunningham, "Non-state Governance and Climate Policy:
The Fossil Fuel Divestment Movement," *Climate Policy* 17, no. 2 (2017): 131–149.

81. Task Force on Climate-Related Financial Disclosures homepage, accessed June
30, 2019, https://www.fsb-tcfd.org/.

82. See, for example, Hong Kong Exchanges and Clearing Limited, *Review of the Environmental, Social and Governance Reporting Guide and Related Listing Rules* (Hong Kong:
Hong Kong Exchanges and Clearing Limited, May 2019), https://www.hkex.com.hk
/-/media/HKEX-Market/News/Market-Consultations/2016-Present/May-2019-Review
-of-ESG-Guide/Consultation-Paper/cp201905.pdf; New York Stock Exchange, Nasdaq,
"Introducing Nasdaq's ESG Reporting Guide 2.0." https://www.nasdaq.com/articles
/introducing-nasdaqs-esg-reporting-guide-2.0-2019-05-15 (accessed 07/31/2020), and
London Stock Exchange Group, *Revealing the Full Picture: Your Guide to ESG Reporting*
(London: London Stock Exchange Group, January 2018), https://www.lseg.com/sites
/default/files/content/images/Green_Finance/ESG/2018/February/LSEG_ESG_report
_January_2018.pdf.

83. Forum for Sustainable and Responsible Investment, 2018 Report on US Sustainable, Responsible and Impact Investing Trends (Forum for Sustainable and
Responsible Investment, 2018), https://www.ussif.org/files/2018%20_Trends_OneP-
ager_Overview(3).pdf.

84. "10 Years of Green Bonds: Creating the Blueprint for Sustainability across Capital Markets," World Bank, March 18, 2019, https://www.worldbank.org/en/news
/immersive-story/2019/03/18/10-years-of-green-bonds-creating-the-blueprint-for
-sustainability-across-capital-markets.

85. United Nations Environment, 9.

86. Deborah Lehr, "China's Green Finance Model Shows How Saving the Planet Can
Also Be a Savvy Investment," *South China Morning Post*, February 20, 2019, https://www

.scmp.com/comment/insight-opinion/united-states/article/2186814/chinas-green
-finance-model-shows-how-saving.

87. "Global Green Bonds: 2019 Full Year Review," Asset, January 17, 2020, https://
www.theasset.com/capital-markets/39561/global-green-bonds-2019-full-year-review.

88. "Korea Climate Bond Market Overview and Opportunities," Climate Bonds
Initiative, March 19, 2018, https://www.climatebonds.net/resources/reports/korea
-climate-bond-market-overview-and-opportunities.

89. Sustainable Stock Exchanges Initiative, *2018 Report on Progress* (Sustainable Stock
Exchanges Initiative, 2019), 24.

90. Esarey et al.

91. Roland Irle, "Global BEV & PHEV Sales for 2019," EV-Volumes, accessed May 10,
2010, http://www.ev-volumes.com/country/total-world-plug-in-vehicle-volumes/.

92. International Renewable Energy Agency, *A New World: The Geopolitics of the
Energy Transformation* (International Renewable Energy Agency, 2019), 40–41.

93. Jason Deign, "8 Ways China Is Encouraging Zero-Subsidy Renewables," GTM,
January 17, 2019, https://www.greentechmedia.com/articles/read/china-zero-subsidy
-renewables.

94. Renewable Energy Institute, "Feed-in Tariffs in Japan: Five Years of Achievements" (September 2017), https://www.renewable-ei.org/en/activities/reports/img/pdf
/20170810/REI_Report_20170908_FIT5years_Web_EN.pdf (accessed July 31, 2020).

95. Institute for Sustainable Energy Policies, "Share of Renewable Energy Power in
Japan, 2018 (Preliminary Report)," April 5, 2019, https://www.isep.or.jp/en/717/
(accessed July 30, 2020).

96. "Electrified vehicles account for 40% of overall sales," Yomiuri Shimbun, Aug.
9, 2019, https://the-japan-news.com/news/article/0005929739 (accessed July 31,
2020).

97. Ju-min Park, "South Korea to Provide $3 Billion in Financial Support for Trou-
bled Auto Suppliers," Reuters, December 17, 2018, https://www.reuters.com/article
/us-southkorea-autos-suppliers-idUSKBN1OH0A3.

98. Hyungguen Park and Changhee Kim, "Do Shifts in Renewable Energy Operation
Policy Affect Efficiency: Korea's Shift from FIT to RPS and Its Results," *Sustainability*
10, no. 6 (2018): 1723–1736.

99. Solar Thermal World, "One Million Green Home Programme," https://www.solar
thermalworld.org/content/one-million-green-home-programme (accessed 07/31/2020).

100. Michael Herh, "S. Korea To Build World's Biggest Floating Solar Power Plant
on Saemangeum Lake," Business Korea (July 19, 2019). http://www.businesskorea

.co.kr/news/articleView.html?idxno=34083#:~:text=The%20South%20Korean%20
government%20will,power%20plant%20on%20Saemangeum%20Lake.

101. "Taiwan: Pursuing a New Green Energy Revolution in the East," Power Technology, November 5, 2018, https://www.power-technology.com/features/taiwan
-pursuing-new-green-energy-revolution-east/.

102. "Electric Vehicles in Taiwan," Smart Cities Dive, accessed May 10, 2019, https://
www.smartcitiesdive.com/ex/sustainablecitiescollective/electric-vehicles-taiwan
/23760/.

103. *Taiwan: Key Innovative Industry—Green Energy* (Taiwan: Department of Investment Services, Ministry of Economic Affairs, and InvesTaiwan Service Center, 2017),
https://www.roc-taiwan.org/uploads/sites/30/2018/03/Green-Energy.pdf.

104. "Taiwan to Introduce Emissions Trading System," International Carbon Action
Partnership, accessed May 10, 2019, https://icapcarbonaction.com/en/news-archive
/456-taiwan-to-introduce-national-emissions-trading-system.

105. Jin Zhen, "Current Status of Emission Trading Schemes (ETSs) in Japan, China
and South Korea" (presentation, 2017), https://www.iges.or.jp/isap/2017/files/TT3
/pdf/TT3_Zhen_Jin_rev.pdf.

106. Marcello Rossi, "Taiwan Has One of the Highest Recycling Rates in the World.
Here's How That Happened," Ensia, December 18, 2018, https://ensia.com/features
/taiwan-recycling-upcycling/.

107. State Forestry Administration of China, *Forestry in China*, (State Forestry Administration of China, 2014).

108. Jianguo Liu et al., "Ecological and Socioeconomic Effects of China's Policies
for Ecosystem Services," *Proceedings of the National Academy of Sciences* 105, no. 28
(2008): 9477–9482; Jodi S. Brandt et al., "The Relative Effectiveness of Protected
Areas, a Logging Ban, and Sacred Areas for Old-Growth Forest Protection in Southwest China," *Biological Conservation* 181 (2015): 1–8.

109. Graeme Lang and Cathy Hiu Wan Chan, "China's Impact on Forests in Southeast Asia," *Journal of Contemporary Asia* 36, no. 2 (2006): 167–194; Alicia S. T. Robbins
and Stevan Harrell, "Paradoxes and Challenges for China's Forests in the Reform
Era," *China Quarterly* 218 (2014): 381–403.

110. Asian Institute of Technology, A-6, 11.

111. Auld and Gulbrandsen; Cheung; Gordon L. Clark, Andreas Feiner, and Michael
Viehs, *From the Stockholder to the Stakeholder: How Sustainability Can Drive Financial
Outperformance* (Swiss Sustainable Finance, 2015); Björn Fasterling, "Development of
Norms through Compliance Disclosure," *Journal of Business Ethics* 106 (2012): 73–87;
Fung, Graham, and Weil; Institute of Public and Environmental Affairs and Natural

Resources Defense Council; Jiang; Kraft, Stephan, and Abel; Mitchell; Prakash and Potoski, *Voluntary Environmentalists*.

112. Agrawal and Lemos; Daniel Arenas, Josep Lozano, and Laura Albareda, "The Role of NGOs in CSR: Mutual Perceptions among Stakeholders," *Journal of Business Ethics* 88, no. 1 (August 2009): 175–197; Auld and Gulbrandsen; Haddad, "Increasing Environmental Performance"; Clodia Vurro, Angeloantonio Russo, and Francesco Perrini, "Shaping Sustainable Value Chains: Network Determinants of Supply Chain Governance Models," *Journal of Business Ethics* 90, no. 4 (2009): 607–621; Morton Winston, "NGO Strategies for Promoting Corporate Social Responsibility," *Ethics and International Affairs* 16, no. 2 (2002): 71–87; Young.

113. Murdie and Urpelainen; Ayling and Gunningham; Ewing-Chow and Soh.

114. For a good example, see "Mandarin Oriental: We're Not Fans of Your Rainforest Destruction," Forest Heroes, May 18, 2015, http://www.forestheroes.org/mandarin -oriental-were-not-fans-of-your-rainforest-destruction/.

115. Haddad, "Increasing Environmental Performance"; Friends of Nature et al., *Apple Opens Up: IT Industry Supply Chain Investigative Report* (Beijing: Institute for Public and Environmental Affairs, 2013); Agrawal and Lemos; Auld and Gulbrandsen; Kraft, Stephan, and Abel; Young.

Chapter 8

1. Edelman, p. 2.

2. Thomas Vernon Reed, *The Art of Protest: Culture and Activism from the Civil Rights Movement to the Present* (Minneapolis: University of Minnesota Press, 2019); Manabe; Jacqueline Adams, "Art in Social Movements: Shantytown Women's Protest in Pino-chet's Chile," *Sociological Forum* 17, no. 2 (2002): 21–56; Peter Weibel et al., *Global Activism: Art and Conflict in the 21st Century* (Cambridge, MA: MIT Press, 2015).

3. Erika Doss, "'Revolutionary Art Is a Tool for Liberation': Emory Douglas and Pro-test Aesthetics at *The Black Panther*," *New Political Science* 21, no. 2 (1999): 245–259; Reed; Weibel et al.; Claudia Mesch, *Art and Politics: A Small History of Art for Social Change since 1945* (London: I. B. Tauris, 2013); William Walling, "'Art' and 'Protest': Ralph Ellison's *Invisible Man* Twenty Years After," *Phylon* 34, no. 2 (1973): 120–134.

4. Anthony Downey, *Art and Politics Now* (New York: Thames and Hudson, 2014), 14.

5. Downey, 22.

6. Mesch; Reed.

7. Adams; Doss; Walling.

8. Edelman.

9. Mark A. Graham, "Art, Ecology and Art Education: Locating Art Education in a Critical Place-Based Pedagogy," *Studies in Art Education* 48, no. 4 (2007): 375–391; Barbara Bickel et al., "Richgate: Transforming Public Spaces through Community-Engaged Art," *Amerasia Journal* 33, no. 2 (2007): 115–124.

10. "Photo Exhibition: A Body in Fukushima (2014–)," Eiko & Koma, accessed August 5, 2019, http://eikoandkoma.org/abodyinfukushimaexhibit.

11. George.

12. Ben Maddow and W. Eugene Smith, *Let Truth Be the Prejudice: W. Eugene Smith, His Life and Photographs* (New York: Aperture, 1985).

13. For a detailed timeline of the lawsuits and related events, see "Timeline of Minamata Disease History (1889–2007)," Minamata Disease Municipal Museum, accessed May 29, 2019, https://minamata195651.jp/pdf/timeline_en.pdf.

14. Morris Low, "Eco-cities in Japan: Past and Future," *Journal of Urban Technology* 20, no. 1 (2013): 7–22.

15. Thomas Lennon and Ruby Yang, "The Warriors of Qiugang: A Chinese Village Fights Back," *Yale Environment 360*, January 10, 2011, https://e360.yale.edu/features/the_warriors_of_qiugang_a_chinese_village_fights_back.

16. Tree Planet Brand Book 2015, posted on Issuu February 5, 2017, pp. 95–101, https://issuu.com/treeplanet/docs/treeplanet_brandbook_print.

17. Treeplanetvideo, "[Tree Planet] sTREEt Campaign," September 28, 2014, YouTube video, 2:14, https://www.youtube.com/watch?v=FBX_uJ8jDvY.

18. Tree Planet Brand Book, 98.

19. Nosheen Iqbal, "That's Not Just a Water Bottle—It's a Status Symbol," *Guardian*, November 18, 2018, https://www.theguardian.com/media/2018/nov/18/water-bottle-status-symbol-reusable-fashion-statement.

20. Gordon and Hill.

21. Institute of Public and Environmental Affairs, Lvse Jiangnan, National Resources Defense Council, Zhaolu Environmental, and EnviroFriends, *No Excuses: Taking Full Responsibility for Pollution from Manufacturing*, Green Choice Alliance Phase 4 Textile Industry Report (December 2014), http://wwwoa.ipe.org.cn//Upload/IPE-Reports/Report-Textiles-Phase-IV-EN.pdf; https://www.eileenfisher.com/ (accessed May 20, 2019)

22. WildAid Hong Kong (@WildAidHK), "Spotted in Hong Kong: 'Shark Fin Makes Your Penis Small,'" Twitter, February 16, 2018, https://twitter.com/wildaidhk/status/964500276743782401; WildAid Hong Kong (@WildAidHK), "Spotted in Sheung Wan, Hong Kong's #shark fin trade area: 'SHARK FIN MAKES YOUR PENIS SMALL. VERY VERY SMALL,'" Twitter, December 13, 2018, https://twitter.com/WildAidHK/status/1073479398169276417;

"About Shark Fin and Penis Size, Hong Kong," EGS Photography, accessed May 21, 2019, https://www.egsphotography.com/Portfolio/Others/i-hM87BXp.

23. Christopher Jobson, "Green Pedestrian Crossing in China Creates Leaves from Footprints," Colossal, August 25, 2012, https://www.thisiscolossal.com/2012/08/green -pedestrian-crossing-in-china-creates-leaves-from-footprints/.

24. Matt Taylor, "From Luo Ta-You to Wang Leehom—A Crash Course on Environmental Activism in Taiwanese Music," Asian Pop Weekly, (March 7, 2018), https:// asianpopweekly.com/features/thinkpieces/from-luo-ta-you-to-wang-leehom-a-crash -course-on-environmental-activism-in-taiwanese-music/ (accessed Aug. 1, 2020).

25. Taylor, 2018, https://www.asianpopweekly.com/features/from-luo-ta-yu-to-wang -leehom-a-crash-course-on-environmental-activism-in-taiwanese-music (accessed May 20, 2019).

26. Manabe.

27. Weller and Hsiao, 94–95; Ming-sho Ho, "Resisting Naphtha Crackers," 8.

28. Pamela Gossin, "Animated Nature: Aesthetics, Ethics, and Empathy in Miyazaki Hayao's Ecophilosophy," *Mechademia* 10 (2015): 209–234.

29. "Kedogawa River Art Project: Moving Water Days: Backpacking Water for the Future," Ichi Ikeda Art Projects and Network, accessed May 29, 2019, http:// ikedawater.com/06Kedogawa/06KedogawaE.html.

30. "Moving Water Days 2007 for Spreading Ecological Action: Water-in-Water: Bamboo Raft with a Length of 100 Meters," Ichi Ikeda Art Projects and Network, accessed May 30, 2019, http://ikedawater.com/07Kedogawa/07Kedogawa.html.

31. "United Nations Seminar and Exhibition: Unlearning Intolarance [*sic*]: Art Changing Attitudes toward the Environment," Ichi Ikeda Art Projects and Network, accessed May 30, 2019, http://ikedawater.com/08UnitedNations.html.

32. Liam Morgan, "Art Exhibition on Pyeongchang Opens in Seoul to Help Raise Awareness of Winter Olympic and Paralympic Games," Inside the Games, December 2, 2017, https://www.insidethegames.biz/articles/1058623/art-exhibition-on-pyeongchang -opens-in-seoul-to-help-raise-awareness-of-winter-olympic-and-paralympic-games.

33. Under the Dome (with English Subtitles), https://www.youtube.com/watch ?v=V5bHb3ljjbc (accessed July 31, 2020).

34. Shuqin Cui, "Chai Jing's *Under the Dome*: A Multimedia Documentary in the Digital Age," *Journal of Chinese Cinemas* 11, no. 1 (2017): 40.

35. King, Pan, and Roberts, 326.

36. Deborah Seligsohn and Angel Hsu, "How China's 13th Five-Year Plan Addresses Energy and the Environment," China File, March 10, 2016, https://www.chinafile

.com/reporting-opinion/environment/how-chinas-13th-five-year-plan-addresses
-energy-and-environment.

37. "How the Art Industry Is Benefiting from the Pop-Up Trend: A Success Story You Can Learn From," Storefront, last updated March 22, 2019, https://www.thestorefront .com/mag/how-the-art-industry-is-benefiting-from-the-pop-up-trend/.

38. Daylily Art Circus 2011 homepage [in Japanese], accessed May 31, 2019, http:// daylily-art-circus-2011.blogspot.com/; Daylily Art Circus homepage, accessed May 31, 2019, http://www.shiro1000.jp/history/circus/circus.html.

39. Graham.

40. Richard Florida, *Cities and the Creative Class* (New York: Routledge, 2005).

41. Nancy Duxbury, Heather Campbell, and Elizabeth Keurvorst, "Developing and Revitalizing Rural Communities through Arts and Culture," Small Cities Imprint 3: 111–122; Chris Gibson, Gordon Waitt, Jim Walmsley, and John Connell, "Cultural Festivals and Economic Development in Nonmetropolitan Australia," *Journal of Planning Education and Research* 29: (2011): 280–293.

42. Bernadette Quinn, "Arts Festivals and the City," *Urban Studies* 42, no. 5–6 (2005): 927–943.

43. Phone interview with James Jack, February 27, 2017.

44. Jessica Stewart, "97-Year-Old Grandpa Saves Village by Painting Buildings with Colorful Art," My Modern Met, May 22, 2019, https://mymodernmet.com/painted -rainbow-village-taiwan/.

45. Eliot Stein, "The 96-Year-Old Painter Who Saved a Village," BBC, November 29, 2018, http://www.bbc.com/travel/gallery/20181128-the-96-year-old-painter-who-saved -a-village.

46. "96-Year-Old Taiwanese Man Saves His Village from Demolition by Turning It into a 'Rainbow Village,'" Bored Panda, 2019, https://www.boredpanda.com/rainbow -village-taichung-taiwan-huang-yung-fu/?utm_source=google&utm_medium=organic &utm_campaign=organic.

47. "The 'Rainbow Grandpa' Saving a Taiwan Village with Art," NDTV, last updated August 27, 2015, https://www.ndtv.com/offbeat/the-rainbow-grandpa-saving-a-taiwan -village-with-art-1211466.

48. The Artro, "A Review of Se Art Festival of Busan 2017." (October 27, 2017). https://www.theartro.kr:440/eng/features/features_view.asp?idx=918&b_code=12 (accessed Aug. 1, 2020).

49. "2017: Floride," Sea Art Festival, 2017, http://www.busanbiennale.org/eng/index .php?pCode=author&pg=3&mode=view&idx=2331.

50. Naoko Takahasi, "The Role of Arts Festivals in Contemporary Japan" (Discussion Papers in Arts and Festivals Management 15/2, December 2015), https://www.academia.edu/22293167/The_Role_of_Arts_Festivals_in_Contemporary_Japan.

51. "About ETAT," Echigo-Tsumari Art Field, accessed June 3, 2019, http://www.echigo-tsumari.jp/en/about/.

52. "About ETAT."

53. Ysabelle Cheung, "Revitalizing a Dying Region of Rural Japan with Art," Hyperallergic, September 4, 2015, https://hyperallergic.com/234480/revitalizing-a-dying-region-of-rural-japan-with-art/.

54. Art Setouchi homepage, accessed June 3, 2019, https://setouchi-artfest.jp/en/.

55. Lucy Dayman, "11 Best Art Festivals in Japan You Should Visit in 2019," Japan Objects, March 29, 2019, https://japanobjects.com/features/japan-festivals (accessed June 3, 2019).

56. Adrian Favell, "Socially Engaged Art in Japan: Mapping the Pioneers," Field 7 (Spring 2017), http://field-journal.com/issue-7/socially-engaged-art-in-japan-mapping-the-pioneers.

57. "North China Oil Spill Threatens Yellow River," Phys.org, January 3, 2010, https://phys.org/news/2010-01-north-china-oil-threatens-yellow.html (accessed June 4, 2019).

58. Lena Fritsch, "'Tree,' Ai Weiwei, 2010," Tate, August 2015, https://www.tate.org.uk/art/artworks/ai-tree-t14630.

59. Ai Weiwei, "Ai Weiwei: The Artwork That Made Me the Most Dangerous Person in China," Guardian, February 25, 2018, https://www.theguardian.com/artanddesign/2018/feb/15/ai-weiwei-remembering-sichuan-earthquake.

60. "Ai Weiwei Beijing Studio Demolished 'without Warning,'" BBC, August 4, 2018, https://www.bbc.com/news/world-asia-china-45070214.

61. Kat Barandy, "This Is the Largest Ever Exhibition Staged by Chinese Artist Ai Weiwei," Designboom, November 14, 2018, https://www.designboom.com/art/ai-weiwei-raiz-sao-paulo-brazil-11-14-18/.

62. http://www.bbc.com/culture/story/20140401-the-most-important-artist-alive (accessed June 4, 2019); Alastair Sooke, "Ai Weiwei: The Most Important Artist Alive?" April 1, 2014. BBC Culture [weblink no longer available].

Chapter 9

1. For the searchable database, see "Records," Institute of Public and Environmental Affairs, accessed April 14, 2020, http://wwwen.ipe.org.cn/IndustryRecord/Regulatory_DD.aspx. For the additional reports, see "Reports," Institute of Public

and Environmental Affairs, accessed April 14, 2020, http://wwwen.ipe.org.cn /reports/NewsReport.html.

2. Dexter Roberts and Pete Engardio, "Secrets, Lies, and Sweatshops," *Businessweek*, November 26, 2006.

3. Jiang, 86–87.

4. While Panasonic and Sanyo were initially responsive, subsequent reports show that they became less so.

5. "Green Supply Chain CITI Evaluation," Institute of Public and Environmental Affairs, accessed April 14, 2020, http://wwwen.ipe.org.cn/GreenSupplyChain/CITI.html.

6. Interview with Ma Jun, Beijing, April 2011.

7. Motoko Aizawa and Chaofei Yang, "Green Credit, Green Stimulus, Green Revolution? China's Mobilization of Banks for Environmental Cleanup," *Journal of Environment Development* 19, no. 2 (2010): 119–144; Cheung.

8. Interview with Ma Jun, Beijing, April 2011.

9. Securities Times Corporate Social Responsibility Research Center and Institute of Public and Environmental Affairs, *Online Pollutant Data Index of Publicly-Listed A-Share Companies: Annual Report* (Securities Times Corporate Social Responsibility Research Center and Institute of Public and Environmental Affairs, February 2, 2016), 1, http:// wwwoa.ipe.org.cn//Upload/201609010413027665.pdf.

10. "Records."

11. "Green Stocks," Institute of Public and Environmental Affairs, accessed April 14, 2020, http://wwwen.ipe.org.cn/GreenSecurities/Securities.html.

12. Institute of Public and Environmental Affairs, *Green Finance: Blue Map Solution by IPE* (Beijing: Institute of Public and Environmental Affairs, accessed July 21, 2019), http://wwwoa.ipe.org.cn//Upload/201704051040530172.pdf.

13. Rob Schmitz, "China's New Weapon against Water Pollution: Its People," Marketplace, May 2, 2016, https://www.marketplace.org/2016/05/02/chinas-new-weapon -against-water-pollution-its-people/.

14. Interview with Ma Jun, Beijing, November 16, 2016.

15. Haddad, "Increasing Environmental Performance."

16. Jessie Steinhauer, "From Cairo to Tokyo," *Cairo Review of Global Affairs*, Summer 2017, https://www.thecairoreview.com/midan/from-cairo-to-tokyo/.

17. Yoshifumi Nakashima, "Climate Change Policies in Japan: What Are Cool Biz and Warm Biz?," *Japan Environment Quarterly* 3 (October 2013), https://www.env.go.jp /en/focus/jeq/issue/vol03/feature.html.

18. "Air-Con Disease or 'Cooler-Byo,'" *Education in Japan Community Blog*, accessed July 28, 2019, https://educationinjapan.wordpress.com/httpeducationinjapan-wordpress -comkids-health-safety-issuesnational-institute-of-infectious-diseases-no-of-cases-of-rs -virus-respiratory-syncytial-virus-infections-which-cause-serious-pneumonia-i/air-con -disease-or-cooler-byo.

19. "The Japanese Cool Biz Campaign: Increasing Comfort in the Workplace," Environmental and Energy Study Institute, September 30, 2015, https://www.eesi.org /articles/view/the-japanese-cool-biz-campaign-increasing-comfort-in-the-workplace.

20. "Lion Heart—Message from Prime Minister Junichiro Koizumi (Provisional Translation)," *Koizumi Cabinet E-mail Magazine*, no. 192 (June 16, 2005), https://japan.kantei .go.jp/m-magazine/backnumber/koizumi/2005/0616.html.

21. Andrew Morse and Miho Inada, "Japanese Men Dress Down to Cut Summer's Energy Costs," *Wall Street Journal*, June 9, 2006, https://www.wsj.com/articles/SB114 979837550375306.

22. Cheng Herng Shinn, "'Super Cool Biz' Springs Sales Boost," *Japan Real Time* (blog), *Wall Street Journal*, August 16, 2011, https://blogs.wsj.com/japanrealtime/2011/08/16 /super-cool-biz-springs-sales-boost/.

23. Mutsuko Murakami, "Cool Dudes," *South China Morning Post*, June 11, 2005, https://www.scmp.com/article/503914/cool-dudes (accessed, July 27, 2019).

24. "'Cool Biz' Dress Code Spreads through Halls of Promotion," *Japan Times*, July 9, 2005, https://www.japantimes.co.jp/news/2005/07/09/national/cool-biz-dress-code -spreads-through-halls-of-promotion/#.XTyY1ZNKhmA.

25. Morse and Inada, "Japanese Men Dress Down"; "Workers Eagerly Embrace 'Cool Biz,'" *Japan Times*, August 19, 2005, https://www.japantimes.co.jp/news/2005/08/19 /business/workers-eagerly-embrace-cool-biz/#.XT2DaZNKhmA.

26. "Japan Promotes 'Super Cool Biz' Energy Saving Campaign," BBC, June 1, 2011, https://www.bbc.com/news/business-13620900.

27. "Arab Envoys Support Japan's 'Cool Biz' Campaign: Koizumi," Kuwait News Agency, June 16, 2005, https://www.kuna.net.kw/ArticleDetails.aspx?id=1568912 &language=en.

28. Erix Prideaux, "Dress-Fest for a Warming World Thaws Political Chill," *Japan Times*, June 18, 2006, https://www.japantimes.co.jp/life/2006/06/18/to-be-sorted /dress-fest-for-a-warming-world-thaws-political-chill/#.XT2G8ZNKhmA.

29. "Let's Cut CO2 Emission! Lose the Tie and Keep the Summer and the Earth Cool," Ministry of Environment Republic of Korea, June 7, 2006, https://web .archive.org/web/20070311145432/http://eng.me.go.kr/docs/news/press_view.html ?seq=335.

30. Samrana Hussain, "TUC Calls for Dress-Down Summer," *Guardian*, July 18, 2006, https://www.theguardian.com/environment/2006/jul/18/weather.climatechange1.

31. Chun Knee Tan, "Cool United Nations," Our World, August 26, 2008, https://ourworld.unu.edu/en/cool-united-nations.

32. Tomoyuki Tachikawa, "Hot! Also in North Korea, Cool Biz Already Takes Root," Kyodo News, July 27, 2018, https://english.kyodonews.net/news/2018/07/4a965b5a6de0-feature-hot-also-in-north-korea-cool-biz-already-takes-root.html?phrase=neyagawa&words=.

33. "Minister Koike Created the 'Mottainai Furoshiki,'" Ministry of the Environment, Government of Japan, April 3, 2006, https://www.env.go.jp/en/focus/060403.html.

34. Tools of Change, "Cool Biz, Japan," https://toolsofchange.com/en/case-studies/detail/662 (accessed Aug. 1, 2020).

35. Nakashima, "Climate Change Policies in Japan."

36. Memorandum of understanding between the City of London Corporation and Tokyo Metropolitan Government, signed December 4, 2017, http://www.metro.tokyo.jp/english/topics/2017/documents/171204.pdf.

37. "Tokyo Declaration on Realization of Clean Cities & Clear Skies," Tokyo Metropolitan Government, May 23, 2018, http://www.metro.tokyo.jp/english/topics/2018/180528_01.html.

38. "Urban 20 Group of Cities Meet in Tokyo and Urge G20 to Act Urgently on Climate Change, Social Inclusion and Sustainable Economic Growth," C40 Cities, May 22, 2019, https://www.c40.org/press_releases/urban-20-group-of-cities-meet-in-tokyo-and-urge-g20-to-act-urgently-on-climate-change-social-inclusion-and-sustainable-economic-growth.

39. "Results of Tokyo Cap-and-Trade Program in the 8th Fiscal Year: Covered Facilities Continue Reducing Emissions in Second Compliance Period," Tokyo Metropolitan Government, Bureau of Environment, February 19, 2019, http://www.kankyo.metro.tokyo.jp/en/climate/cap_and_trade/index.files/8thYearResult.pdf.

40. https://www.bccjapan.com/news/2017/12/governor-koike-achieving-sustainable-tokyo/ (accessed July 29, 2019). Sterling Content, "Governor Koike on Achieving a Sustainable Tokyo," Dec. 15, 2017, https://bccjapan.com/news/governor-koike-on-achieving-a-sustainable-tokyo/,.

41. Yachi Chen and Suiki Park, "Interview with the Founders of Tree Planets, Hyungsoo Kim and Min Choel Jeong (1)," Cultural Entrepreneurship, March 17, 2017, https://blogs.warwick.ac.uk/cultent/entry/interview_with_the_1_2_3/; interview with Jeong Mincheol, Seoul, June 19, 2019.

42. Social Enterprise Promotion Act, Act No. 8217, January 3, 2007, http://www
.ilo.org/dyn/natlex/docs/ELECTRONIC/78610/84122/F-684569511/KOR78610%20
Eng%202012.pdf.

43. Seoul Legal Administration Services, Amendment of Other Laws 5208. https://
legal.seoul.go.kr/legal/front/page/existing.html?pAct=lawComparison&pPromNo=1307
&type=en (accessed July 23, 2019).

44. "Social Innovation," Seoul Metropolitan Government, accessed July 23, 2019,
http://english.seoul.go.kr/policy-information/key-policies/city-initiatives/4-social
-innovation/.

45. Kyujin Jung, Hee Soun Jang, and Inseok Seo, "Government-Driven Social Enterprises in South Korea: Lessons from the Social Enterprise Promotion Program in the
Seoul Metropolitan Government," *International Review of Administrative Sciences* 82,
no. 3 (2016): 607.

46. StartBigger, "Tree Planet," Medium, August 30, 2018, https://medium.com/@
startbigger/tree-planet-869b51b77712.

47. Kathleen Emerson, "Tree Planet: Solving Environmental Problems through Planting, Gaming, and Commerce," Ubique, January 13, 2016, https://ubique.americangeo
.org/company-and-not-for-profit-spotlights/tree-planet-solving-environmental
-problems-through-planting-gaming-and-commerce/.

48. "Hanwha Solar Forest Helps Make the World Greener," Hanwha, accessed July
23, 2019, https://www.hanwha.com/en/news_and_media/press_release/hanwha-solar
-forest-helps-make-the-world-greener.html.

49. Treeplanetvideo, "[Tree Planet] About Tree Planet ENG Sub," October 22, 2014,
YouTube video, 3:39, https://www.youtube.com/watch?v=10W1lQ-FobY.

50. Emerson, "Tree Planet."

51. "K-POP Star's Forests in Seoul," Trazy, accessed June 22, 2020, https://www.trazy
.com/theme/kpop_forest.

52. Tree Planet Brand Book 2015, posted on Issuu February 5, 2017, p. 32, https://
issuu.com/treeplanet/docs/treeplanet_brandbook_print.

53. "Yellow Ribbon Memorial Forest Near Site of Sewol Sinking Completed," Han-
kyoreh, April 11, 2016, http://english.hani.co.kr/arti/english_edition/e_international
/739163.html (accessed July 24, 2019).

54. Interview with Jeong Mincheol, Seoul, June 19, 2019.

55. "Yellow Ribbon Memorial Forest."

56. Peace Park in Seoul: Tree Planet Blog. "Forest for victims of military sexual slav-
ery by Japan," Oct. 24, 2015, https://tpmembers.blog.me/220523271196 (Korean,

accessed July 24, 2019); China: Tree Planet "foRest in Peace" https://youtu.be /YG1yLi7mYkQ (Korean, accessed July 24, 2019).

57. Interview with Jeong Mincheol, Seoul, June 19, 2019.

58. Tree Planet Brand Book, 40.

59. "S. Korea's Air Quality, Worst in World," *Seoul Times*, accessed July 25, 2019, http://theseoultimes.com/ST/?url=/ST/db/read.php?idx=13085.

60. "Air Purifier Market Soars on Yellow Dust Concerns," *Korea Herald*, April 24, 2016, http://www.koreaherald.com/view.php?ud=20160424000217.

61. Interview with Jeong Mincheol, Seoul, June 19, 2019.

62. "Air Purification to Become Mandatory at Schools for Young and Special Needs Children," Korea Bizwire, April 6, 2018, http://koreabizwire.com/air-purification-to -become-mandatory-at-schools-for-young-and-special-needs-children/116362.

63. Kim Arin, "Government Funds Mandatory Air Purifiers at Schools," *Korea Herald*, March 27, 2019, http://www.koreaherald.com/view.php?ud=20190327000802.

64. Elizabeth Choi, "Air Purifiers Are Not Always Effective, and Some May Be Harmful," *South China Morning Post*, June 16, 2014, https://www.scmp.com/lifestyle/health /article/1531695/air-purifiers-are-not-always-effective-and-some-may-be-harmful.

65. Interview with Jeong Mincheol, Seoul, June 19, 2019.

66. Interview with Jeong; "Sudokwon Landfill Site Management Corp. to Deliver Purification Plants to Elementary Schools," Korea Bizwire, April 4, 2019, http://koreabizwire .com/sudokwon-landfill-site-management-corp-to-deliver-purification-plants-to -elementary-schools/135401.

67. Interview with Jeong Mincheol, Seoul, June 19, 2019. See also Choi Ye-jin, "Tree Planet Takes a Step towards Saving the Planet by Promoting 'Funation,'" Ewha Voice, April 15, 2019, https://evoice.ewha.ac.kr/news/articleView.html?idxno=5791; and Tree Planet Blog, Nepal Forests, https://project.treepla.net/project/nepal_forest (Korean, accessed July 25, 2019).

68. Bloomberg New Economy, "The Chinese Gaming App That Lets Users Plant Real Trees," June 25, 2019, YouTube video, 3:43, https://www.youtube.com/watch ?v=DeGK7WObbQU.

69. Xu Keyue, "Ant Forest Users Plant 55m Trees in 507 Square Kilometers," *Global Times*, February 18, 2019, http://www.globaltimes.cn/content/1139299.shtml.

70. Jeong Tae Kim, "Korean Social Enterprises Go Global," Devex, March 13, 2015, https://www.devex.com/news/korean-social-enterprises-go-global-85637.

71. "Buying Plant Gifts for People in Seoul South Korea," FlowerGift Korea, accessed July 25, 2019, https://flowergiftkorea.com/buying-plant-gifts-people-seoul-south-korea/.

72. Jisun Choi and Dong-A-Ilbo, "Reforestation in South Korea to Prevent Fine Dust," *Hindu*, April 22, 2019, https://www.thehindu.com/sci-tech/energy-and-environment /reforestation-in-south-korea-to-prevent-fine-dust/article26903426.ece.

73. Treeplanetvideo, "[Tree Planet] sTREEt Campaign," September 28, 2014, YouTube video, 2:14, https://www.youtube.com/watch?v=FBX_uJ8jDvY.

Conclusion

1. I would like to acknowledge my son Reja for helping me realize that "replenishing the commons" is a suitable antonym for "tragedy of the commons."

2. Alagappa; Broadbent and Brockman; Dalton, Recchia, and Rohrschneider; Harris and Lang; Peter Ho and Edmonds; Schreurs, *Environmental Politics*.

3. Agrawal and Ostrom; Bosso; Dalton; Fuentes-George; Giugni; Gunningham, Kagan, and Thornton; Miller; O'Neill; Rodger A. Payne, "Freedom and the Environment," *Journal of Democracy* 6, no. 3 (1995): 41–55; Pekkanen, Smith, and Tsujinaka; Miranda Schreurs, "Democratic Transition and Environmental Civil Society: Japan and South Korea Compared," *Good Society* 11, no. 2 (2002): 57–64; Teets, *Civil Society under Authoritarianism*.

4. Hugh Heclo, *Modern Social Politics in Britain and Sweden: From Relief to Income Maintenance* (New Haven, CT: Yale University Press, 1974); North; Sabatier and Jenkins-Smith.

5. R. Edward Freeman, *Strategic Management: A Stakeholder Approach* (Cambridge: Cambridge University Press, 2010), 53.

6. Freeman, 53.

7. Yiqi Zhang et al.

8. Pekkanen, Smith, and Tsujinaka.

9. Steven Rathgeb Smith and Lipsky.

10. Berry and Arons; Salamon; Portney and Berry.

11. Jens Newig and Oliver Fritsch, "Environmental Governance: Participatory, Multilevel—and Effective?," *Environmental Policy and Governance* 19, no. 3 (2009): 197–214.

12. Daly.

13. Eisner; John; Kraft, Stephan, and Abel; Prakash and Gugerty.

14. Helmut K. Anheier, Marlies Glasius, and Mary Kaldor, *Global Civil Society 2001* (New York: Oxford University Press, 2001); Sheri Berman, "Civil Society and the Collapse of the Weimar Republic," *World Politics* 49, no. 3 (1997): 401–429; Nancy Bermeo and Philip Nord, eds., *Civil Society before Democracy: Lessons from Nineteenth-Century Europe* (New York: Rowman and Littlefield, 2000); Michael Bernhard and

Ekrem Karakoç, "Civil Society and the Legacies of Dictatorship," *World Politics* 59, no. 4 (2007): 539–567; Blaney and Pasha; Larry Diamond, "Rethinking Civil Society: Toward Democratic Consolidation," *Journal of Democracy* 5, no. 3 (July 1994): 4–17; B. Michael Frolic, "State-Led Civil Society," in *Civil Society in China*, ed. Timothy Brook and B. Michael Frolic (New York: M. E. Sharpe, 1997); Makoto Iokibe, "Japan's Civil Society: An Historical Overview," in *Deciding the Public Good: Governance and Civil Society in Japan*, ed. Tadashi Yamamoto (New York: Japan Center for International Exchange, 1999); Norton; Lester M. Salamon et al., eds., *Global Civil Society: Dimensions of the Nonprofit Sector* (Baltimore: Johns Hopkins Center for Civil Society Studies, 1999); Teets, *Civil Society under Authoritarianism*; Wiktorowicz.

15. Andrew B. Whitford, Vicky M. Wilkins, and Mercedes G. Ball, "Descriptive Representation and Policymaking Authority: Evidence from Women in Cabinets and Bureaucracies," *Governance* 20, no. 4 (2007): 559–580.

16. Sheryl Sandberg, *Lean In: Women, Work, and the Will to Lead* (New York: Knopf, 2013); Lois Frankel, *Nice Girls Don't Get the Corner Office: 101 Mistakes Women Make That Sabotage Their Careers* (New York: Business Plus, 2004); Linda Babcock and Sara Laschever, *Women Don't Ask: Negotiation and the Gender Divide* (Princeton, NJ: Princeton University Press, 2003); Jodi Kantor, "Harvard Business School Case Study: Gender Equity," *New York Times*, September 7, 2013; R. Kage, F. Rosenbluth, and S. Tanaka, "What Explains Low Female Political Representation? Evidence from Survey Experiments in Japan," *Politics & Gender* 15, no. 2 (2019): 285–309.

17. Garrett Hardin, "The Tragedy of the Commons," *Science* 162, no. 3859 (1968): 1243–1248; Ostrom.

18. "Elinor Ostrom: Facts," Nobel Prize, accessed June 23, 2020, https://www.nobel prize.org/prizes/economic-sciences/2009/ostrom/facts/.

Appendix B

1. The US groups were created by searching the IRS's cumulative list of environmental organizations found here: Internal Revenue Service, Tax Exempt Search, https://www.irs .gov/charities-non-profits/tax-exempt-organization-search (search conducted between July 1–7, 2012). We first selected 501(c)(3) organizations that had missions related to the environment (all the C codes, and then D20, D30–D34, and K25), which generated 29,498 organizations. We then randomly selected 100 organizations and added them to the database. Organizations for which we could find no information were eliminated, resulting in a total of 105 organizations.

2. NPO Hiroba [in Japanese], accessed January 12, 2012, http://www.npo-hiroba.or .jp/search/.

3. Korean Ministry of the Environment, List of Environmental Organizations, http:// www.me.go.kr/home/web/policy_data/read.do?pagerOffset=0&maxPageItems=10

&maxIndexPages=10&searchKey=&searchValue=&menuId=10260&orgCd=&condition
.code=A1&seq=6330 (accessed October 4–25, 2014; website no longer available).

4. David G. Ortiz et al., "Where Do We Stand with Newspaper Data?," *Mobilization: An International Quarterly* 10, no. 3 (2005): 397–419; Ruud Koopmans and Dieter Rucht, "Protest Event Analysis," in *Methods of Social Movement Research*, ed. Bert Klandermans and Suzanne Staggenborg, Social Movements, Protest, and Contention 16 (Minneapolis: University of Minnesota Press, 2002).

5. Pekkanen, Smith, and Tsujinaka.

6. Steinhoff; Sarat and Scheingold.

7. R Development Core Team, "A Language and Environment for Statistical Computing" (R Foundation for Statistical Computing, Vienna, 2012); Torsten Hothorn et al., "Survival Ensembles," *Biostatistics* 7, no. 3 (2006): 355–373.

8. Leo Breiman, "Random Forests," *Machine Learning* 45, no. 1 (2001): 5–32.

9. Carolin Strobl, Anne-Laure Boulesteix, Thomas Kneib, Thomas Augustin, and Achim Zeileis, "Conditional Variable Importance for Random Forests," *BMC Bioinformatics* 9 (2008).

References

Adams, Jacqueline. "Art in Social Movements: Shantytown Women's Protest in Pinochet's Chile." *Sociological Forum* 17, no. 2 (2002): 21–56.

Adeola, Francis O. "Cross-national Environmental Injustice and Human Rights Issues: A Review of Evidence in the Developing World." *American Behavioral Scientist* 43, no. 4 (2000): 686–706.

Agrawal, Arun, and Maria Carmen Lemos. "A Greener Revolution in the Making? Environmental Governance in the 21st Century." *Environment: Science and Policy for Sustainable Development* 49, no. 5 (2007): 36–45.

Agrawal, Arun, and Elinor Ostrom. "Collective Action, Property Rights, and Decentralization in Resource Use in India and Nepal." *Politics and Society* 29, no. 4 (2001): 485–514.

Aivazian, Varouj A., Ying Ge, and Jiaping Qiu. "Can Corporatization Improve the Performance of State-Owned Enterprises Even without Privatization?" *Journal of Corporate Finance* 11, no. 5 (2005): 791–808.

Aizawa, Motoko, and Chaofei Yang. "Green Credit, Green Stimulus, Green Revolution? China's Mobilization of Banks for Environmental Cleanup." *Journal of Environment Development* 19, no. 2 (2010): 119–144.

Alagappa, Muthiah, ed. *Civil Society and Political Change in Asia: Expanding and Contracting Democratic Space.* Stanford, CA: Stanford University Press, 2004.

Aldrich, Daniel P. "It's Who You Know: Factors Driving Recovery from Japan's 11 March 2011 Disaster." *Public Administration* 94, no. 2 (2016): 399–413.

Aldrich, Daniel P. *Site Fights: Divisive Facilities and Civil Society in Japan and the West.* Ithaca, NY: Cornell University Press, 2008.

Anderson, Christina L., and Rebecca L. Bieniaszewska. "The Role of Corporate Social Responsibility in an Oil Company's Expansion into New Territories." *Corporate Social Responsibility and Environmental Management* 12, no. 1 (2005): 1–9.

Anheier, Helmut K., Marlies Glasius, and Mary Kaldor. *Global Civil Society 2001*. New York: Oxford University Press, 2001.

Arenas, Daniel, Josep Lozano, and Laura Albareda. "The Role of NGOs in CSR: Mutual Perceptions among Stakeholders." *Journal of Business Ethics* 88, no. 1 (August 2009): 175–197.

Armstrong, Charles, ed. *Korean Society: Civil Society, Democracy and the State*. New York: Taylor and Francis, 2002.

Asian Institute of Technology. *Regional Collaboration on Ecolabelling—Asia Pacific*. United Nations Environment Programme and European Union, 2016.

Auld, Graeme, and Lars H. Gulbrandsen. "Transparency in Nonstate Certification: Consequences for Accountability and Legitimacy." *Global Environmental Politics* 10, no. 3 (August 2010): 97–119.

Avenell, Simon. "Legal Experts and Environmental Activism in Japan: Fighting for 'Environmental Rights.'" In *Greening East Asia: The Rise of the Eco-developmental State*, edited by Ashley Esarey, Mary Alice Haddad, Stevan Harrell, and Joanna Lewis. Seattle: University of Washington Press, forthcoming.

Avenell, Simon. *Transnational Japan in the Global Environmental Movement*. Honolulu: University of Hawai'i Press, 2017.

Ayling, Julie, and Neil Gunningham. "Non-state Governance and Climate Policy: The Fossil Fuel Divestment Movement." *Climate Policy* 17, no. 2 (2017): 131–149.

Babcock, Linda, and Sara Laschever. *Women Don't Ask: Negotiation and the Gender Divide*. Princeton, NJ: Princeton University Press, 2003.

Bader, Christine. *The Evolution of a Corporate Idealist: When Girl Meets Oil*. New York: Bibliomotion, 2014.

Baek, Kyung-Min, Young-Kyo Seo, Jun-Young Kim, and Sung-Ok Baek. "Monitoring of Particulate Hazardous Air Pollutants and Affecting Factors in the Largest Industrial Area in South Korea: The Sihwa-Banwol Complex." *Environmental Engineering Research*. Published ahead of print, December 17, 2019. https://doi.org/10.4491/eer.2019.419.

Balla, Steven J., and John R. Wright. "Interest Groups, Advisory Committees, and Congressional Control of the Bureaucracy." *American Journal of Political Science* 45, no. 4 (2001): 799–812.

Bamber, Greg J., and Chris J. Leggett. "Changing Employment Relations in the Asia-Pacific Region." *International Journal of Manpower* 22, no. 4 (2001): 300–317.

Barabási, Albert-László, and Márton Pósfai. *Network Science*. New York: Cambridge University Press, 2016.

Bardach, Eugene. *A Practical Guide for Policy Analysis: The Eightfold Path to More Effective Problem Solving*. Washington, DC: Sage, 2012.

Baumgartner, Frank R., and Bryan D. Jones. *Agendas and Instability in American Politics*. Chicago: University of Chicago Press, 1993.

Berger, Suzanne, and Ronald Dore, eds. *National Diversity and Global Capitalism*. Ithaca, NY: Cornell University Press, 1996.

Berman, Sheri. "Civil Society and the Collapse of the Weimar Republic." *World Politics* 49, no. 3 (1997): 401–429.

Bermeo, Nancy, and Philip Nord, eds. *Civil Society before Democracy: Lessons from Nineteenth-Century Europe*. New York: Rowman and Littlefield, 2000.

Bernhard, Michael, and Ekrem Karakoç. "Civil Society and the Legacies of Dictatorship." *World Politics* 59, no. 4 (2007): 539–567.

Berry, Jeffrey, and David F. Arons. *A Voice for Nonprofits*. Washington, DC: Brookings Institution, 2003.

Berry, Jeffrey, and Kent Portney. "The Group Basis of City Politics." In *Nonprofits and Advocacy: Engaging Community and Government in an Era of Retrenchment*, edited by Robert J. Pekkanen, Steven Rathgeb Smith, and Yutaka Tsujinaka, 21–46. Baltimore: Johns Hopkins University Press, 2014.

Berry, Jeffrey, Kent Portney, Robin Liss, Jessica Simoncelli, and Lisa Berger. "Power and Interest Groups in City Politics." Working paper, Rappaport Institute for Greater Boston, Kennedy School of Government, Harvard University, Cambridge, MA, 2006.

Bickel, Barbara, Valerie Triggs, Stephanie Springgay, Rita Irwin, Kit Grauer, Gu Xiong, Ruth Beer, and Pauline Sameshima. "Richgate: Transforming Public Spaces through Community-Engaged Art." *Amerasia Journal* 33, no. 2 (2007): 115–124.

Blaney, David L., and Mustapha Kamal Pasha. "Civil Society and Democracy in the Third World: Ambiguities and Historical Possibilities." *Studies in Comparative International Development* 28, no. 1 (1993): 3–24.

Bosso, Christopher. *Environment, Inc.* Studies in Government and Public Policy. Lawrence: University Press of Kansas, 2005.

Bouteligier, Sofie. *Cities, Networks, and Global Environmental Governance: Spaces of Innovation, Places of Leadership*. New York: Routledge, 2013.

Brandt, Jodi S., Van Butsic, Benjamin Schwab, Tobias Kuemmerle, and Volker C. Radeloff. "The Relative Effectiveness of Protected Areas, a Logging Ban, and Sacred Areas for Old-Growth Forest Protection in Southwest China." *Biological Conservation* 181 (2015): 1–8.

Brecher, W. Puck. *An Investigation of Japan's Relationship to Nature and Environment*. Lewiston, NY: Edwin Mellen, 2000.

Breiman, Leo. "Random Forests." *Machine Learning* 45, no. 1 (2001): 5–32.

Broadbent, Jeffrey. *Environmental Politics in Japan: Networks of Power and Protest*. New York: Cambridge University Press, 1998.

Broadbent, Jeffrey, and Vicky Brockman, eds. *East Asian Social Movements: Power, Protest, and Change in a Dynamic Region*. New York: Springer, 2010.

Bryant, William E. *Japanese Private Economic Diplomacy: An Analysis of Business-Government Linkages*. New York: Praeger, 1975.

Bryce, Herrington. *Players in the Public Policy Process: Nonprofits as Social Capital and Agents*. New York: Palgrave Macmillan, 2005.

Bryson, John M., Barbara C. Crosby, and Melissa Middleton Stone. "The Design and Implementation of Cross-sector Collaborations: Propositions from the Literature." In "Collaborative Public Management." Special issue, *Public Administration Review* 66 (2006): 44–55.

Burstein, Paul. "Policy Domains: Organization, Culture, and Policy Outcomes." *Annual Review of Sociology* 17 (1991): 327–350.

Busch, Per-Olof, Helge Jörgens, and Kerstin Tews. "The Global Diffusion of Regulatory Instruments: The Making of a New International Environmental Regime." *Annals of the American Academy of Political and Social Science* 598 (March 2005): 146–167.

Campbell, John Creighton. *How Policies Change: The Japanese Government and the Aging Society*. Princeton, NJ: Princeton University Press, 1992.

Carter, Neil, and Arthur Mol. *Environmental Governance in China*. New York: Routledge, 2008.

Chang, Young-Bae, Jae-Kak Han, and Woo-Hyun Kim. "Green Growth and Green New Deal Policies in Korea: Are They Creating Decent Green Jobs?" Paper presented at the International Trade Union Conference / Global Unions Research Network workshop "A Green Economy That Works for Social Progress," Brussels, Belgium, 2011.

Chawla, Louise. "Research Priorities in Environmental Education." *Children's Environments* 9, no. 1 (1992): 68–71.

Cheng, Tun-jen. "Democratizing the Quasi-Leninist Regime in Taiwan." *World Politics* 61, no. 4 (1989): 471–499.

Cheol, Kim Yong. "The Shadow of the Gwangju Uprising in the Democratization of Korean Politics." *New Political Science* 25, no. 2 (2003): 225–240.

Cheung, Ray. "Making Green from Green—How Improving the Environmental Performance of Supply Chains Can Be a Win-Win for China and the World." Brief, China Environment Forum, 2011.

Chi, Chun-Chieh. "Capitalist Expansion and Indigenous Land Rights: Emerging Environmental Justice Issues in Taiwan." *Asia Pacific Journal of Anthropology* 2, no. 2 (2001): 135–153.

China Development Brief. *A Report on the Policy Environment for Chinese NGOs.* Hong Kong: China Development Brief, 2018.

Cho, Myung-Rae. "Emergence and Evolution of Environmental Discourses in South Korea." *Korea Journal* 44, no. 3 (2004): 138–164.

Cho, Myung-Rae. "The Politics of Urban Nature Restoration: The Case of Cheong-gyecheon Restoration in Seoul, Korea." *International Development Planning Review* 32, no. 2 (2010): 145–165.

Cho, Myung-Rae. "A Progressive City in the Making? The Seoul Experience." In *The Rise of Progressive Cities East and West*, edited by Mike Douglass, Romain Garbaye, and K. C. Ho, 47–64. Singapore: Springer, 2019.

Choi, Jang Jip. "The Democratic State Engulfing Civil Society: The Ironies of Korean Democracy." *Korean Studies* 34 (2010): 1–24.

Clark, Gordon L., Andreas Feiner, and Michael Viehs. *From the Stockholder to the Stakeholder: How Sustainability Can Drive Financial Outperformance.* Swiss Sustainable Finance, 2015.

Cui, Shuqin. "Chai Jing's *Under the Dome*: A Multimedia Documentary in the Digital Age." *Journal of Chinese Cinemas* 11, no. 1 (2017): 30–45.

Dahl, Robert A. *Polyarchy: Participation and Opposition.* New Haven, CT: Yale University Press, 1971.

Dalton, Russell J. *The Green Rainbow: Environmental Interest Groups in Western Europe.* New Haven, CT: Yale University Press, 1994.

Dalton, Russell J., Steve Recchia, and Robert Rohrschneider. "The Environmental Movement and the Modes of Political Action." *Comparative Political Studies* 36, no. 7 (2003): 743–771.

Daly, John A. *Advocacy: Championing Ideas and Influencing Others.* New Haven, CT: Yale University Press, 2012.

Dauvergne, Peter. *Shadows in the Forest: Japan and the Politics of Timber in Southeast Asia.* Cambridge, MA: MIT Press, 1997.

Dauvergne, Peter, and Jane Lister. "Big Brand Sustainability: Governance Prospects and Environmental Limits." *Global Environmental Change* 22, no. 1 (2012): 36–45.

Detomasi, David Antony. "The Multinational Corporation and Global Governance: Modelling Global Public Policy." *Journal of Business Ethics* 71, no. 3 (2007): 321–334.

Diamond, Larry. "Rethinking Civil Society: Toward Democratic Consolidation." *Journal of Democracy* 5, no. 3 (July 1994): 4–17.

Dominick, Raymond. *The Environmental Movement in Germany: Prophets and Pioneers, 1871–1971.* Bloomington: Indiana University Press, 1992.

Dong, Baomin, Jiong Gong, and Xin Zhao. "FDI and Environmental Regulation: Pollution Haven or a Race to the Top?" *Journal of Regulatory Economics* 41, no. 2 (2012): 216–237.

Dore, Ronald. *British Factory, Japanese Factory: The Origins of National Diversity in Industrial Relations*. Berkeley: University of California Press, 1973.

Doss, Erika. "'Revolutionary Art Is a Tool for Liberation': Emory Douglas and Protest Aesthetics at *The Black Panther*." *New Political Science* 21, no. 2 (1999): 245–259.

Downey, Anthony. *Art and Politics Now*. New York: Thames and Hudson, 2014.

Easterly, William. *The Tyranny of Experts: Economists, Dictators, and the Forgotten Rights of the Poor*. New York: Basic Books, 2014.

Economist. "Business in the Blood: Family Firms." November 1, 2014.

Economy, Elizabeth. *The River Runs Black: The Environmental Challenge to China's Future*. Ithaca, NY: Cornell University Press, 2004.

Edelman, Murray. *From Art to Politics*. Chicago: University of Chicago Press, 1995.

Eder, Norman. *Poisoned Prosperity: Development, Modernization, and the Environment in South Korea*. New York: M. E. Sharpe, 1997.

Edmondson, Robert. "The February 28 Incident and National Identity." In *Memories of the Future: National Identity Issues and the Search for a New Taiwan*, edited by Stéphane Corcuff, 25–46. Armonk, NY: M. E. Sharpe, 2002.

Efird, Rob. "Closing the Green Gap: Policy and Practice in Chinese Environmental Education." In *Schooling for Sustainable Development across the Pacific*, edited by John Chi-Kin Lee and Rob Efird, 279–292. Dordrecht: Springer, 2014.

Eisner, Marc. *Governing the Environment*. New York: Lynne Rienner, 2006.

Elder, Mark. "Regional Governance for Environmental Sustainability in Asia in the Context of Sustainable Development: A Survey of Regional Cooperation Frameworks." In *Routledge Handbook of Sustainable Development in Asia*, edited by Sara Hsu, 468–494. New York: Routledge, 2018.

Elver, Hilal. "International Environmental Law, Water and the Future." *Third World Quarterly* 27, no. 5 (2006): 885–901.

Epstein, Marc, and Marie-Josée Roy. "Managing Corporate Environmental Performance: A Multinational Perspective." *European Management Journal* 16, no. 3 (1998): 284–296.

Esarey, Ashley, Mary Alice Haddad, Stevan Harrell, and Joanna Lewis, eds. *Greening East Asia: The Rise of the Eco-developmental State*. Seattle: University of Washington Press, forthcoming.

Estévez-Abe, Margarita. *Welfare and Capitalism in Postwar Japan.* New York: Cambridge University Press, 2008.

Evans, Peter. *Embedded Autonomy: States and Industrial Transformation.* Princeton, NJ: Princeton University Press, 1995.

Evans, Peter, ed. *State-Society Synergy: Government and Social Capital in Development.* Berkeley: University of California Press, 1997.

Ewing-Chow, Michael, and Darryl Soh. "Pain, Gain, or Shame: The Evolution of Environmental Law and the Role of Multinational Corporations." *Indiana Journal of Global Legal Studies* 16, no. 1 (2009): 195–222.

Fasterling, Björn. "Development of Norms through Compliance Disclosure." *Journal of Business Ethics* 106 (2012): 73–87.

Feiock, Richard C. "The Institutional Collective Action Framework." *Policy Studies Journal* 41, no. 3 (2013): 397–425.

Feiock, Richard C., Ruowen Shen, Liming Suo, Jiasheng Zhang, and Hongtao Yi. "Local Level Collaborations on Environmental Issues in China through the Lens of Institutional Collective Action." Paper presented at the Annual Meeting of the Association of Public Policy Analysis and Management, Miami, 2015.

Fell, Dafydd. "The Evolution of the Anti-nuclear Movement in Taiwan since 2008." In *Taiwan's Social Movements under Ma Ying-Jeou: From the Wild Strawberries to the Sunflowers*, edited by Dafydd Fell, 170–192. London: Routledge, 2017.

Fischer, Frank. *Citizens, Experts, and the Environment: The Politics of Local Knowledge.* Durham, NC: Duke University Press, 2000.

Fischer, Frank. *Technocracy and the Politics of Expertise.* Newbury Park, CA: Sage, 1989.

Fisher, Dana R. "The Broader Importance of #Fridaysforfuture." *Nature Climate Change* 9, no. 6 (2019): 430–431.

Florida, Richard. *Cities and the Creative Class.* New York: Routledge, 2005.

Frankel, Lois. *Nice Girls Don't Get the Corner Office: 101 Mistakes Women Make That Sabotage Their Careers.* New York: Business Plus, 2004.

Freeman, Linton. *The Development of Social Network Analysis: A Study in the Sociology of Science.* Vancouver, BC: Empirical, 2004.

Freeman, R. Edward. *Strategic Management: A Stakeholder Approach.* Cambridge: Cambridge University Press, 2010.

Friends of Nature, Institute of Public and Environmental Affairs, Envirofriends, Nature University, and Nanjing Greenstone. *Apple Opens Up: IT Industry Supply Chain Investigative Report.* Beijing: Institute for Public and Environmental Affairs, 2013.

Frolic, B. Michael. "State-Led Civil Society." In *Civil Society in China*, edited by Timothy Brook and B. Michael Frolic, 46–67. New York: M. E. Sharpe, 1997.

Fuentes-George, Kemi. *Between Preservation and Exploitation: Transnational Advocacy Networks and Conservation in Developing Countries*. Cambridge, MA: MIT Press, 2016.

Fung, Archon, Mary Graham, and David Weil. *Full Disclosure: The Perils and Promise of Transparency*. New York: Cambridge University Press, 2007.

George, Timothy. *Minamata: Pollution and the Struggle for Democracy in Postwar Japan*. Cambridge, MA: Harvard University Press, 2002.

Gilby, Simon, Matthew Hengesbaugh, Premakumara Jagath Dickella Gamaralage, Kazunobu Onogawa, Eddy Soedjono, and Nurina Fitriani. *Planning and Implementation of Integrated Solid Waste Management Strategies at Local Level: The Case of Surabaya City*. Osaka: UN Environment; Hayama, Kanagawa: Institute for Global Environmental Strategies, 2017.

Gillion, Daniel. *The Political Power of Protest: Minority Activism and Shifts in Public Policy*. Cambridge Studies in Contentious Politics. New York: Cambridge University Press, 2013.

Giugni, Marco. *Social Protest and Policy Change: Ecology, Antinuclear, and Peace Movements in Comparative Perspective*. New York: Rowman and Littlefield, 2004.

Goodman, Roger, and Ito Peng. "The East Asian Welfare States: Peripatetic Learning, Adaptive Change, and Nation-Building." In *Welfare States in Transition: National Adaptations in Global Economies*, edited by Gosta Esping-Andersen, 192–224. London: Sage, 1996.

Goodman, Roger, Gordon White, and Huck-ju Kwong, eds. *The East Asian Welfare Model: Welfare Orientalism and the State*. New York: Routledge, 1998.

Gordon, Jennifer Farley, and Colleen Hill. *Sustainable Fashion: Past, Present and Future*. New York: Bloomsbury Academic, 2015.

Gossin, Pamela. "Animated Nature: Aesthetics, Ethics, and Empathy in Miyazaki Hayao's Ecophilosophy." *Mechademia* 10 (2015): 209–234.

Graham, Mark A. "Art, Ecology and Art Education: Locating Art Education in a Critical Place-Based Pedagogy." *Studies in Art Education* 48, no. 4 (2007): 375–391.

Grano, Simona A. *Environmental Governance in Taiwan: A New Generation of Activists and Stakeholders*. New York: Routledge, 2015.

Groothuis, Peter A., and Gail Miller. "Locating Hazardous Waste Facilities: The Influence of NIMBY Beliefs." *American Journal of Economics and Sociology* 53, no. 3 (1994): 335–346.

Gunningham, Neil, Robert Kagan, and Dorothy Thornton. *Shades of Green: Business, Regulation, and Environment*. Stanford, CA: Stanford Law and Politics, 2003.

Haas, Peter M. "Banning Chlorofluorocarbons: Epistemic Community Efforts to Protect Stratospheric Ozone." *International Organization* 46, no. 1 (Winter 1992): 187–224.

Haas, Peter M. "Introduction: Epistemic Communities and International Policy Coordination." *International Organization* 46, no. 1 (1992): 1–35.

Haddad, Mary Alice. *Building Democracy in Japan.* New York: Cambridge University Press, 2012.

Haddad, Mary Alice. "From Backyard Environmental Advocacy to National Democratisation: The Cases of South Korea and Taiwan." In *NIMBY Is Beautiful: Cases of Local Activism and Environmental Innovation around the World,* edited by Carol Hager and Mary Alice Haddad, 179–199. New York: Berghahn Books, 2015.

Haddad, Mary Alice. "Increasing Environmental Performance in a Context of Low Governmental Enforcement: Evidence from China." *Journal of Environment and Development* 24, no. 1 (March 2015): 3–25.

Haddad, Mary Alice. "Paradoxes of Democratization: Environmental Politics in East Asia." In *Routledge Handbook of East Asia and the Environment,* edited by Paul G. Harris and Graeme Lang, 86–104. New York: Routledge, 2015.

Haddad, Mary Alice. "Working with and around Strong States: Environmental Networks in East Asia." In *Civil Society and the State in Democratic East Asia: Between Entanglement and Contention in Post High Growth,* edited by David Chiavacci, Simona A. Grano, and Julia Obinger, 59–84. Amsterdam: Amsterdam University Press, 2020.

Hadden, Jennifer. "Explaining Variation in Transnational Climate Change Activism: The Role of Inter-movement Spillover." *Global Environmental Politics* 14, no. 2 (2014): 7–25.

Hager, Carol. *Technological Democracy: Bureaucracy and Citizenry in the German Energy Debate.* Ann Arbor: University of Michigan Press, 1995.

Hager, Carol, and Mary Alice Haddad, eds. *NIMBY Is Beautiful: Cases of Local Activism and Environmental Innovation around the World.* New York: Berghahn Books, 2015.

Haggard, Stephan. *Pathways from the Periphery: The Politics of Growth in the Newly Industrializing Countries.* Ithaca, NY: Cornell University Press, 1990.

Haggard, Stephan. "The Political Economy of Regionalism in Asia and the Americas." In *The Political Economy of Regionalism,* edited by Edward D. Mansfield and Helen V. Milner, 20–49. New York: Columbia University Press, 1996.

Hakelberg, Lukas. "Governance by Diffusion: Transnational Municipal Networks and the Spread of Local Climate Strategies in Europe." *Global Environmental Politics* 14, no. 1 (February 2014): 107–129.

Hao, Zhidong, and Henry A. Giroux, eds. *Intellectuals at a Crossroads: The Changing Politics of China's Knowledge Workers*. Albany: State University of New York Press, 2003.

Hardin, Garrett. "The Tragedy of the Commons." *Science* 162, no. 3859 (1968): 1243–1248.

Harris, Paul G., and Graeme Lang, eds. *Routledge Handbook of Environment and Society in Asia*. New York: Routledge, 2015.

Harris, Paul G., and Chihiro Udagawa. "Defusing the Bombshell? Agenda 21 and Economic Development in China." *Review of International Political Economy* 11, no. 3 (2004): 618–640.

Hasmath, Reza, Timothy Hildebrandt, and Jennifer Y. J. Hsu. "Conceptualizing Government-Organized Non-governmental Organizations." Paper presented at the Development Studies Association Annual Meeting, Oxford, UK, 2016.

Hassard, John, Jonathan Morris, Jackie Sheehan, and Xiao Yuxin. "China's State-Owned Enterprises: Economic Reform and Organizational Restructuring." *Journal of Organizational Change Management* 23, no. 5 (2010): 500–516.

Hatch, Walter, and Kozo Yamamura. *Asia in Japan's Embrace: Building a Regional Production Alliance*. New York: Cambridge University Press, 1996.

Heclo, Hugh. "Issue Networks and the Executive Establishment." In *The New American Political System*, edited by Anthony King, 46–57. Washington, DC: American Enterprise Institute, 1978.

Heclo, Hugh. *Modern Social Politics in Britain and Sweden: From Relief to Income Maintenance*. New Haven, CT: Yale University Press, 1974.

Hildebrandt, Timothy, and Jennifer Turner. "Green Activism? Reassessing the Role of Environmental NGOs in China." In *State and Society Responses to Social Welfare Needs in China: Serving the People*, edited by Jonathan Schwartz and Shawn Shieh, 88–110. New York: Routledge, 2009.

Hirata, Keiko. "Whither the Developmental State? The Growing Role of NGOs in Japanese Aid Policymaking." *Journal of Comparative Policy Analysis: Research and Practice* 4 (2002): 165–188.

Ho, Ming-sho. *Challenging Beijing's Mandate of Heaven: Taiwan's Sunflower Movement and Hong Kong's Umbrella Movement*. Philadelphia: Temple University Press, 2019.

Ho, Ming-sho. "Environmental Movement in Democratizing Taiwan (1980–2004): A Political Opportunity Structure Perspective." In *East Asian Social Movements: Power, Protest, and Change in a Dynamic Region*, edited by Jeffrey Broadbent and Vicky Brockman, 283–314. New York: Springer, 2010.

Ho, Ming-sho. "Lukang Anti-Dupont Movement (Taiwan)." In *The Wiley-Blackwell Encyclopedia of Social and Political Movements*, edited by David Snow, Donatella della Porta, Bert Klandermans, and Doug McAdam. Malden, MA: Wiley-Blackwell, 2013. http://homepage.ntu.edu.tw/~msho/book.files/BC/B34.pdf.

Ho, Ming-sho. "Resisting Naphtha Crackers: A Historical Survey of Environmental Politics in Taiwan." *China Perspectives* 2014, no. 3 (September 2014): 5–14.

Ho, Ming-sho, and Chen-Shuo Hong. "Challenging New Conservative Regimes in South Korea and Taiwan." *Asian Survey* 52, no. 4 (July/August 2012): 643–665.

Ho, Peter. "Embedded Activism and Political Change in a Semiauthoritarian Context." *China Information* 21 (2007): 187–209.

Ho, Peter, and Richard Edmonds, eds. *China's Embedded Activism: Opportunities and Constraints of a Social Movement*. New York: Routledge, 2007.

Hoshi, Takeo, and Anil Kashyap. *Corporate Financing and Governance in Japan: The Road to the Future*. Cambridge, MA: MIT Press, 2004.

Hoshi, Takeo, Anil Kashyap, and David Scharfstein. "Corporate Structure, Liquidity, and Investment: Evidence from Japanese Industrial Groups." *Quarterly Journal of Economics* 106, no. 1 (1991): 33–60.

Hothorn, Torsten, Peter Bühlmann, Sandrine Dudoit, Annette Molinaro, and Mark J. Van Der Laan. "Survival Ensembles." *Biostatistics* 7, no. 3 (2006): 355–373.

Hsiao, Hsin-Huang Michael. "Environmental Movements in Taiwan." In *Asia's Environmental Movements: Comparative Perspectives*, edited by Yok-shiu F. Lee and Alvin Y. So, 31–54. New York: M. E. Sharpe, 1999.

Hsu, Sara, ed. *Routledge Handbook of Sustainable Development in Asia*. New York: Routledge, 2018.

Huntington, Samuel P. "Democracy's Third Wave." *Journal of Democracy* 2, no. 2 (Spring 1991): 12–34.

Ibrahim, Saad Eddin. "Civil Society and Prospects of Democratization in the Arab World." In *Civil Society in the Middle East*, edited by Augustus Richard Norton, 1:27–54. New York: E. J. Brill, 1995.

Imada, Makoto. "The Voluntary Response to the Hanshin Awaji Earthquake: A Trigger for the Development of the Voluntary and Non-profit Sector in Japan." In *The Voluntary and Non-profit Sector in Japan: The Challenge of Change*, edited by Stephen Osborne, 40–50. Nissan Institute / RoutledgeCurzon Japanese Studies Series. New York: RoutledgeCurzon, 2003.

Imura, Hidefumi, and Miranda Schreurs, eds. *Environmental Policy in Japan*. Northampton, MA: Edward Elgar, 2005.

Institute for Global Environmental Strategies. *Towards a Sustainable Asia and the Pacific: Report of Eco Asia Long-Term Perspective Project Phase II*. Hayama, Kanagawa: Institute for Global Environmental Strategies, 2001.

Institute of Public and Environmental Affairs and Natural Resources Defense Council. *Greening the Global Supply Chain*. CITI Index and Annual Report. Beijing: Institute of Public and Environmental Affairs and Natural Resources Defense Council, 2014.

International Institute of Environment and Development and World Business Council for Sustainable Development. *Breaking New Ground: Mining, Minerals, and Sustainable Development—Report of the Mining, Minerals and Sustainable Development Project*. London: International Institute of Environment and Development and World Business Council for Sustainable Development, 2002.

International Renewable Energy Agency. *A New World: The Geopolitics of the Energy Transformation*. International Renewable Energy Agency, 2019.

Iokibe, Makoto. "Japan's Civil Society: An Historical Overview." In *Deciding the Public Good: Governance and Civil Society in Japan*, edited by Tadashi Yamamoto, 51–96. New York: Japan Center for International Exchange, 1999.

Ito, Hiroshi, and Nobuo Kawazoe. "A Review of Toyota City's Eco-policy: Changes in Citizens' Awareness between 2012 and 2015." *Urban Research and Practice* 11, no. 1 (2018): 19–36.

Jakobson, Linda, and Dean Knox. "New Foreign Policy Actors in China." Policy paper, Stockholm International Peace Research Institute, Solna, Sweden, 2010.

Jenkins-Smith, Hank C., and Paul A. Sabatier. "Evaluating the Advocacy Coalition Framework." *Journal of Public Policy* 14, no. 2 (1994): 175–203.

Jiang, Bin. "Implementing Supplier Codes of Conduct in Global Supply Chains: Process Explanations from Theoretic and Empirical Perspectives." *Journal of Business Ethics* 85 (2009): 77–92.

Johansen, Morgen, and Kelly LeRoux. "Managerial Networking in Nonprofit Organizations: The Impact of Networking on Organizational and Advocacy Effectiveness." *Public Administration Review* 73, no. 2 (2013): 355–363.

John, DeWitt. *Civic Environmentalism: Alternatives to Regulation in States and Communities*. Washington, DC: Congressional Quarterly Press, 1994.

Johnson, Chalmers. *MITI and the Japanese Miracle: The Growth of Industrial Policy, 1925–1975*. Stanford, CA: Stanford University Press, 1982.

Johnson, Edward, and Michael Mappin, eds. *Environmental Education and Advocacy: Changing Perspectives of Ecology and Education*. New York: Cambridge University Press, 2009.

Johnson, McKenzie F., Corrie Hannah, Leslie Acton, Ruxandra Popovici, Krithi K. Karanth, and Erika Weinthal. "Network Environmentalism: Citizen Scientists as Agents for Environmental Advocacy." *Global Environmental Change* 29 (2014): 235–245.

Jones, Bryan D. *Reconceiving Decision-Making in Democratic Politics: Attention, Choice, and Public Policy.* Chicago: University of Chicago Press, 1994.

Joo, Sungsoo, Seonmi Lee, and Youngjae Jo. "The Explosion of CSOs and Citizen Participation: An Assessment of Civil Society in South Korea 2004." CIVICUS http://www.civicus.org/media/CSI_South_Korea_Country_Report.pdf (accessed July 28, 2020).

Jung, Kyujin, Hee Soun Jang, and Inseok Seo. "Government-Driven Social Enterprises in South Korea: Lessons from the Social Enterprise Promotion Program in the Seoul Metropolitan Government." *International Review of Administrative Sciences* 82, no. 3 (2016): 598–616.

Kagan, Robert, Dorothy Thornton, and Neil Gunningham. "Explaining Corporate Environmental Performance: How Does Regulation Matter?" *Law and Society Review* 37, no. 1 (March 2003): 51–90.

Kage, Rieko, Frances McCall Rosenbluth, and Seiki Tanaka. "What Explains Low Female Political Representation? Evidence from Survey Experiments in Japan." *Politics & Gender* 15, no. 2 (2019): 285–309.

Kalish, Debbie, Susan Burek, Amy Costello, Lawrence Schwartz, and John Taylor. "Integrating Sustainability into New Product Development." *Research-Technology Management* 61, no. 2 (2018): 37–46.

Kalland, Arne and Gerard Persoon, eds. *Environmental Movements in Asia.* Man and Nature, vol. 4. New York: RoutledgeCurzon, 1999.

Kang, Xiaoguang, and Qun Wang, eds. "Nonprofit Policymaking in China." Special issue, *Nonprofit Policy Forum* 9, no. 1 (2018).

Kantor, Jodi. "Harvard Business School Case Study: Gender Equity." *New York Times,* September 7, 2013.

Kataoka, Yatsuka, Shiko Hayashi, Junko Akagi, Kohei Hibino, Junko Ota, Fritz Akhmad Nuzir, and Shino Horizonoa. *Actions for a Sustainable Society: Collaboration between Asia and the City of Kitakyushu.* Kitakyushu City: Institute for Global Environmental Strategies, 2018.

Katsiaficas, Georgy. *South Korean Democracy: Legacy of the Gwangju Uprising.* New York: Routledge, 2013.

Kauffman, Craig M. *Grassroots Global Governance: Local Watershed Management Experiments and the Evolution of Sustainable Development.* New York: Oxford University Press, 2017.

Keck, Margaret, and Kathryn Sikkink. *Activists beyond Borders: Advocacy Networks in International Politics*. Ithaca, NY: Cornell University Press, 1998.

Kern, Kristine, and Harriet Bulkeley. "Cities, Europeanization and Multi-level Governance: Governing Climate Change through Transnational Municipal Networks." *JCMS: Journal of Common Market Studies* 47, no. 2 (2009): 309–332.

Ketchum, Robert Glenn. *The Tongass: Alaska's Vanishing Rain Forest*. New York: Aperture, 1987.

Kijek, Tomasz. "Modelling of Eco-innovation Diffusion: The EU Eco-label." *Comparative Economic Research* 18, no. 1 (2015): 65–79.

Kim, Hyuk-Rae. "Dilemmas in the Making of Civil Society in Korean Political Reform." *Journal of Contemporary Asia* 34, no. 1 (2004): 55–69.

Kim, Hyuk-Rae. "The State and Civil Society in Transition: The Role of Non-governmental Organizations in South Korea." *Pacific Review* 13, no. 4 (2010): 595–613.

Kim, Sung-Young, and Elizabeth Thurbon. "Developmental Environmentalism: Explaining South Korea's Ambitious Pursuit of Green Growth." *Politics and Society* 43, no. 2 (2015): 213–240.

Kim, Sunhyuk. "Civic Engagement and Democracy in South Korea." *Korean Observer* 40, no. 1 (Spring 2009): 1–26.

Kim, Sunhyuk. *Politics of Democratization in Korea: The Role of Civil Society*. Pittsburgh: University of Pittsburgh Press, 2000.

King, Gary, Jennifer Pan, and Margaret E. Roberts. "How Censorship in China Allows Government Criticism but Silences Collective Expression." *American Political Science Review* 107, no. 2 (2013): 326–343.

Kingdon, John W. *Agendas, Alternatives, and Public Policies*. New York: HarperCollins, 1984.

Kingston, Jeff, ed. *Natural Disaster and Nuclear Crisis in Japan: Response and Recovery after Japan's 3/11*. Nissan Institute / Routledge Japan Studies. New York: Routledge, 2012.

Kirk, Emerson, Tina Nabatchi, and Stephen Balogh. "An Integrative Framework for Collaborative Governance." *Journal of Public Administration Research and Theory* 22 (2011): 1–29.

Knoke, David, Franz Urban Pappi, Jeffrey Broadbent, and Yutaka Tsujinaka. *Comparing Policy Networks: Labor Politics in the U.S., Germany, and Japan*. Cambridge Studies in Comparative Politics. Cambridge: Cambridge University Press, 1996.

Konisky, David M. "Regulatory Competition and Environmental Enforcement: Is There a Race to the Bottom?" *American Journal of Political Science* 51, no. 4 (2007): 853–872.

Koopmans, Ruud, and Dieter Rucht. "Protest Event Analysis." In *Methods of Social Movement Research*, edited by Bert Klandermans and Suzanne Staggenborg, 231–259. Social Movements, Protest, and Contention 16. Minneapolis: University of Minnesota Press, 2002.

Kostka, Genia, and Arthur P. J. Mol. "Implementation and Participation in China's Local Environmental Politics: Challenges and Innovations." *Journal of Environmental Policy and Planning* 15, no. 1 (2013): 3–16.

Kraft, Michael, Mark Stephan, and Troy Abel. *Coming Clean: Information Disclosure and Environmental Performance.* Cambridge, MA: MIT Press, 2011.

Krauss, Ellis, and Bradford Simcock. "Citizens' Movements: The Growth and Impact of Environmental Protest in Japan." In *Political Opposition and Local Politics in Japan*, edited by Kurt Steiner, Ellis Krauss, and Scott Flanagan, 187–227. Princeton, NJ: Princeton University Press, 1980.

Ku, Dowan. "The Korean Environmental Movement: Green Politics through Social Movement." In *East Asian Social Movements: Power, Protest, and Change in a Dynamic Region*, edited by Jeffrey Broadbent and Vicky Brockman, 205–235. New York: Springer, 2011.

Ku, Do-Wan. "The Structural Change of the Korean Environmental Movement." *Korea Journal of Population and Development* 25, no. 1 (1996): 155–180.

Ku, Duwan. "Environmental Movement and Policies during High Economic Growth in Korea." In *Environment and Our Sustainability in the 21st Century: Understanding and Cooperation between Developed and Developing Countries*, edited by Yuko Arayama, 65–87. Nagoya, Japan: Nagoya University, 2002.

Lake, David A., and Matthew A. Baum. "The Invisible Hand of Democracy: Political Control and the Provision of Public Services." *Comparative Political Studies* 34, no. 6 (2001): 587–621.

Lang, Graeme, and Cathy Hiu Wan Chan. "China's Impact on Forests in Southeast Asia." *Journal of Contemporary Asia* 36, no. 2 (2006): 167–194.

Lau, Kai-Hon, Wai Man Wu, Jimmy C. H. Fung, and Bill Barron. *Significant Marine Source for SO₂ Levels in Hong Kong.* Hong Kong: Civic Exchange, 2005.

Lee, John Chi-Kin, and Rob Efird. "Introduction: Schooling and Education for Sustainable Development (ESD) across the Pacific." In *Schooling for Sustainable Development across the Pacific*, edited by John Chi-Kin Lee and Rob Efird, 3–36. Dordrecht: Springer, 2014.

Lee, Jong Youl. "Theory and Application of Urban Governance: The Case of Seoul." *Journal of Urban Technology* 10, no. 2 (2003): 69–86.

Lee, See-Jae. "The Environmental Movement and Its Political Empowerment." *Korea Journal* 40, no. 3 (2000): 131–160.

Lee, Su-Hoon. "Environmental Movements in Korea." In *Asia's Environmental Movements: Comparative Perspectives*, edited by Yok-shiu F. Lee and Alvin Y. So, 90–119. New York: M. E. Sharpe, 1999.

Lee, Taedong. *Global Cities and Climate Change: The Translocal Relations of Environmental Governance*. Cities and Global Governance. New York: Routledge, 2015.

Lee, Yok-shiu F., and Alvin Y. So, eds. *Asia's Environmental Movements: Comparative Perspectives*. New York: M. E. Sharpe, 1999.

Lejano, Raul, Mrill Ingram, and Helen Ingram. *The Power of Narrative in Environmental Networks*. Cambridge, MA: MIT Press, 2013.

Lemos, Maria Carmen, and Arun Agrawal. "Environmental Governance." *Annual Review of Environmental Resources* 31 (2006): 297–325.

Li, Cheng, ed. *Bridging Minds across the Pacific: U.S.-China Educational Exchanges, 1978–2003*. Lexington, MA: Lexington Books, 2005.

Libby, Patricia J. *The Lobby Strategy Handbook: 10 Steps to Advancing Any Cause Effectively*. New York: Sage, 2011.

Lieberthal, Kenneth, and Michel Oksenberg. *Policy Making in China: Leaders, Structures, and Process*. Princeton, NJ: Princeton University Press, 1988.

Lin, Chin-Huang, Ho-Li Yang, and Dian-Yan Liou. "The Impact of Corporate Social Responsibility on Financial Performance: Evidence from Business in Taiwan." *Technology in Society* 31, no. 1 (2009): 56–63.

Lindenberg, Nannette. *Definition of Green Finance*. Bonn: Deutsches Institut für Entwicklungspolitik, 2014.

Litfin, Karen T. "Advocacy Coalitions along the Domestic-Foreign Frontier: Globalization and Canadian Climate Change Policy." *Policy Studies Journal* 28, no. 1 (Spring 2000): 236–252.

Litzinger, Ralph. "In Search of the Grassroots: Hydroelectric Politics in Northwest Yunnan." In *Grassroots Political Reform in Contemporary China*, edited by Elizabeth Perry and Merle Goldman, 282–299. Cambridge, MA: Harvard University Press, 2007.

Liu, Jianguo, Shuxin Li, Zhiyun Ouyang, Christine Tam, and Xiaodong Chen. "Ecological and Socioeconomic Effects of China's Policies for Ecosystem Services." *Proceedings of the National Academy of Sciences* 105, no. 28 (2008): 9477–9482.

Lobel, Orly. "Sustainable Capitalism or Ethical Transnationalism: Offshore Production and Economic Development." *Journal of Asian Economics* 17, no. 1 (2006): 56–62.

Lora-Wainwright, Anna. *Resigned Activism: Living with Pollution in Rural China*. Cambridge, MA: MIT Press, 2017.

Lorentzen, Peter, Pierre Landry, and John Yasuda. "Undermining Authoritarian Innovation: The Power of China's Industrial Giants." *Journal of Politics* 76, no. 1 (January 2014): 182–194.

Low, Morris. "Eco-cities in Japan: Past and Future." *Journal of Urban Technology* 20, no. 1 (2013): 7–22.

Ma, Jun, Ray Cheung, Wang Jingjing, and Ruan Qingyuan. "Greening Supply Chains in China: Practical Lessons from China-Based Suppliers in Achieving Environmental Performance." Working paper, World Resources Institute, Washington, DC, 2010.

Maddow, Ben, and W. Eugene Smith. *Let Truth Be the Prejudice: W. Eugene Smith, His Life and Photographs*. New York: Aperture, 1985.

Maeda, Toshizo. "Reducing Waste through the Promotion of Composting and Active Involvement of Various Stakeholders: Replicating Surabaya's Solid Waste Management Model." Policy brief, Institute for Global Environmental Strategies, Kitakyushu, Japan, 2009.

Mahoney, James, and Kathleen Thelen. *Explaining Institutional Change: Ambiguity, Agency, and Power*. New York: Cambridge University Press, 2009.

Manabe, Noriko. "The No Nukes 2012 Concert and the Role of Musicians in the Anti-nuclear Movement." *Asia-Pacific Journal* 10, no. 29 (2012): article 2. https://apjjf.org/-Noriko-Manabe/3799/article.pdf.

Manion, Melanie. "Chinese Democratization in Perspective: Electorates and Selectorates at the Township Level." *China Quarterly* 163 (2000): 764–782.

McAdam, Doug, John D. McCarthy, and Mayer N. Zald, eds. *Comparative Perspectives on Social Movements: Political Opportunities, Mobilizing Structures, and Cultural Framings*. New York: Cambridge University Press, 1996.

McBeth, Gerald, and Tse-Kang Leng. *Governance of Biodiversity Conservation in China and Taiwan*. Environmental Governance in Asia. Northampton, MA: Edward Elgar, 2006.

McGurty, Eileen Maura. "From NIMBY to Civil Rights: The Origins of the Environmental Justice Movement." *Environmental History* 2, no. 3 (July 1997): 301–323.

McKean, Margaret. *Environmental Protest and Citizen Politics in Japan*. Berkeley: University of California Press, 1981.

Mertha, Andrew. *China's Water Warriors: Citizen Action and Policy Change*. Ithaca, NY: Cornell University Press, 2008.

Mertha, Andrew. "'Fragmented Authoritarianism 2.0': Political Pluralization in the Chinese Policy Process." *China Quarterly* 200, no. 1 (2009): 995–1012.

Mesch, Claudia. *Art and Politics: A Small History of Art for Social Change since 1945*. London: I. B. Tauris, 2013.

Migdal, Joel S. "The State in Society: An Approach to Struggles for Domination." In *State Power and Social Forces: Domination and Transformation in the Third World*, edited by Joel S. Migdal, Atul Kohli, and Vivienne Shue, 7–34. New York: Cambridge University Press, 1994.

Migdal, Joel S. *State in Society: Studying How States and Societies Transform and Constitute One Another.* Cambridge Studies in Comparative Politics. New York: Cambridge University Press, 2001.

Miller, Norman. *Environmental Politics: Interest Groups, the Media, and the Making of Policy.* New York: Taylor and Francis, 2002.

Min, Seo-Hyeon, Seul-Ye Lim, and Seung-Hoon Yoo. "Consumers' Willingness to Pay a Premium for Eco-labeled LED TVs in Korea: A Contingent Valuation Study." *Sustainability* 9, no. 5 (2017): 814–825.

Ministry of the Environment. "White Paper on the Environment 2017." Ministry of the Environment, Tokyo, 2018.

Ministry of the Environment Korea. *Ecorea: Environmental Review 2015, Korea.* Seoul: Ministry of the Environment, 2017.

Minkoff, Debra C., Silke Aisenbrey, and Jon Agone. "Organizational Diversity in the U.S. Advocacy Sector." *Social Problems* 55, no. 4 (2008): 525–548.

Mitchell, Ronald. "Transparency for Governance: The Mechanisms and Effectiveness of Disclosure-Based and Education-Based Transparency Policies." *Ecological Economics* 70, no. 11 (2011): 1882–1890.

Monden, Yasuhiro. *Toyota Production System: An Integrated Approach to Just-in-Time.* Portland, OR: Productivity, 2011.

Morikawa, Jun. *Whaling in Japan: Power, Politics, and Diplomacy.* New York: Columbia University Press, 2009.

Muller-Rommel, Ferdinand, and Thomas Poquntke. *Green Parties in National Government* New York: Routledge, 2002.

Murdie, Amanda, and Johannes Urpelainen. "Why Pick on Us? Environmental INGOs and State Shaming as a Strategic Substitute." *Political Studies* 63, no. 2 (2015): 353–372.

Nakamura, Eri. "Does Environmental Investment Really Contribute to Firm Performance? An Empirical Analysis Using Japanese Firms." *Eurasian Business Review* 1, no. 2 (2011): 91–111.

Newig, Jens, and Oliver Fritsch. "Environmental Governance: Participatory, Multilevel—and Effective?" *Environmental Policy and Governance* 19, no. 3 (2009): 197–214.

Ng, Simon. *Fair Winds Charter: How Civic Exchange Influenced Policymaking to Reduce Ship Emissions in Hong Kong 2006–2015.* Hong Kong: Civic Exchange, 2018.

Nissani, Moti. "Media Coverage of the Greenhouse Effect." *Population and Environment* 21, no. 1 (1999): 27–43.

Noble, Gregory W., and John Ravenhill. *The Asian Financial Crisis and the Architecture of Global Finance*. New York: Cambridge University Press, 2000.

North, Douglass C. *Institutions, Institutional Change and Economic Performance*. New York: Cambridge University Press, 1990.

Norton, Augustus Richard, ed. *Civil Society in the Middle East*. Vol. 1. New York: E. J. Brill, 1995.

Nuzir, Fritz Akhmad. *Development Model of Takakura Composting Method (TCM) as an Appropriate Environmental Technology (AET) for Urban Waste Management*. Hiroshima: IGES Kitakyushu Urban Centre, 2018.

O'Neill, Michael. *Green Parties and Political Change in Contemporary Europe: New Politics, Old Predicaments*. Burlington, VT: Ashgate, 1997.

OECD (Organisation for Economic Co-operation and Development). *Environmental Labelling and Information Schemes*. OECD, 2016.

Oh, Jennifer S. "Strong State and Strong Civil Society in Contemporary South Korea: Challenges to Democratic Governance." *Asian Survey* 52, no. 3 (May–June 2012): 528–549.

Oh, Won Yong, Young Kyun Chang, and Aleksey Martynov. "The Effect of Ownership Structure on Corporate Social Responsibility: Empirical Evidence from Korea." *Journal of Business Ethics* 104, no. 2 (2011): 283–297.

Okimoto, Daniel I. *Between MITI and the Market: Japanese Industrial Policy for High Technology*. Stanford, CA: Stanford University Press, 1989.

Onis, Ziya. "The Logic of the Developmental State." *Comparative Politics* 24, no. 1 (1991): 109–126.

Ortiz, David G., Daniel J. Myers, Eugene N. Walls, and Maria-Elena D. Diaz. "Where Do We Stand with Newspaper Data?" *Mobilization: An International Quarterly* 10, no. 3 (2005): 397–419.

Osborne, Stephen, ed. *The Voluntary and Non-profit Sector in Japan*. Nissan Institute / RoutledgeCurzon Japanese Studies Series. New York: RoutledgeCurzon, 2003.

Ostrom, Elinor. *Governing the Commons: The Evolution of Institutions for Collective Action*. Cambridge: Cambridge University Press, 1990.

Ozawa, Connie P. "Science in Environmental Conflicts." *Sociological Perspectives* 39, no. 2 (Summer 1996): 219–230.

Park, Hyungguen, and Changhee Kim. "Do Shifts in Renewable Energy Operation Policy Affect Efficiency: Korea's Shift from FIT to RPS and Its Results." *Sustainability* 10, no. 6 (2018): 1723–1736.

Payne, Rodger A. "Freedom and the Environment." *Journal of Democracy* 6, no. 3 (1995): 41–55.

Pekkanen, Robert Joseph. "Japan's Dual Civil Society: Members without Advocates." PhD diss., Harvard University, 2002.

Pekkanen, Robert Joseph. *Japan's Dual Civil Society: Members without Advocates.* Stanford, CA: Stanford University Press, 2006.

Pekkanen, Robert, Steven Rathgeb Smith, and Yutaka Tsujinaka, eds. *Nonprofits and Advocacy: Engaging Community and Government in an Era of Retrenchment.* Baltimore: Johns Hopkins University Press, 2014.

Perry, Elizabeth, and Merle Goldman, eds. *Grassroots Political Reform in Contemporary China.* Cambridge, MA: Harvard University Press, 2007.

Pierson, Paul. "When Effect Becomes Cause: Policy Feedback and Political Change." *World Politics* 45, no. 4 (1993): 595–628.

Portney, Kent E. *Taking Sustainable Cities Seriously: Economic Development, the Environment, and Quality of Life in American Cities.* Cambridge, MA: MIT Press, 2013.

Portney, Kent E., and Jeffrey Berry. "Civil Society and Sustainable Cities." *Comparative Political Studies* 47, no. 3 (2014): 395–419.

Poulos, Helen M. "The Media and NIMBY: How Do Grassroots Environmental Protests Incite Innovation?" In *NIMBY Is Beautiful: Cases of Local Activism and Environmental Innovation around the World*, edited by Carol J. Hager and Mary Alice Haddad, 15–32. New York: Berghahn Books, 2015.

Poulos, Helen M., and Mary Alice Haddad. "Violent Repression of Environmental Protests." *SpringerPlus* 5, no. 230 (2016): 1–12.

Prakash, Aseem, and Mary Kay Gugerty, eds. *Advocacy Organizations and Collective Action.* New York: Cambridge University Press, 2010.

Prakash, Aseem, and Matthew Potoski. "Racing to the Bottom? Trade, Environmental Governance, and ISO 14001." *American Journal of Political Science* 50, no. 2 (2006): 350–364.

Prakash, Aseem, and Matthew Potoski. *The Voluntary Environmentalists: Green Clubs, ISO 14001, and Voluntary Environmental Regulations.* New York: Cambridge University Press, 2006.

Pralle, Sarah. "Shopping Around: Environmental Organizations and the Search for Policy Venues." In *Advocacy Organizations and Collective Action*, edited by Aseem Prakash and Mary Kay Gugerty, 177–201. New York: Cambridge University Press, 2010.

Premakumara, D. G. J. "Kitakyushu City's International Cooperation for Organic Waste Management in Surabaya City, Indonesia and Its Replication in Asian Cities."

Discussion paper, Kitakyushu Urban Centre, Institute for Global Environmental Strategies, Kitakyushu City, Japan, 2012.

Purdy, Jedediah. "The Politics of Nature: Climate Change, Environmental Law, and Democracy." *Yale Law Journal* 119, no. 6 (2010): 1122–1209.

Qiao, Liming, and Peng Wang. "The 26 Degree Campaign: Saving Energy." In *The China Environment Yearbook (2005)*, edited by Congjie Liang and Dongping Yang, 331–339 London: Brill, 2005.

Quinn, Bernadette. "Arts Festivals and the City." *Urban Studies* 42, no. 5–6 (2005): 927–943.

Quinn, James Brian, and Frederick G. Hilmer. "Strategic Outsourcing." *MIT Sloan Management Review* 35, no. 4 (1994): 43.

Rabe, Barry. *Beyond NIMBY: Hazardous Waste Siting in Canada and the United States*. Washington, DC: Brookings Institution, 1994.

Ran, Ran. "Perverse Incentive Structure and Policy Implementation Gap in China's Local Environmental Politics." *Journal of Environmental Policy and Planning* 15, no. 1 (2013): 17–39.

R Development Core Team. "A Language and Environment for Statistical Computing." R Foundation for Statistical Computing, Vienna, 2012.

Reardon-Anderson, James. *Pollution, Politics, and Foreign Investment in Taiwan: The Lukang Rebellion*. New York: M. E. Sharpe, 1997.

Reed, Thomas Vernon. *The Art of Protest: Culture and Activism from the Civil Rights Movement to the Present*. Minneapolis: University of Minnesota Press, 2019.

Reimann, Kim. "Building Global Civil Society from the Outside In? Japanese International Development NGOs, the State, and International Norms." In *The State of Civil Society in Japan*, edited by Frank Schwartz and Susan Pharr, 298–315. New York: Cambridge University Press, 2003.

Reimann, Kim. *The Rise of Japanese NGOs*. New York: Routledge, 2009.

Robbins, Alicia S. T., and Stevan Harrell. "Paradoxes and Challenges for China's Forests in the Reform Era." *China Quarterly* 218 (2014): 381–403.

Roberts, Dexter, and Pete Engardio. "Secrets, Lies, and Sweatshops." *Businessweek*, November 26, 2006.

Rodrigues, Maria Guadalupe Moog. *Global Environmentalism and Local Politics: Transnational Advocacy Networks in Brazil, Ecuador, and India*. Albany: State University of New York Press, 2003.

Rohrschneider, Robert, and Russell J. Dalton. "A Global Network? Transnational Cooperation among Environmental Groups." *Journal of Politics* 64, no. 2 (2003): 510–533.

Rowe, William T. "The Problem of 'Civil Society' in Late Imperial China." *Modern China* 19, no. 2 (April 1993): 139–157.

Rubik, Frieder, and Paolo Frankl. *The Future of Eco-labelling: Making Environmental Product Information Systems Effective*. London: Routledge, 2017.

Rubik, Frieder, Dirk Scheer, and Fabio Iraldo. "Eco-labelling and Product Development: Potentials and Experiences." *International Journal of Product Development* 6, no. 3–4 (2008): 393–419.

Sabatier, Paul. "An Advocacy Coalition Framework of Policy Change and the Role of Policy-Oriented Learning Therein." *Policy Sciences* 21, no. 2/3 (1988): 129–168.

Sabatier, Paul, and Hank C. Jenkins-Smith, eds. *Policy Change and Learning: An Advocacy Coalition Approach*. Boulder, CO: Westview, 1993.

Saikawa, Eri. "Policy Diffusion of Emission Standards: Is There a Race to the Top?" *World Politics* 65, no. 1 (2013): 1–33.

Salamon, Lester M. *Partners in Public Service: Government-Nonprofit Relations in the Modern Welfare State*. Baltimore: Johns Hopkins University Press, 1995.

Salamon, Lester M., Helmut K. Anheier, Regina List, Stefan Toepler, and S. Wojciech Skolowski, eds. *Global Civil Society: Dimensions of the Nonprofit Sector*. Baltimore: Johns Hopkins Center for Civil Society Studies, 1999.

Samuels, Richard J. *The Business of the Japanese State: Energy Markets in Comparative and Historical Perspective*. Ithaca, NY: Cornell University Press, 1987.

Samuels, Richard J. *The Politics of Regional Policy in Japan: Localities Incorporated?* Princeton, NJ: Princeton University Press, 1983.

Samuels, Richard J. *"Rich Nation, Strong Army": National Security and the Technological Transformation of Japan*. Ithaca, NY: Cornell University Press, 1994.

Sandberg, Sheryl. *Lean In: Women, Work, and the Will to Lead*. New York: Knopf, 2013.

Sarat, Austin, and Stuart Scheingold, eds. *Cause Lawyers and Social Movements*. Stanford, CA: Stanford Law and Politics, 2006.

Schlesinger, Jacob M. *Shadow Shoguns: The Rise and Fall of Japan's Postwar Political Machine*. 2nd ed. Stanford, CA: Stanford University Press, 1999.

Schlosberg, David. *Defining Environmental Justice: Theories, Movements, and Nature*. Oxford: Oxford University Press, 2009.

Schreurs, Miranda. "Democratic Transition and Environmental Civil Society: Japan and South Korea Compared." *Good Society* 11, no. 2 (2002): 57–64.

Schreurs, Miranda. *Environmental Politics in Japan, Germany, and the United States*. New York: Cambridge University Press, 2002.

Schwartz, Frank J. *Advice and Consent: The Politics of Consultation in Japan*. New York: Cambridge University Press, 1998.

Schwartz, Jonathan. "Environmental NGOs in China: Roles and Limits." *Pacific Affairs* 77, no. 1 (2004): 28–49.

Scott, John. *Social Network Analysis: A Handbook*. New York: Sage, 2000.

Shapiro, Judith. *Mao's War against Nature: Politics and the Environment in Revolutionary China*. New York: Cambridge University Press, 2001.

Shen, Bo, Barbara Finamore, and Mona Yew. "Promoting Energy Efficiency as a Cost-Effective Resource in China: A Review of Jiangsu's Efficiency Power Plant Pilot." *ACEEE Summer Study on Energy Efficiency in Industry* 4 (2009): 114–125.

Sherman, Daniel. *Not Here, Not There, Not Anywhere: Politics, Social Movements, and the Disposal of Low-Level Radioactive Waste*. Washington, DC: RFF, 2011.

Shimizu, Teruyuki, Yohei Tsukushi, Kei Hasegawa, Manabu Ihara, Tatsuya Okubo, and Yasunori Kikuchi. "A Region-Specific Analysis of Technology Implementation of Hydrogen Energy in Japan." *International Journal of Hydrogen Energy*. Published ahead of print, December 15, 2017. https://doi.org/10.1016/j.ijhydene.2017.11.128.

Shinkawa, Toshimitsu, and T. J. Pempel. "Occupational Welfare and the Japanese Experience." In *The Privatization of Social Policy? Occupational Welfare and the Welfare State in America, Scandinavia and Japan*, edited by Michael Shalev, 280–326. New York: Routledge, 1996.

Shirk, Susan L. "China in Xi's 'New Era': The Return to Personalistic Rule." *Journal of Democracy* 29, no. 2 (2018): 22–36.

Simão, Lídia, and Ana Lisboa. "Green Marketing and Green Brand—the Toyota Case." *Procedia Manufacturing* 12 (2017): 183–194.

Simon, Karla. *Civil Society in China: The Legal Framework from Ancient Times to the "New Reform Era."* New York: Oxford University Press, 2013.

Siniawer, Eiko Maruko. *Waste: Consuming Postwar Japan*. Ithaca, NY: Cornell University Press, 2018.

Smith, David Horton. *Grassroots Associations*. Thousand Oaks, CA: Sage, 2000.

Smith, Steven Rathgeb, and Michael Lipsky. *Nonprofits for Hire: The Welfare State in the Age of Contracting*. Cambridge, MA: Harvard University Press, 1993.

Sörqvist, Patrik, Andreas Haga, Linda Langeborg, Mattias Holmgren, Maria Wallinder, Anatole Nöstl, Paul B. Seager, and John E. Marsh. "The Green Halo: Mechanisms and Limits of the Eco-label Effect." *Food Quality and Preference* 43 (2015): 1–9.

State Forestry Administration of China. *Forestry in China*. State Forestry Administration of China, 2014.

Steiner, Kurt. "Toward a Framework for the Study of Local Opposition." In *Political Opposition and Local Politics in Japan*, edited by Kurt Steiner, Ellis Krauss, and Scott Flanagan, 3–32. Princeton, NJ: Princeton University Press, 1980.

Steiner, Kurt, Ellis Krauss, and Scott Flanagan, eds. *Political Opposition and Local Politics in Japan*. Princeton, NJ: Princeton University Press, 1980.

Steinhoff, Patricia, ed. *Going to Court to Change Japan*. Ann Arbor, MI: Center for Japanese Studies, 2014.

Sustainable Stock Exchanges Initiative. *2018 Report on Progress*. Sustainable Stock Exchanges Initiative, 2019.

Strobl, Carolin, Anne-Laure Boulesteix, Thomas Kneib, Thomas Augustin, and Achim Zeileis. "Conditional Variable Importance for Random Forests." *BMC Bioinformatics* 9 (2008).

Switzer, Jacqueline Vaughn. *Environmental Activism: A Reference Handbook*. Santa Barbara, CA: ABC-CLIO, 2003.

Tan, Yeling. "Transparency without Democracy: The Unexpected Effects of China's Environmental Disclosure Policy." *Governance* 27, no. 1 (2014): 37–62.

Tang, Shui-Yan, and Ching-Ping Tang. "Democratization and Environmental Politics in Taiwan." *Asian Survey* 37, no. 3 (March 1997): 281–294.

Tarrow, Sidney. *Power in Movement: Social Movements and Contentious Politics*. 2nd ed. Ithaca, NY: Cornell University Press, 1998.

Teets, Jessica C. *Civil Society under Authoritarianism: The China Model*. New York: Cambridge University Press, 2014.

Teets, Jessica C. "Let Many Civil Societies Bloom: The Rise of Consultative Authoritarianism in China." *China Quarterly* 213 (March 2013): 19–38.

Teets, Jessica C. "Post-earthquake Relief and Reconstruction Efforts: The Emergence of Civil Society in China?" *China Quarterly* 198 (June 2009): 330–347.

Teets, Jessica C. "The Power of Policy Networks in Authoritarian Regimes: Changing Environmental Policy in China." *Governance* 31, no. 1 (2017): 125–141.

Teets, Jessica C. "Reforming Service Delivery in China: The Emergence of a Social Innovation Model." *Journal of Chinese Political Science* 17 (2012): 15–32.

Teets, Jessica C., and William Hurst, eds. *Local Governance Innovation in China: Experimentation, Diffusion, and Defiance*. Routledge Contemporary China Series. New York: Routledge, 2015.

Terao, Tadayoshi, and Kenji Otsuka, eds. *Development of Environmental Policy in Japan and Asian Countries*. New York: Palgrave Macmillan and IDE-JETRO, 2007.

Terazono, Atsushi, Yuichi Moriguchi, Yuko Sato Yamamoto, Shin-ichi Sakai, Bulent Inanc, Jianxin Yang, Stephen Siu, et al. "Waste Management and Recycling in Asia." *International Review of Environmental Strategies* 5, no. 2 (2005): 477–498.

Tiberghien, Yves. "The Battle for the Global Governance of Genetically Modified Organisms: The Roles of the European Union, Japan, Korea, and China in a Comparative Context." *Les Etudes du CERI*, no. 124 (June 2006): 1–49.

Tiberghien, Yves. "The Global Governance of Biotechnology: Mediating Chinese and Canadian Interests." *China Papers*, no. 13 (July 2010): 111–127.

Tilly, Charles, and Sidney Tarrow. *Contentious Politics*. Boulder, CO: Paradigm, 2006.

Tokunaga, Shojiro. "Japan's FDI-Promoting Systems and Intra-Asian Networks: New Investment and Trade Systems Created by the Borderless Economy." In *Japan's Foreign Investment and Asian Economic Interdependence: Production, Trade, and Financial Systems*, edited by Shojiro Tokunaga, 5–48. Tokyo: University of Tokyo Press, 1992.

Toyota Corporation. *Environmental Report 2018*. Toyota Corporation, 2018.

Tsai, Lily. *Accountability without Democracy: How Solidary Groups Provide Public Goods in Rural China*. New York: Cambridge University Press, 2007.

United Nations Department of Economic and Social Affairs. *People Matter: Civic Engagement in Public Governance—World Public Sector Report*. New York: United Nations, 2009.

United Nations Environment. *Sustainable Finance Progress Report*. Geneva: United Nations, 2019.

Upham, Frank. "Litigation and Moral Consciousness in Japan: An Interpretive Analysis of Four Japanese Pollution Suits." *Law and Society Review* 10, no. 4 (Summer 1976): 579–619.

Vanderheiden, Steve. "Eco-terrorism or Justified Resistance? Radical Environmentalism and the 'War on Terror.'" *Politics and Society* 33, no. 3 (2005): 425–447.

Van Rooij, Benjamin, Gerald E. Fryxell, Carlos Wing-Hung Lo, and Wei Wang. "From Support to Pressure: The Dynamics of Social and Governmental Influences on Environmental Law Enforcement in Guangzhou City, China." *Regulation and Governance* 7, no. 3 (2013): 321–347.

Van Waarden, Frans. "Dimensions and Types of Policy Networks." *European Journal of Political Research* 21, no. 1–2 (1992): 29–52.

Vig, Norman, and Michael Kraft, eds. *Environmental Policy: New Directions for the Twenty-First Century*. 8th ed. Washington, DC: SAGE, 2013.

Vitalis, Vangelis. "Private Voluntary Eco-labels: Trade Distorting, Discriminatory and Environmentally Disappointing." Background paper for the Round Table on Sustainable Development meeting "Eco-labelling and Sustainable Development,"

Organisation for Economic Co-operation and Development Headquarters, Paris, December 6, 2002. .

Vogel, Steven K. *Freer Markets, More Rules: Regulatory Reform in Advanced Industrial Countries*. Ithaca, NY: Cornell University Press, 1996.

Von Hippel, David, Sun-Jin Yun, and Myung-Rae Cho. "The Current Status of Green Growth in Korea: Energy and Urban Security." *Asia-Pacific Journal* 9, no. 44 (2011): article 4. https://apjjf.org/-Myung-Rae-Cho--Sun-Jin-YUN--David-von-Hippel/3628 /article.pdf.

Vurro, Clodia, Angeloantonio Russo, and Francesco Perrini. "Shaping Sustainable Value Chains: Network Determinants of Supply Chain Governance Models." *Journal of Business Ethics* 90, no. 4 (2009): 607–621.

Wade, Robert. *Governing the Market: Economic Theory and the Role of Government in East Asian Industrialization*. Princeton, NJ: Princeton University Press, 1990.

Waley, Paul. "Ruining and Restoring Rivers: The State and Civil Society in Japan." *Pacific Affairs* 78, no. 2 (2005): 195–215.

Walker, Brett. *Toxic Archipelago: A History of Industrial Disease in Japan*. Weyerhaeuser Environmental Books. Seattle: University of Washington Press, 2011.

Walling, William. "'Art' and 'Protest': Ralph Ellison's *Invisible Man* Twenty Years After." *Phylon* 34, no. 2 (1973): 120–134.

Walmart. *2018 Global Responsibility Report*. Bentonville, AR: Walmart, 2019.

Wang, Haikun, Yanxu Zhang, Hongyan Zhao, Xi Lu, Yanxia Zhang, Weimo Zhu, Chris P. Nielsen, et al. "Trade-Driven Relocation of Air Pollution and Health Impacts in China." *Nature Communications* 8, no. 1 (2017): article 738.

Wasserman, Stanley, and Katherine Faust. *Social Network Analysis: Methods and Applications*. Structural Analysis in the Social Sciences 8. Cambridge: Cambridge University Press, 1994.

Watanabe, Hiroyuki. *Japan's Whaling: The Politics of Culture in Historical Perspective*. Victoria, Australia: Trans Pacific, 2008.

Watanabe, Takehiro. "Talking Sulfur Dioxide: Air Pollution and the Politics of Science in Late Meiji Japan." In *Japan at Nature's Edge: The Environmental Context of a Global Power*, edited by Ian Jared Miller, Julia Adeney Thomas, and Brett Walker, 73–89. Honolulu: University of Hawai'i Press, 2013.

Weibel, Peter, Can Altay, Sruti Bala, Tatiana Bazzichelli, Olaf Bertram-Nothangel, Ángela Bonadies, Juan José Olavarría, Bruno Latour, and Peter Sloterdijk. *Global Activism: Art and Conflict in the 21st Century*. Cambridge, MA: MIT Press, 2015.

Weller, Robert Paul. *Alternate Civilities: Democracy and Culture in China and Taiwan*. Boulder, CO: Westview, 1999.

Weller, Robert Paul. *Discovering Nature: Globalization and Environmental Culture in China and Taiwan*. Cambridge: Cambridge University Press, 2006.

Weller, Robert Paul, and Hsin-Huang Michael Hsiao. "Culture, Gender and Community in Taiwan's Environmental Movement." In *Environmental Movements in Asia*, edited by Arne Kalland and Gerard Persoon, 83–109. Man and Nature, vol. 4. New York: RoutledgeCurzon, 1999.

Wells-Dang, Andrew. *Civil Society Networks in China and Vietnam: Informal Pathbreakers in Health and the Environment* (New York: Palgrave Macmillan, 2012).

White, Jenny B. *Islamist Mobilization in Turkey: A Study in Vernacular Politics*. Seattle: University of Washington Press, 2002.

Whitford, Andrew B., Vicky M. Wilkins, and Mercedes G. Ball. "Descriptive Representation and Policymaking Authority: Evidence from Women in Cabinets and Bureaucracies." *Governance* 20, no. 4 (2007): 559–580.

Wiktorowicz, Quintan. "Civil Society as Social Control: State Power in Jordan." *Comparative Politics* 33, no. 1 (2000): 43–62.

Wilkening, Kenneth. *Acid Rain Science and Politics in Japan: A History of Knowledge and Action toward Sustainability*. Cambridge, MA: MIT Press, 2004.

Winston, Morton. "NGO Strategies for Promoting Corporate Social Responsibility." *Ethics and International Affairs* 16, no. 2 (2002): 71–87.

Wong, Joseph. "The Adaptive Developmental State in East Asia." *Journal of East Asian Studies* 4, no. 3 (September–December 2004): 345–362.

Wong, Rebecca W. Y. *The Illegal Wildlife Trade in China*. Cham: Springer, 2019.

Woo-Cumings, Meredith, ed. *The Developmental State*. Ithaca, NY: Cornell University Press, 1999.

Woods, Neal D. "Interstate Competition and Environmental Regulation: A Test of the Race-to-the-Bottom Thesis." *Social Science Quarterly* 87, no. 1 (2006): 174–189.

Wu, Fengshi. "New Partners or Old Brothers? GONGOs in Transnational Environmental Advocacy in China." *China Environment Review*, no. 5 (2002): 45–58.

Wuthnow, Robert, ed. *Between States and Markets: The Voluntary Sector in Comparative Perspective*. Princeton, NJ: Princeton University Press, 1991.

Xie, Lei. *Environmental Activism in China*. New York: Routledge, 2009.

Xie, Lei, and Peter Ho. "Urban Environmentalism and Activists' Networks in China: The Cases of Xiangfan and Shanghai." *Conservation and Society* 6 (2008): 141–153.

Xin, Katherine R., and Jone L. Pearce. "Guanxi: Connections as Substitutes for Formal Institutional Support." *Academy of Management Journal* 39, no. 6 (1996): 1641–1658.

Xu, Bin. *The Politics of Compassion: The Sichuan Earthquake and Civic Engagement in China*. Stanford, CA: Stanford University Press, 2017.

Yamamoto, Tadashi, ed. *The Nonprofit Sector in Japan*. New York: Manchester University Press, 1998.

Yang, Guobin. *Power of the Internet in China*. New York: Columbia University Press, 2009.

Yeh, Yin-hua, Tsun-siou Lee, and Tracie Woidtke. "Family Control and Corporate Governance: Evidence from Taiwan." *International Review of Finance* 2, no. 1–2 (2001): 21–48.

Yoo, Sangjin, and Sang M. Lee. "Management Style and Practice of Korean Chaebols." *California Management Review* 29, no. 4 (1987): 95–110.

Young, Oran. "Governance for Sustainable Development in a World of Rising Interdependencies." In *Governance for the Environment: New Perspectives*, edited by Magali Delmas and Oran R. Young, 12–40. New York: Cambridge University Press, 2009.

Zelenovskaya, Ekaterina. *Green Growth Policy in Korea: A Case Study*. International Center for Climate Governance, 2012.

Zhan, Xueyong, Carlos Wing-Hung Lo, and Shui-Yan Tang. "Contextual Changes and Environmental Policy Implementation: A Longitudinal Study of Street-Level Bureaucrats in Guangzhou, China." *Journal of Public Administration Research and Theory* 24, no. 4 (2014): 1005–1035.

Zhang, Qiang, Yixuan Zheng, Dan Tong, Min Shao, Shuxiao Wang, Yuanhang Zhang, Xiangde Xu, et al. "Drivers of Improved PM2.5 Air Quality in China from 2013 to 2017." *Proceedings of the National Academy of Sciences* 116, no. 49 (2019): 24463–24469.

Zhang, Yiqi, Christine Loh, Peter K. K. Louie, Huan Liu, and Alexis K. H. Lau. "The Roles of Scientific Research and Stakeholder Engagement for Evidence-Based Policy Formulation on Shipping Emissions Control in Hong Kong." *Journal of Environmental Management* 223 (October 2018): 49–56.

Zhou, Yuezhi. "Watchdogs on Party Leashes? Contexts and Implications of Investigative Journalism in Post-Deng China." *Journalism Studies* 1, no. 2 (2000): 577–595.

Zhou, Yuezhi, and Sun Wusan. "Public Opinion Supervision: Possibilities and Limits of the Media in Constraining Local Officials." In *Grassroots Political Reform in Contemporary China*, edited by Elizabeth Perry and Merle Goldman, 300–324. Cambridge, MA: Harvard University Press, 2007.

Zissis, Carin, and Jayshree Bajoria. "China's Environmental Crisis." Council on Foreign Relations, 2008. https://www.cfr.org/backgrounder/chinas-environmental-crisis.

Index

Figures and tables are indicated by "f" and "t" following the page number. Note material is indicated by an "n" and note number following the page number.

American and Comparative Environmental Policy

Sheldon Kamieniecki and Michael E. Kraft, series editors

Russell J. Dalton, Paula Garb, Nicholas P. Lovrich, John C. Pierce, and John M. Whiteley, *Critical Masses: Citizens, Nuclear Weapons Production, and Environmental Destruction in the United States and Russia*

Daniel A. Mazmanian and Michael E. Kraft, editors, *Toward Sustainable Communities: Transition and Transformations in Environmental Policy*

Elizabeth R. DeSombre, *Domestic Sources of International Environmental Policy: Industry, Environmentalists, and U.S. Power*

Kate O'Neill, *Waste Trading among Rich Nations: Building a New Theory of Environmental Regulation*

Joachim Blatter and Helen Ingram, editors, *Reflections on Water: New Approaches to Transboundary Conflicts and Cooperation*

Paul F. Steinberg, *Environmental Leadership in Developing Countries: Transnational Relations and Biodiversity Policy in Costa Rica and Bolivia*

Uday Desai, editor, *Environmental Politics and Policy in Industrialized Countries*

Kent Portney, *Taking Sustainable Cities Seriously: Economic Development, the Environment, and Quality of Life in American Cities*

Edward P. Weber, *Bringing Society Back In: Grassroots Ecosystem Management, Accountability, and Sustainable Communities*

Norman J. Vig and Michael G. Faure, editors, *Green Giants? Environmental Policies of the United States and the European Union*

Robert F. Durant, Daniel J. Fiorino, and Rosemary O'Leary, editors, *Environmental Governance Reconsidered: Challenges, Choices, and Opportunities*

Paul A. Sabatier, Will Focht, Mark Lubell, Zev Trachtenberg, Arnold Vedlitz, and Marty Matlock, editors, *Swimming Upstream: Collaborative Approaches to Watershed Management*

Sally K. Fairfax, Lauren Gwin, Mary Ann King, Leigh S. Raymond, and Laura Watt, *Buying Nature: The Limits of Land Acquisition as a Conservation Strategy, 1780–2004*

Steven Cohen, Sheldon Kamieniecki, and Matthew A. Cahn, *Strategic Planning in Environmental Regulation: A Policy Approach That Works*

Michael E. Kraft and Sheldon Kamieniecki, editors, *Business and Environmental Policy: Corporate Interests in the American Political System*

Joseph F. C. DiMento and Pamela Doughman, editors, *Climate Change: What It Means for Us, Our Children, and Our Grandchildren*

Christopher McGrory Klyza and David J. Sousa, *American Environmental Policy, 1990–2006: Beyond Gridlock*

John M. Whiteley, Helen Ingram, and Richard Perry, editors, *Water, Place, and Equity*

Judith A. Layzer, *Natural Experiments: Ecosystem-Based Management and the Environment*

Daniel A. Mazmanian and Michael E. Kraft, editors, *Toward Sustainable Communities: Transition and Transformations in Environmental Policy*, second edition

Henrik Selin and Stacy D. VanDeveer, editors, *Changing Climates in North American Politics: Institutions, Policymaking, and Multilevel Governance*